CZECHOSLOVAK ACADEMY OF SCIENCES

Helminths of Fish-Eating Birds of the Palaearctic Region I

CZECHOSLOVAK ACADEMY OF SCIENCES

Scientific Editor
Academician Bohumír Rosický

Scientific Adviser
Dr. Jozef Macko, CSc.

Language Editor
Dr. Sheila Willmott

Helminths of Fish-Eating Birds of the Palaearctic Region I

Nematoda

V. Baruš, T. P. Sergeeva, M. D. Sonin and K. M. Ryzhikov

B. Ryšavý and K. M. Ryzhikov - Editors

DR. W. JUNK, b. v., Publishers
The Hague

ACADEMIA,
Publishing House of the
Czechoslovak Academy
of Sciences,
Prague
1978

Distribution throughout the world with the exception of Socialist countries
Dr. W. Junk, b. v., Publishers, P.O.Box 13713, 2501 ES The Hague, The Netherlands

ISBN 90 6193 551 2

© ACADEMIA, Publishing House of the Czechoslovak Academy of Sciences, Prague 1978

Translation © Marie Dašková, 1978

All rights reserved. No part of this book may be reproduced in any form, by photostat, microfilm, retrieval system, or by any other means, without written permission from the publishers
Printed in Czechoslovakia

Contents

Preface 8
Introduction 11
General account of nematodes 13
Systematic part 17
Subclass *Adenophorea* (Linstow, 1905; Chitwood, 1950) 18
Suborder *Trichocephalata* Railliet & Henry, 1913 18
Family *Capillariidae* Neveu-Lemaire, 1936 19
Genus *Capillaria* Zeder, 1800 19
Genus *Thominx* Dujardin, 1845 26
Suborder *Dioctophymata* Skrjabin, 1927 32
Family *Dioctophymidae* Railliet, 1915 32
Genus *Eustrongylides* Jägerskiöld, 1909 32
Genus *Hystrichis* Dujardin, 1845 40
Subclass *Secernentea* (Linstow, 1905) Dougherty, 1958 43
Suborder *Rhabditata* Chitwood, 1933 43
Family *Strongyloididae* Chitwood & McIntosh, 1934 43
Genus *Strongyloides* Grassi, 1879 44
Suborder *Strongylata* Railliet & Henry, 1913 45
Family *Amidostomidae* Baylis & Daubney, 1926 46
Genus *Amidostomum* Railliet & Henry, 1909 46
Family *Trichostrongylidae* Leiper, 1912 53
Genus *Epomidiostomum* Skrjabin, 1915 53
Family *Syngamidae* Leiper, 1912 55
Genus *Syngamus* Siebold, 1836 56
Genus *Cyathostoma* Blanchard, 1849 59
Genus *Hovorkonema* Ryzhikov, 1967 66
Suborder *Ascaridata* Skrjabin, 1915 68
Family *Anisakidae* Skrjabin & Karokhin, 1945 68
Genus *Contracaecum* Railliet & Henry, 1912 69
Genus *Porrocaecum* Railliet & Henry, 1912 91
Suborder *Oxyurata* Skrjabin, 1923 104
Family *Heterakidae* Railliet & Henry, 1914 105
Genus *Heterakis* Dujardin, 1845 105
Family *Subuluridae* Yorke & Maplestone, 1926 110
Genus *Subulura* Molin, 1860 110
Suborder *Spirurata* Railliet, 1914 112
Family *Spiruridae* Oerley, 1885 113
Genus *Cyrnea* Seurat, 1914 114
Genus *Excisa* Gendre, 1928 118

Family *Acuariidae* (Railliet, Henry & Sisoff, 1912 subfam.) Seurat, 1913 120
Genus *Acuaria* Bremser, 1841 122
Genus *Aviculariella* Wehr, 1931 123
Genus *Chevreuxia* Seurat, 1918 125
Genus *Cosmocephalus* Molin, 1858 126
Genus *Decorataria* Skrjabin, Sobolev & Ivashkin, 1965 133
Genus *Desportesius* Skrjabin, Sobolev & Ivashkin, 1965 135
Genus *Dispharynx* Railliet, Henry & Sisoff, 1912 144
Genus *Paracuaria* Rao, 1951 145
Genus *Pectinospirura* Wehr, 1939 149
Genus *Rusguniella* Seurat, 1919 151
Genus *Sexansocara* Sobolev & Sudarikov, 1939 153
Genus *Skrjabinocara* Kurashvili, 1941 155
Genus *Skrjabinocerca* Shikhobalova, 1930 158
Genus *Syncuaria* Gilbert, 1930 160
Genus *Synhimantus* Railliet, Henry & Sisoff, 1912 162
Genus *Echinuria* Soloviev, 1912 165
Genus *Cordonema* Schmidt & Kuntz, 1972 168
Genus *Skrjabinoclava* Sobolev, 1943 170
Family *Desmidocercidae* Cram, 1927 174
Genus *Desmidocerca* Skrjabin, 1915 174
Genus *Desmidocercella* Yorke & Maplestone, 1926 175
Family *Physalopteridae* Leiper, 1908 179
Genus *Physaloptera* Rudolphi, 1819 179
Family *Schistorophidae* Skrjabin, 1941 180
Genus *Schistorophus* Railliet, 1916 181
Genus *Sciadiocara* Skrjabin, 1916 186
Genus *Sobolevicephalus* Parukhin, 1964 188
Genus *Stellocaronema* Gilbert, 1930 190
Genus *Viktorocara* Gushanskaya, 1950 191
Family *Streptocaridae* Skrjabin, Sobolev & Ivashkin, 1965 195
Genus *Streptocara* Railliet, Henry & Sisoff, 1912 196
Genus *Ingliseria* Gibson, 1968 200
Genus *Proyseria* Petter, 1959 201
Genus *Stegophorus* Wehr, 1934 203
Genus *Seuratia* Skrjabin, 1916 206
Family *Tetrameridae* Travassos, 1914 208
Genus *Tetrameres* Creplin, 1846 209
Genus *Microtetrameres* Cram, 1927 219
Family *Thelaziidae* Skrjabin, 1915 221
Genus *Thelaziella* Skrjabin, Sobolev & Ivashkin, 1967 222
Suborder *Filariata* Skrjabin, 1915 224
Family *Aproctidae* Skrjabin & Shikhobalova, 1945 224
Genus *Aprocta* Linstow, 1883 225
Family *Splendidofilariidae* Sonin, 1962 226
Genus *Eufilaria* Seurat, 1921 227
Genus *Parornithofilaria* Sonin, 1965 228
Genus *Skrjabinocta* Chertkova, 1946 229
Family *Diplotriaenidae* Anderson, 1958 230
Genus *Dicheilonema* Diesing, 1861 230

Genus *Monopetalonema* Diesing, 1861 232
Family *Oswaldofilariidae* Sonin, 1968 234
Genus *Lemdana* Seurat, 1917 234
Genus *Heterospiculoides* Shigin, 1957 237
Genus *Heterospiculum* Shigin, 1951 238
Genus *Paronchocerca* Peters, 1936 240
Genus *Pelecitus* Railliet & Henry, 1910 243
Suborder *Camallanata* Chitwood, 1936 245
Family *Dracunculidae* Leiper, 1912 246
Genus *Avioserpens* Wehr & Chitwood, 1934 246
Family *Robertdollfusidae* Chabaud & Campana, 1950 250
Genus *Robertdollfusa* Chabaud & Campana, 1950 250

List of fish-eating birds and their nematodes recorded in the Palaearctic Region . . . 252

References 274

Index to Nematode Taxa 305

Preface

This monograph is the first attempt in the world literature to prepare a complete survey of helminths parasitizing the fish-eating birds of one of the largest and most interesting zoogeographic territories — the Palaearctic Region. This first volume deals with the *Nematoda*.

The work is based on the wide experience of Soviet parasitologists and is the result of many years' cooperation between them and Czechoslovak parasitologists.

Helminthology as a modern scientific discipline has developed rapidly in recent years. Thanks to the ingenious work of one of the founders of world helminthology, Academician Konstantin Ivanovich Skrjabin, member of the USSR Academy of Sciences and Czechoslovak Academy of Sciences, it has been possible to appreciate the importance of this discipline in public health and in agriculture as well as in studies of the nature of living matter and theories of animal evolution.

I vividly remember the friendly meetings with Academician K. I. Skrjabin, then director of the Helminthological Laboratory of the USSR Academy of Sciences and the great friend and teacher of Czechoslovak parasitologists. We decided that the joint work of teams from the Helminthological Laboratory, USSR Academy of Sciences and from the Institute of Parasitology of the Czechoslovak Academy of Sciences should be published as monographs prepared by outstanding scientists from both institutes. Having discussed this suggestion with other editors, we agreed that the helminths of fish-eating birds should be the first of these and should be published in three separate volumes. The first volume dealing with the class Nematoda is now prepared for publication. We regret very much that Academician Skrjabin has not lived to see the manuscript...

We chose fish-eating birds as the subject of the first monograph because, as hosts of helminths, they are a very numerous and wide-spread group, including as many as 176 species in the Palaearctic Region. This ecological group is not only interesting scientifically, but is also of great practical importance. Fish-eating birds are definitive hosts of some helminth species of which the larval stages cause great losses in fish breeding (e. g. infections by members of the genera *Diplostomulum*, *Ligula*, *Contracaecum* and others). They are also reservoir hosts of many helminth species causing serious diseases of poultry, (hystrichosis, streptocarosis and polymorphosis). Some fish-eating birds are also known to serve as definitive hosts of parasites dangerous to man, for example cestodes of the genus *Diphyllobothrium*. This host group is not uniform from a systematic point of view but represents an ecological group, the common feature of which is the dependence on an aquatic environment. Up to the present, many data on the helminth fauna of fish-eating birds from the Palaearctic Region have been recorded, especially in the vast territory of the U.S.S.R. and in central Europe in Czechoslovakia. Unfortunately, the papers are dispersed through a large number of journals and many of them are not easily available. Having regard to the present state and rapid development of helminthology it appeared necessary to assemble and evaluate all data on this subject.

At first we intended to publish this work in the form of a key to the helminths of fish-eating birds of the Palaearctic Region but in the course of preparation it became obvious that the extent of the work was far in excess of that for a key. It therefore became a critical synthesis of current knowledge of the helminth fauna of these hosts. The authors evaluated not only their own original material, but also numerous data from the literature, consulting more than 600 original scientific papers dealing with the subject. This gives some evidence of the extent of this work.

We were very glad that the initial work on this monograph could be guided by the wise advice of Academician K. I. Skrjabin. His pupils, from both the U.S.S.R. and Czechoslovakia, undertook the work, the Soviet helminthologists being headed by K. M. Ryzhikov and the Czechoslovak helminthologists working under the guidance of B. Ryšavý. K. M. Ryzhikov is a corresponding member of the USSR Academy of Sciences, the present director of the Helminthological Laboratory and one of the close co-workers of Academician Skrjabin and B. Ryšavý is a corresponding member of the Czechoslovak Academy of Sciences, Head of the Laboratory of Experimental Helminthology of the Institute of Parasitology and one of the first Czechoslovak pupils of Academician Skrjabin.

As I have already mentioned, the first volume is devoted to the class *Nematoda*. Parasitic worms of this class were found in 97 species of fish-eating birds of the Palaearctic Region. The nematodes belonged to 148 species of 62 genera, 23 families and 9 suborders. Each species is briefly described and figured, its definitive hosts in the group of fish-eating birds are listed and its distribution in the Palaearctic Region (or in other zoogeographical regions) is given. For those in which the life-cycle has been studied, brief data on this aspect are also attached. The notes of the authors reflecting their views on the systematic position of the respective species are of importance in systematics and taxonomy. Undoubtedly these will initiate further studies. The book also contains keys to genera and species of nematodes parazitizing fish-eating birds in the Palaearctic Region, which will provide a rapid and exact determination of the species.

The participation of outstanding specialists in nematodology, namely K. M. Ryzhikov, V. Baruš, M. D. Sonin and T. P. Sergeeva, has made it possible to prepare an original publication, interesting and useful for helminthologists working in various institutions. This book is the first comprehensive work written by Czechoslovak and Soviet helminthologists and I believe that this scientific co-operation will lead to further valuable results. This volume, as well as the subsequent ones devoted to cestodes, trematodes and acanthocephalans parasitizing fish-eating birds in the Palaearctic Region, will certainly be of interest to parasitologists throughout the world. The authors have prepared an excellent monograph, the content and critical approach of which represent the advances in helminthology in two Socialist countries. It is a compendium indispensable for every institution and worker engaged in the study of parasitic worms of fish-eating birds, a group of great economic importance and theoretical interest.

Academician Bohumír Rosický
Director of the Institute of Parasitology, Czechoslovak Academy of Sciences, Vice-President of the Czechoslovak Academy of Sciences

Introduction

This book is the first part of the key to helminths of fish-eating birds from the Palaearctic Region. It contains the data on the nematodes. The trematodes, cestodes and acanthocephalans will be dealt with separately in the following volumes.

The key has been compiled by several authors including Soviet and Czechoslovak helminthologists from the Helminthological Laboratory, Academy of Sciences of the U.S.S.R. (Moscow) and Institute of Parasitology, Czechoslovak Academy of Sciences (Prague).

Data from the relevant literature served as a basis for the key. The authors used all available monographs on helminthology and papers dealing with this subject and in many cases, the published data were supplemented by the results of the authors' own studies of original material.

Some difficulties were encountered when compiling the list of fish-eating birds of the Palaearctic Region, because many birds are not exclusively fish eaters. Ecological characteristics of many birds, especially the nature of their food, are known to be rather variable and may change with the season, habitat and other related biological factors. Therefore, when deciding whether or not a bird belongs to the fish-eating group the authors could not avoid assigning some conditionally.

Of the bird species living in the territory of the Palaearctic Region, 176 species were included in the fish-eating group. These species are the representatives of 10 orders. Within some orders, all species are included in the list, i.e., the survey covers all their representatives known from the Palaearctic Region. These are: *Gaviiformes* (5 species), *Podicipediformes* (5 species), *Pelecaniformes* (20 species), *Procellariformes* (26 species) and *Ciconiiformes* (32 species). On the other hand, for the remaining five orders only some of the taxa are dealt with, as follows: *Anseriformes*, 4 species of the genus *Mergus; Falconiformes*, one species of the genus *Pandion* and 4 species of *Haliaetus; Charadriiformes*, all species of the suborders *Lari* (51) and *Alcae* (19); *Strigiformes*, the 3 species of the genus *Ketupa;* and *Coraciiformes*, the 7 species of the family *Alcedinidae*.

The nematodes were reported from 97 of the 176 species of fish-eating birds from the Palaearctic Region. In the second part of the book there is a list of these bird species and a survey of nematodes recorded from them. The list of birds was compiled after Vaurie (1965), who published the most complete compendium on the fauna of birds of the Palaearctic Region yet available.

The first basic part of the book comprises a brief account of the nematodes, characterization of individual taxonomic classes, and keys and data on each species of parasite under consideration.

The Key contains data on 148 species of nematodes belonging to 62 genera, 23 families and 9 suborders.

The data on the individual parasite species include a list of hosts (mainly fish-eating birds, the occurrence in other groups of birds is mentioned only briefly), an indication of the localization of the parasite within the body of the bird and the geographical distribution (if possible in detail when the territory of the Palaearctic Region is concerned and briefly when dealing with other regions). A short morphological characterization of the species follows. For these in which the biology of the parasite has been studied, the main results of these findings are also given.

The list of references pertaining to each species includes those authors who have described the parasite (marked by asterisk), authors of monographs and compendia in which the species have been mentioned (also marked by asterisk) and, in addition, authors of those papers, which contain data on the respective species but are not mentioned in the bibliography of monographs and compendia.

Supplementary information on the taxonomic position, synonymy and some other topics is added to the description of each species in the form of notes.

Drawings of nearly all species included in the Key are attached.

The separate parts of the Key were prepared by the following authors:

General account of nematodes	K. M. Ryzhikov
Suborder *Trichocephalata*	V. Baruš
Suborder *Dioctophymata*	M. D. Sonin
Suborder *Rhabditata*	V. Baruš
Suborder *Strongylata*	V. Baruš, K. M. Ryzhikov
Suborder *Ascaridata*	V. Baruš, M. D. Sonin
Suborder *Oxyurata*	V. Baruš, M. D. Sonin
Suborder *Spirurata*	T. P. Sergeeva, M. D. Sonin, V. Baruš
Suborder *Filariata*	M. D. Sonin
Suborder *Camallanata*	M. D. Sonin

The work was directed and co-ordinated by B. Ryšavý and K. M. Ryzhikov, who also edited the whole text.

Our thanks are due to Dr. Sheila Willmott for careful revision of the whole manuscript and valuable advice and criticism.

As far as we know, no publication on helminths of any group of birds from the territory of the Palaearctic Region has appeared till now. Our book is the first of a series of publications in this field.

We hope that the Key will meet with the approval of a wide range of helminthologists and parasitologists, as well as of specialists engaged in some way in the problems of raising the biological productivity of water reservoirs.

The authors will be grateful for any critical comments on the contents and arrangement of the present book.

General Account of Nematodes

The class Nematoda Rudolphi, 1808 comprises a large group of worms. The basic character of the representatives of this class is the presence of a primary body cavity (schizocoel) which is the free space between the internal organs of the worm and the body wall, and is filled with schizocoelic fluid.

The characteristic properties of nematodes are as follows: elongated, usually fusiform body covered with well developed cuticle, absence of ciliated epithelium and protonephridia, dioecism, absence of organs of respiration and blood circulation.

The nematodes show a very wide compass of ecological adaptation. Among them are the free-living forms (occurring in fresh water, sea water and soil) and species parasitizing both animals and plants. The overwhelming majority of nematode species are free-living.

The species parasitizing vertebrates, especially birds, usually have a long, filiform, cylindrical, or fusiform body, round in transverse section. Sometimes the body is sac-shaped (e.g. females of *Tetrameres*).

The length of the body has a wide range, from tenths of millimetre to several metres. The longest of all known nematodes is *Placentonema gigantissima* which parasitizes whales: the female specimens of this nematode are sometimes more than 8 metres long. The length of nematodes parasitizing birds does not often exceed a few centimetres.

On the outside, the nematode body is covered with a thick cuticle which may be smooth but usually bears transverse and often also longitudinal striations. Sometimes the longitudinal striations are sharply distinct giving the appearance of cuticular ridges.

There may be different formations on the cuticle, such as spines, cordons, tubercles, ridges, and verrucae. On the posterior end of some nematodes *(Strongylata)* is a special cuticular outgrowth, the bursa copulatrix. In many species the cuticle forms alae on the lateral sides of the body which, when they extend nearly the entire length of body, are termed lateral alae, and, when they are confined to the anterior or posterior parts of the body, cervical or caudal alae, respectively. Papillae, representing cuticular elevations of different forms and sizes, are found on the surface of the cuticle in different parts of the body. The nerve endings come up to the papillae, some of which have a tactile function and some are chemoreceptors. According to their position they are divided into cephalic, cervical and genital papillae. A pointed cuticular appendage or a group of spines arranged in the form of a rosette is sometimes present on the tip of the tail end of the female.

Under the cuticle is a layer of hypodermis or subcuticle composed of cells which produce the cuticle. Under the hypodermis is a layer of musculature composed of longitudinal muscular elements. All these layers together, including the cuticle, hypodermis and musculature, form the so-called musculo-cutaneous sac, which borders the primary body cavity. The other side of the primary cavity is bordered by the internal organs of the worm.

Nematodes possess digestive, excretory, nervous and reproductive systems. The digestive system is formed of a straight tube beginning at the anterior end of the worm in the mouth opening and terminating at the cloaca or anus. There are three parts of the digestive tube: anterior (composed of vestibule, pharynx and oesophagus), middle (mid-intestine) and posterior (rectum). The border between the anterior (oesophagus) and middle (intestine) sections are usually distinctly marked but that between the mid-intestine and rectum is practically indistinguishable.

The mouth opening is usually surrounded by two, three or six lips but these are absent in some forms. The lips often bear different cuticular structures, e.g. denticles, papillae etc. The mouth opening leads into the buccal cavity, the walls of which may be thickened, in which case it is termed a buccal capsule (vestibule) and which may contain chitinized structures in the form of denticles or plates.

The pharynx, which is lacking in most nematode species although characteristic of Spirurata, follows immediately behind the stoma and represents a small piece of oesophagus (some authors consider it to be only a part of the oesophagus and do not separate it as an independent region of the digestive tube). The pharynx often has distinct transverse striations. When the pharynx is absent, the buccal cavity opens directly into the oesophagus.

In different groups of nematodes the oesophagus takes on diverse forms. For example, in the *Strongylata* the posterior part is swollen and club-shaped; in the *Oxyurata* the oesophagus ends in a globular swelling which is a bulb containing a chitinized device for grinding the food (the so-called valvular apparatus). In contrast with the *Strongylata* and *Oxyurata*, the *Spirurata* and *Filariata* have an oesophagus which is longer and divided into two distinct sections, of which the anterior is muscular and the posterior glandular. In *Trichocephalata* the oesophagus is a long hollow tube along which extends a single row of large cells (stichocytes) giving the oesophagus a beaded appearance. There is also one known nematode species in which both the oesophagus and the intestine atrophy. This is *Robertodollfusa longimicrofilaria*, the parasite of the brain of ospreys. In mature females of this species the digestive tract is completely absent.

In some species of *Ascaridata* a ventriculus is present between the oesophagus and the intestine; the ventriculus may be elongated and bent into the shape of a figure 3 or it may have one or several blind appendages. The form of the oesophagus, as well as that of the entire anterior part of the digestive tube, is of great importance in the systematics of nematodes. The middle and the posterior intestine, which is usually a short rectum, form a straight tube. In males both the intestine and the male duct open into the cloaca, which itself opens on to the ventral side of the tail. In females the intestine opens on to the ventral side of the tail through the anus which may be terminal.

The excretory system of nematodes consists of two symmetrical canals, which originate in the posterior part of the body and run along lateral sides. These two canals unite to form one which opens through the excretory pore on the ventral surface near the anterior end of the body. In some forms, mainly those which are free-living, the excretory system consists of a single glandular cell, the duct of which opens to the outside through a ventral pore.

The nervous system is composed of nerve trunks running in the body and circular commissures which bind the longitudinal trunks into a single system. The most typical circular nerve commissure surrounds the oesophagus and is the so-called "nerve ring". Anterior and posterior to the nerve ring is a complicated system of ganglial cells. Their clusters, or ganglia, are connected by numerous nerve cords to the external sense organs, i.e., the cephalic, cervical and caudal papillae.

As stated previously, nematodes are dioecious and the males and females may usually be easily differentiated by their appearance, the females normally being larger than the males.

The male genital organs are usually unpaired and consist of a thin, gradually widening tube, of which the anterior filiform part is the testis. This leads into the vas deferens which is considerably wider in diameter. The vas deferens widens in its conical part and forms the seminal vesicle, which runs into a short muscular ejaculatory duct. This joins into the ventral part of the rectum, forming the cloaca. Near the opening of the cloaca are found the accessory genital organs which aid in copulation, namely, the spicules, gubernaculum, telamon, copulatory bursa, cuticular alae, genital sucker, papillae etc. The presence or absence and form of these accessory structures are of great importance in nematode systematics.

The spicules are elongated chitinized structures located in the cloaca and partly in the terminal part of the male genital tube. By means of protractor and retractor muscles they may be moved up and down and their main purpose is to help to keep the vulva open during copulation. Two spicules are present in most groups, e.g. *Spirurata, Filariata, Strongylata* and *Ascaridata* and one in others, e.g. *Trichocephalata* and *Dioctophymata*. Some species possess, in addition to the spicules, another accessory chitinized organ, the gubernaculum which lies dorsal to the spicules and directs their movements.

The copulatory bursa is characteristic of the *Strongylata*. It is a cuticular membrane which envelops the tail end of the male. The bursa is divided into two large lateral lobes and a small dorsal one which is absent in some forms. The bursa is supported by rays, the arrangement and configuration of which are characteristic of individual genera and species of *Strongylata*.

The tail end of the male of *Capillaria* is usually provided with a cuticular membrane supported by two lateral processes.

The caudal alae are situated on the lateral sides of the posterior end of the male and are characteristic of *Spirurata*, some *Oxyurata* and *Ascaridata*. Between the caudal alae on the ventral side of the body lie the genital papillae. These are usually arranged in two rows and may be symmetrical or asymmetrical. The papillae may be sessile, subsessile or pedunculate depending on the species. Pedunculate papillae support the caudal alae.

The species of the subclass *Secernentea* have pore-shaped openings of one pair of sense organs—the phasmids—on the tail of both sexes. Phasmids are similar in structure to the paired amphids on the anterior end, while their nerve endings are also connected with a gland cell. In most of the parasitic forms of the subclass *Adenophorea* these caudal glands are lacking. The presence or absence of cervical papillae (deirids) is also of importance.

The female reproductive system consists of two long coiled tubes of which one may be reduced in some species. The anterior part of each tube is an ovary; this is followed by the oviduct which runs into the uterus. The two uteri unite in one muscular canal, termed the vagina, which opens on the ventral side of the body via the female genital opening, or vulva which may be situated in the mid-body or near the posterior or anterior end, at varying levels either projecting beyond the surface or in a depression. Although most nematodes are oviparous some are viviparous, giving birth to larvae. The eggs usually have a many-layered shell with either a smooth or rough external surface and the poles of the eggs may be covered by a characteristic operculum or plug.

There are great differences in the life-cycles of nematodes. Some develop without changing the host (homoxenous nematodes or geohelminths), while others have an obligatory change of host during their life-cycle (heteroxenous nematodes or biohelminths). The heteroxenous nematodes are divided into two groups, according to their life-cycle, one including those species which employ one intermediate host (e.g., *Spirurata, Filariata)* and the other including those which utilize two intermediate hosts (some *Dioctophymata* and *Ascaridata*). Insects and crustaceans serve as intermediate hosts for most nematodes.

Sometimes, in addition to the intermediate hosts, the reservoir (paratenic) hosts participate in the life-cycle and these may be *Oligochaeta*, arthropods or vertebrates—especially fishes. In *Syngamidae* the life-cycle may be direct or indirect—with or without changing the host. When direct, the reservoir (paratenic) hosts are employed (mostly soil *Oligochaeta*). Most nematodes parasitizing fish-eating birds are heteroxenous.

As far as the scheme of classification of the class *Nematoda* is concerned, the present authors of various opinions. The schemes proposed by different authors are not consistent, differing both in the number of taxonomic units and in their grouping.

In the present work, for division of the class into higher taxa we have followed the scheme proposed by Shults & Gvozdev (1970) in their monograph "Osnovy obshchey gelmintologii" (Vol. I). For the lower taxa we have used the data included in the respective volumes of the monograph "Osnovy nematodologii" (Vol. 1—22, 1949—1969, Izd. AN SSSR "Nauka"), the compendium by Yamaguti ("Systema helminthum", III, 1961; ,,The nematodes of vertebrates", New York—London) and papers published more recently than these two monographs which deal with the systematics of individual small groups of nematodes.

In the book by R. S. Shults and E.V. Gvozdev, referred to above, the following scheme of Nematoda was accepted (we mention only the taxa which include forms parasitizing animals).

Class *Nematoda*
Subclass *Adenophorea* (= *Aphasmidia*)
   Order *Trichocephalida*
    Suborder *Trichocephalata*
   Order *Dioctophymida*
    Suborder *Dioctophymata*

Subclass *Secernentea* (= *Phasmidia*)
   Order *Rhabditida*
    Suborder *Rhabditata*
    Suborder *Strongylata*
   Order *Ascaridida*
    Suborder *Ascaridata*
    Suborder *Oxyurata*
   Order *Spirurida*
    Suborder *Spirurata*
    Suborder *Filariata*
    Suborder *Camallanata*

The fish-eating birds of the Palaearctic Region have been recorded as being parasitized by representatives of all those systematic groups (suborders) of nematodes mentioned above

Systematic Part

According to available data, the fish-eating birds of the Palaearctic Region are parasitized by 148 species of nematodes, representatives of 62 genera and 23 families, belonging to all the orders and suborders mentioned above.

KEY TO THE SUBCLASSES, ORDERS AND SUBORDERS OF NEMATODES

1 Phasmids (lateral tail glands) and cervical papillae absent. Males with one spicule . . . . (Subclass *Adenophorea* = *Aphasmidia*) . . 2
— Phasmids and cervical papillae present (the latter indistinct in some species). Males with two spicules . . . . . . (Subclass *Sercernenta* = *Phasmidia*) 3
2 Body thin, filiform. Male tail with cuticular membrane at end. Vulva near end of oesophagus . . . . Order *Trichocephalida*; suborder *Trichocephalata*
— Body thick, massive. Male tail with muscular bursa. Vulva near anus . . . . Order *Dioctophymida*; suborder *Dioctophymata*
3 Mouth opening surrounded by two lateral lips or without lips. Cuticular shields present around mouth in some forms. Oesophagus divided into an anterior muscular and a posterior glandular part . . . (Order *Spirurida*) 4
— Mouth opening surrounded by three or six lips, buccal capsule present in some species. Oesophagus simple, not divided into muscular and glandular parts 6
4 Mouth opening with two lips. Pharynx present. Vulva in middle or posterior part of body, distinct. Parasites of digestive tract . . . . Suborder *Spirurata*
— Mouth opening without lips or surrounded by cuticular shield. Pharynx absent. Vulva in anterior part of body. Parasites of tissues or tissue spaces 5
5 Mouth opening without lips. Glandular part of oesophagus without swelling. Tail tip blunt and rounded in both sexes. Males somewhat smaller than females . . . . Suborder *Filariata*
— Mouth opening surrounded by cuticular shields. Glandular part of oesophagus with distinct swelling. Tail tip conical in both sexes. Males less than one-tenth the size of females . . . . Suborder *Camallanata*
6 Six, 3 or 2 lips usually present but sometimes lacking. Corona radiata present or absent. Oesophagus club-shaped, rhabditiform or filariform. Tail end of male with or without cuticular bursa copulatrix. Parasites of plants, invertebrates and digestive tract and respiratory organs of vertebrates, also free-living forms . . . (Order *Rhabditida*) . . 8
— Three lips usually present (sometimes also interlabia). Corona radiata absent. Oesophagus with or without bulb. Typical bursa copulatrix of males not developed. Parasites of digestive tract of vertebrates and arthropods . . . (Order *Ascaridida*) 7
7 Oesophagus cylindrical, without distinct bulb and valvular apparatus in posterior part. Ventriculus between oesophagus and intestine present in some forms. Ventriculus and

intestine may be provided with caecal appendages. Parasitic mainly in small intestine
  . . . . Suborder *Ascaridata*
— Oesophagus with bulb and valvular apparatus in posterior part. Ventriculus between oesophagus and intestine absent. Parasitic mainly in caecum and large intestine . . . .
  . . . . Suborder *Oxyurata*
8 Heterogonic, parasitic forms parthenogenetic, hermaphrodite or dioecious. Oesophagus in parasitic forms long, narrow, somewhat enlarged posteriorly. Males of free-living generation without typical bursa copulatrix . . . . Suborder *Rhabditata*
— Not heterogonic, parasitic forms dioecious. Oesophagus club-shaped. Tail end of male with typical cuticular bursa copulatrix . . . . Suborder *Strongylata*

Subclass *Adenophorea* (Linstow, 1905; Chitwood, 1950)

The main differentiating feature of the representatives of this subclass is the absence of both phasmids (the glands situated on lateral sides of tail end of male and female) and cervical papillae, but there are also other characteristics of various taxa in this subclass.

Most *Adenophorea* are free-living forms. Those species which have adapted to a parasitic life are divided into two orders, *Trichocephalida* and *Dioctophymida*. Each of these orders is represented by a single suborder.

Fish-eating birds of the Palaearctic Region have been reported as hosts of representatives of both suborders *(Trichocephalata* and *Dioctophymata)*.

SUBORDER *TRICHOCEPHALATA* RAILLIET & HENRY, 1913

Nematodes with slender, filiform body, widening gradually posteriorly. Males considerably smaller than females. Mouth small, surrounded by indefinite lips. Oesophagus divided into two portions: shorter anterior muscular part and longer posterior glandular part. Glandular oesophagus surrounded by longitudinally attenuated rows of moniliform cells (stichocytes). Genital apparatus of males and females single. One spicule present or absent, included in long tubular spicule sheath, which may arise some distance from the body of the spicule. Gubernaculum absent. Vulva at the level of border between oesophagus and gut. Eggs with a plug at each pole. Development direct or indirect. When indirect, soil oligochaetes serve as intermediate hosts.

This suborder parasitizes all classes of vertebrates but only one family, the *Capillariidae*, is found in birds.

Family *Capillariidae* Neveu-Lemaire, 1936

Cuticle nearly always with longitudinal bacillary bands. In males, one more or less pseudochitinized spicule may be present, with smooth or spinous spicular sheath, able to invaginate or evaginate. Small membranous alae often present at posterior end, forming a pseudobursa which is supported by small papillae. Female possesses a vulva which may have prominent margins or a vulvar appendage of varying form. Eggs usually barrel-shaped, with operculum at each pole and with smooth or variously ornamented surface. Parasitic in various organs of both cold-blooded and warm-blooded vetebrates. Life-cycle direct or with intermediate host.

Nematodes of this family parasitizing fish-eating birds belong to two genera, *Capillaria* and *Thominx*. In the Palaearctic Region this group of hosts is parasitized by 8 species.

KEY TO THE GENERA OF THE FAMILY *CAPILLARIIDAE*

1	Spicule sheath aspinous	*Capillaria*
—	Spicule sheath spinous	*Thominx*

Genus *Capillaria* Zeder, 1800

Body filiform, with one or more bacillary bands (dorsal, ventral or lateral). Male: anus terminal or subterminal. Spicule thin, long, may be pseudochitinized. Spicule sheath smooth, sometimes with transverse or longitudinal striations, wrinkles or undulations. Posterior end of body usually with pseudobursa. Female: vulva situated near termination of oesophagus, sometimes with cuticular appendages.

Type species *Capillaria obsignata* Madsen, 1945, parasite of *Columbiformes*, *Galliformes* and others.

Only five of the many species of this genus have been found in fish-eating birds in the Palaearctic Region.

KEY TO THE SPECIES OF THE GENUS *CAPILLARIA*

1	Female without vulval appendage	2
—	Female with vulval appendage	*C. mergi*
2	Spicule longer than 2 mm (2.20—2.30 mm)	*C. ryjikovi*
—	Spicule shorter than 1.5 mm	3
3	Spicule shorter than 1 mm (0.63—0.72 mm)	*C. podicipitis*
—	Spicule longer than 1 mm (1.15—1.32 mm)	*C. carbonis* (*C. herodiae*)

Note: The species *C. herodiae* Boyd, 1966 cannot be distinguished unequivocally from *C. carbonis* (Rudolphi, 1819). There is a possibility of synonymy.

Capillaria carbonis (Rudolphi, 1819)

Fig. 1
Hosts: *Gavia arctica, Podiceps nigricollis, P. griseigena, Phalacrocorax carbo, Ph. pygmaeus, Pelecanus onocrotalus, P. crispus, Ardea cinerea, A. purpurea, Egretta alba, E. garzetta, Plegadis falcinellus, Nycticorax nycticorax, Larus canus, L. genei, L. argentatus, Sterna hirundo, Chlidonias nigra* and *Gelochelidon nilotica*.

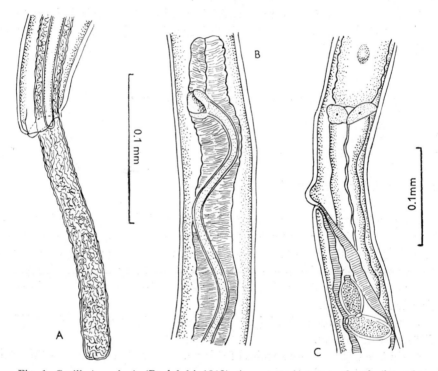

Fig. 1. *Capillaria carbonis* (Rudolphi, 1819). A — posterior part of male (lateral view); B — region of proximal end of spicule; C — region of vulva (lateral view). After Sergeeva (1969).

Localization: intestine.
Distribution: Europe (Austria, U.S.S.R. — Ukraine, Dagestan, Volga delta, Sea of Azov), Asia (U.S.S.R. — Azerbaijan, Kirghizia, East Siberia).

Description: Long thin nematodes, anterior end thinner than posterior. Maximum width at middle of body. Females larger than males.
Male: Body length 8.8—12 mm, body width in head region 0.01—0.014 mm,

at level of termination of oesophagus 0.041—0.042 mm, at tail end 0.035 to 0.036 mm. Oesophagus 5.17—6.16 mm long. Opening of cloaca 0.012 to 0.013 mm from posterior end. Spicule 1.15—1.32 mm long, 0.019—0.02 mm wide. Proximal end of spicule widened, distal end pointed. Spicule sheath 1.0—1.6 mm long, 0.018—0.02 mm wide. Pseudobursa forms four lobes.
Female: Body length 12—23 mm, body width at anterior end 0.01—0.013 mm, at level of termination of oesophagus 0.02—0.04 mm, at posterior end

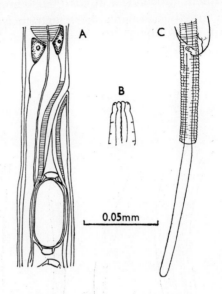

Fig. 2. *Capillaria herodiae* Boyd, 1966. A — region of vulva (lateral view); B — anterior end; C — posterior end of male (lateral view). After Boyd (1966).

0.04—0.06 mm. Oesophagus 7.0—10.2 mm long. Vulva 8.0—12.5 mm from anterior end of body. Eggs 0.042—0.048 × 0.018—0.022 mm.

This species was described as *Trichosomum carbonis* and recovered from *Phalacrocorax* carbo in Central Europe. It was transferred to the genus *Capillaria* by Travassos (1915).

References: Ablasov & Chibichenko (1961, 1962); Ablasov, Iksanov & Chibichenko (1960); Daiya (1967 a); Diesing (1851); Dubinin (1938, 1954); Dubinin & Dubinina (1940)*; Dubinina & Serkova (1951); Ginetsinskaya (1951, 1952); Kosupko (1963); Kulachkova (1950); Nikolskaya (1939); Petrov & Chertkova (1950); Rudolphi (1819)*; Sailov (1965 a,b); Sergeeva (1969)*; Shakhtakhtinskaya (1959 a,b); Shigin (1954); Skrjabin (1915 a, 1923); Smogorzhevskaya (1964); Soloviev (1912); Sonin & Larchenko (1974); Sprehn (1932); Stossich (1890c)*; Travassos (1915); Turemuratov (1963a); Zhatkanbaeva (1964, 1965).

Capillaria herodiae Boyd, 1966

Fig. 2
Host: *Ardea cinerea*.
Localization: small and large intestine.
Distribution: Asia (U.S.S.R. — Far East); outside the Palaearctic Region, in the U.S.A.

Description:
Male: Body length 8.3 mm, body width 0.032—0.040 mm. Nerve ring 0.090 mm from anterior end. Oesophagus 3.7 mm long. Caudal alae small, provided with one small papilla. Spicule sheath without spines, but with delicate annulations. Spicule 1.25 mm long with faint cross striations.
Female: Body length 15.3 mm, body width 0.039 mm. Nerve ring 0.082 mm from anterior end. Oesophagus 6.7 mm long. Anus subterminal, tail 0.010 mm. Vulva without appendage, situated 0.030 mm posterior to junction of oesophagus with intestine. Vagina short, extremely muscular with cutitular lining. Eggs oval, 0.046—0.052 × 0.015—0.026 mm.

The description was based on specimens recovered from *Ardea herodias* in the U.S.A.

References: Boyd (1966)*; Smetanina (1972).

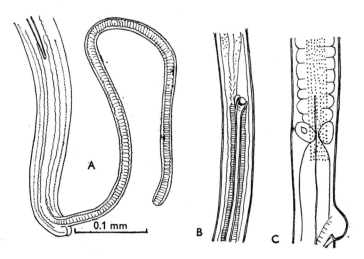

Fig. 3. *Capillaria mergi* Madsen, 1945. A — posterior end of male (lateral view); B — region of proximal end of spicule; C — region of vulva (lateral view). After Czaplinski (1962).

Capillaria mergi Madsen, 1945

Fig. 3
Hosts: *Gavia stellata, Podiceps ruficollis, Ardea cinerea, Mergus serrator, M. merganser, M. albellus, M. squamatus*, and other species of the order *Anseriformes*.
Localization: caeca and rectum, rarely small intestine.
Distribution: Europe (Denmark, Poland, G.D.R., Austria, U.S.S.R. — central regions), Asia (U.S.S.R. — Azerbaijan, Central and East Siberia, Far East).

Description:
Male: Body length 6.82—10.45 mm, maximum body width 0.048—0.057 mm. Oesophagus 3.6—4.9 mm long, with muscular part 0.360—0.440 mm long, 35 to 41 stichocytes. Spicule cylindrical, with enlarged proximal end, 1.1—1.25 mm long; diameter of its proximal end 0.017—0.020 mm, diameter of rest of spicular stem 0.008—0.010 mm. Spicule sheath smooth, slightly folded. Two very indistinct bacillary bands running along body, as in females.
Female: Body length 13.7—16.6 mm, body width 0.055—0.068 mm. Oesophagus 5.3—6.5 mm long, with muscular part 0.380—0.450 mm long, 37 to 44 stichocytes. Vulva situated short distance from oesophagus, provided with characteristic cuticular swelling, 0.050 mm long, 0.020 mm deep. Anus terminal. Eggs 0.048—0.053 × 0.023—0.028 mm, usually slightly asymmetrically oval, shells with small depressions giving the appearance of cross striation. Two bacillary bands run whole length of body, with a maximum width of 0.027 mm.

The description was based on specimens recovered from *Mergus serrator* and other hosts in Denmark.

Notes: In earlier papers some authors connected this species with *Capillaria* (= *Trichosoma*) *brevicole* (Rudolphi, 1819) or *C.* (= *Trichosoma*) *anatis* (Schrank, 1790). Madsen described it as a separate species as late as 1945. This species was referred to by Eberth (1863), in part in papers by Dujardin (1845), Gurlt (1845), Diesing (1851), Linstow (1878, 1909) and most probably also by Mehlis (1831), Creplin (1846), Muehling (1898) and Stossich (1890). These papers dealt with *Capillaria* recovered from the intestine of hosts belonging to the genera *Mergus* and *Clangula*.

References: Alekseev (1970); Belogurov (1965); Creplin (1845—1846); Czaplinski (1962b)*; Daiya (1966*, 1967 a); Diesing (1851); Dujardin (1845)*; Eberth (1863)*; Gubanov & Daiya (1967); Gurlt (1845); Kurochkin, Ryzhikov & Gubanov (1961); Linstow (1878, 1909); Madsen (1945)*; Mehlis (1831); Muehling (1898); Oshmarin (1963); Oshmarin & Parukhin (1963); Ryzhikov (1963 a,b); Ryzhikov & Daiya (1967); Sailov (1966); Shigin (1957, 1959); Smetanina (1972); Sonin & Larchenko (1974); Stossich (1890 c)*; Tolkacheva (1967); Tsimbalyuk (1965); Vaidova (1965).

Capillaria podicipitis Yamaguti, 1941

Fig. 4
Hosts: *Podiceps cristatus, P. auritus, P. nigricollis* and *P. ruficollis*.
Localization: intestine.
Distribution: Asia (U.S.S.R. — Tajikistan, Far East), Japan.

Description: Thin, filiform whitish worms. Cuticle with two longitudinal bacillary bands.
Male: Body length 6.28—8.96 mm, maximum body width 0.044—0.055 mm.

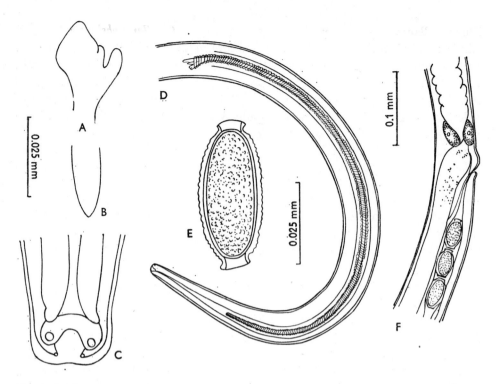

Fig. 4. *Capillaria podicipitis* Yamaguti, 1941. A — proximal end of spicule; B — distal end of spicule; C — posterior end of male (ventral view); D — posterior end of male (lateral view); E — egg; F — region of vulva (lateral view); After Borgarenko & Daiya (1972).

Oesophagus 3.68—4.32 mm long, 0.044—0.048 mm wide. Ratio of oesophageal length to body length 1:1.7—1:2. Spicule massive, well pseudochitinized, 0.63—0.72 mm long, with its proximal end slightly widened, in form of funnel with indentated margin, 0.011—0.016 mm wide; distal end pointed, smooth, 0.003—0.005 mm wide. Spicule sheath 0.01 mm wide, obliquely wrinkled, resembling a spiral encircling the spicule. Small lateral lobes present at posterior end of body, each bearing a stout papilla. Two small sessile papillae supporting pseudobursa. Cloaca subterminal, 0.016—0.017 mm from posterior end of body.

Female: Body length 6.80—10.08 mm, maximum body width 0.044—0.069 mm. Cuticular striations at intervals of 0.014 mm. Oesophagus 3.57—5.17 mm long, 0.044 wide. Ratio of oesophageal to body length 1 : 1.9 Nerve ganglion 0.048—0.072 mm from anterior end. Anus subterminal. Vulva without cuticular appendages, situated 3.57—5.17 mm from anterior end. Eggs 0.042 to 0.048×0.016—0.022 mm, with tuberculate surface, shell 0.0009 mm thick.

The description was based on specimens recovered from *P. ruficollis* in Japan.

Notes: Babaev (1970) reported, under the name *Capillaria obsignata* Madsen, 1945, the capillariids recovered from grebes in Turkmenia. In our opinion, these nematodes are *C. podicipitis*, a common parasite of members of the genus *Podiceps*.

The above description published by Borgarenko & Daiya (1972) differs slightly from the data given by Yamaguti (1941).

References: Alekseev & Smetanina (1972)*; Babaev (1970); Borgarenko (1970); Borgarenko & Daiya (1972)*; Oshmarin & Parukhin (1963); Parukhin (1964c); Smetanina (1972); Smetanina & Alekseev (1967); Yamaguti (1941)*.

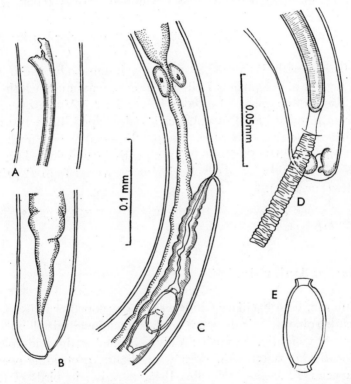

Fig. 5. *Capillaria ryjikovi* Daiya, 1972. A — region of proximal end of spicule; B — posterior end of female (lateral view); C — region of vulva (lateral view); D — posterior end of male (latero-ventral view); E — egg. After Borgarenko & Daiya (1972).

Capillaria ryjikovi Daiya, 1972

Fig. 5
Hosts: *Podiceps nigricollis*, *P. griseigena* and *P. ruficollis*.
Localization: small intestine, caeca.
Distribution: Europe (Czechoslovakia), Asia (U.S.S.R. — Far East).

Description: Slender, filiform, whitish worms. Cuticle with fine transverse striations. Bacillary bands not observed.

Male: Body length 18.43—24.20 mm, maximum body width 0.050—0.060 mm. Length of oesophagus 10.90—13.00 mm, maximum width 0.034 mm. Ratio of oesophageal to body length 1:1.9—2.0. Posterior end of body rounded, forming membranous pseudobursa supported by lateral papillae. Length of pseudobursa 0.011—0.023 mm, maximum width in dorsoventral position 0.026—0.031 mm. Spicule well pseudochitinized, with fine transverse striations, 2.20—2.30 mm long, slightly widened at proximal end which measures 0.023 to 0.026 mm in diameter; distal end rounded, 0.007—0.008 mm wide. Spicule sheath with distinct longitudinal and fine transverse striations, without spines. Cloaca opens subterminally 0.019 mm from posterior border of pseudobursa.

Female: Body length 17.90—32.30 mm, maximum body width 0.049 to 0.065 mm. Length of oesophagus 7.60—11.30 mm, maximum width 0.029 to 0.038 mm. Ratio of oesophageal to body length 1 : 2.7. Vulva opens 0.15 to 0.17 mm from termination of oesophagus and is shaped like a transverse slit with rounded non-elevated borders. Anus 0.015—0.019 mm from posterior end. Eggs 0.040—0.053 × 0.020—0.029 mm, with finely wrinkled surface.

The description was based on specimens recovered from *P. griseigena* in the U.S.S.R. (Far East).

References: Alekseev & Smetanina (1968)*; Baruš & Zajíček (1967)*; Borgarenko & Daiya (1972)*; Smetanina & Alekseev (1967).

Genus *Thominx* Dujardin, 1845

Body filiform, with bacillary bands weakly developed or indistinct. Male: Spicule sheath spinous, spicule more or less pseudochitinized but in some species only very slightly. Female: Vulva situated at the mid-body, sometimes with cuticular outgrowth. Eggs with slightly protruding opercula.

Type species *T. manica* Dujardin, 1845, parasitic in birds of the order *Passeriformes*.

This genus comprises many species of which only three have been found parasitizing fish-eating birds of the Palaearctic Region.

KEY TO THE SPECIES OF THE GENUS *THOMINX*

1 Parasites of buccal cavity and oesophagus 2
— Parasites of intestine . . . . *T. anatis*
2 Spicule sheath less than 0.5 mm long (about 0.208 mm) . . . . *T. spirale*
— Spicule sheath more than 0.5 mm long (0.76—0.98 mm) . . . . *T. contorta*

Thominx anatis (Schrank, 1790)

Fig. 6A-C
Hosts: *Podiceps griseigena, Mergus merganser, M. serrator, Larus argentatus, L. genei, L. minutus, L. ridibundus, Hydroprogne tschegrava, Sterna sandvicensis, S. albifrons* and very often birds of the order *Anseriformes*.
Localization: caecum, small intestine and rectum.
Distribution: Europe (England, U.S.S.R. — Ukraine), Asia (U.S.S.R. — Azerbaijan, Kirghizia). Outside the Palaearctic Region in Cuba.

Description:
Male: Body length 8.2—12.2 mm, body width 0.045—0.051 mm. Oesophagus 4.6—5.38 mm long, with muscular part 0.300—0.360 mm long; 37—45 (average 39) stichocytes between 0.074 and 0.17 mm long (average 0.126 mm). Spicule 0.91—1.59 mm long, with enlarged proximal part 0.015—0.024 mm

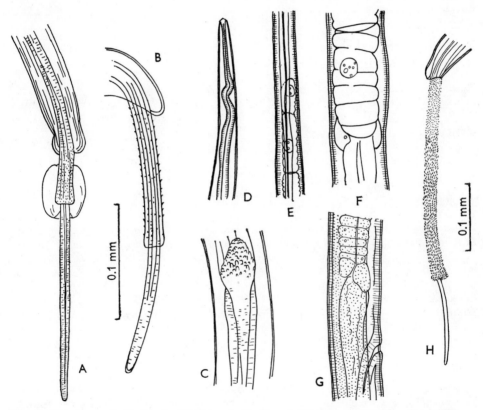

Fig. 6. *Thominx anatis* (Schrank, 1790) — A, B, C and *Thominx contorta* (Creplin, 1839) — D, E, F, G, H. A — posterior end of male (ventral view); B — posterior end of male (lateral view); C — region of proximal end of spicule; D — anterior end; E — junction of muscular and glandular part of oesophagus; F — junction of oesophagus and intestine; G — region of vulva (lateral view); H — posterior end of male (lateral view). After Czaplinski (1962).

wide, median part to distal end 0.007—0.012 mm in diameter. Spicule sheath provided with very small spines, barely visible when the sheath is within the male body.

Female: Body length 18.0—20.7 mm, body width 0.056—0.062 mm. Oesophagus 6.41—7.05 mm long, with 37—44 stichocytes similar in shape and size to those of male. Vulva smooth, provided with very small cuticular swelling, situated 0.026—0.140 mm from termination of oesophagus. Fissure-like genital pore lies transversely. Anus terminal. Eggs measure, without membranous appendages on the poles, 0.055—0.061 × 0.023—0.031 mm; surface of egg shell with numerous small depressions.

The species was described as *Trichocephalus anatis* recovered from *Anas querquedula* in Europe. It was transferred to the genus *Thominx* by Skrjabin & Shikhobalova (1954).

Biology: On the basis of incomplete experiments, Gagarin, 1951 (in Skrjabin, Shikhobalova & Orlov, 1957) presumed indirect development in this species (earthworms may serve as intermediate hosts).

Note: An historical survey and systematic analysis of this species is available in papers by Madsen (1945) and Skrjabin et al. (1957).

References: Ablasov & Chibichenko (1961, 1962); Czaplinski (1962 b)*; Jennings & Soulsby (1956); Leonov (1958); Madsen (1945)*; Sailov (1962, 1965 b, 1970); Shakhtakhtinskaya (1959 a,b); Skrjabin (1923); Skrjabin, Shikhobalova & Orlov (1957)*; Smogorzhevskaya (1964).

Thominx contorta (Creplin, 1839)

Fig. 6 D-H

Hosts: *Podiceps ruficollis, Phalacrocorax carbo, Botaurus stellaris, Larus ridibundus, L. argentatus, L. fuscus, L. canus, L. minutus, L. marinus, L. hyperboreus, L. crassirostris, L. schistisagus, L. genei, L. ichthyaetus, Hydroprogne tschegrava, Sterna aleutica, S. albifrons, S. hirundo, S. sandvicensis, Chlidonias nigra, Ch. leucoptera, Gelochelidon nilotica, Rissa tridactyla, Uria lomvia, Cepphus grylle* and frequently also in birds of the other orders.

Localization: mucous membrane of oral cavity and oesophagus.

Distribution: cosmopolitan, in fish-eating birds in Europe (England, Denmark, G.D.R., Poland, Czechoslovakia, Austria, Sweden, Norway, Rumania, Italy, U.S.S.R.—Estonia, Latvia, Ukraine, north and central regions), Asia (U.S.S.R.—republics of the Transcaucasus region and Middle Asia, Kazakhstan). Outside the Palaearctic Region, in North and South America and Australia.

Description: Cuticle usually clearly cross striated. Two bacillary bands present, often barely visible especially in preserved preparations, running almost the whole length of the body. Width of bacillary bands usually variable, one narrow (about 1/4 of the body diameter), the other broader (about 3/4 of the body diameter) but often of equal width in anterior part of body.

Male: Body length 12.0—16.2 mm, body width 0.052—0.078 mm. Oesophagus 3.9—4.8 mm long, with muscular part 0.42—0.57 mm long. 27—33 stichocytes, 0.095—0.230 mm long. Spicule 1.52—1.8 mm long, indistinct and with only the distal part, which is stronger than in other species, chitinized; diameter of distal end 0.003—0.006 mm. Because the spicule sheath is extensible, it is difficult to determine its exact length which can reach 0.76—0.98 mm. Distal end of everted sheath armed with elongated, thin, hair-like spines, spines of proximal part, which is usually enlarged, are thicker and shorter. Proximal end of sheath smooth.

Female: Body length 17.2—36.8 mm. Body width in vulvar region 0.090 to 0.120 mm. Oesophagus 4.87—7.1 mm long, with muscular part 0.340—0.795 mm long, 28—35 (usually 29—32) stichocytes which are slightly longer than in males and reach 0.105—0.270 mm. Pharyngeal lumen passing excentrically through the stichocytes. Distance between vulva and junction of oesophagus and intestine ranges between 0.030 and 0.099 mm. Cuticular swelling of vulvar region usually very small or absent. Bacillary band covers vulva in some specimens, or passes near vulva in others. Vagina strongly muscular, elongated, with a smooth lumen. Anus terminal. Eggs 0.050—0.069 × 0.025—0.029 mm.

The species was described as *Trichosoma contortum* recovered from *Larus canus* and other birds of different orders in Central Europe. It was transferred to the genus *Thominx* by Skrjabin & Shikhobalova (1954).

Biology: According to Cram (1936) this species has a direct life-cycle. The development of the eggs to the infective stage takes 35 to 40 days and the prepatent period is about 24 days.

Note: According to the literature two species of capillarids have been recovered from the oesophagus of fish-eating birds, especially gulls, namely, *T. contorta* (Creplin, 1839) and *Eucoleus laricola* (Vasilkova, 1930). *T. contorta* has been found more frequently and in a wide spectrum of definitive hosts, while *E. laricola* has been found only sporadically. López-Neyra (1947) assumed that the species *T. contorta* could be divided into separate species according to their relationships with the systematic groups of definitive hosts (e.g., *T. raillieti* for *Anseriformes*, *T. contorta* for *Passeriformes*, *T. laricola* for *Lariformes*). The large size variation in *T. contorta* (see Madsen, 1945, Gagarin, 1951, Cram, 1936, Czaplinski, 1962), as well as the morphological similarity of oesophageal capillarids from different hosts, lead to the conclusion that *T. raillieti* is a synonym of *T. contorta*. The species *E. corvicola* Vasilkova, 1930 was regarded by Baruš et al. (1972) as synonym of *T. contorta*. We agree that *E. laricola* cannot be differentiated from *T. contorta* on the basis of morphological criteria or measurements and it is, therefore, considered a synonym. Recently, Baruš (1974) has indicated that *Trichosoma pachyderma* Linstow, 1877 (= *E. pachyderma*) is also a synonym of *T. contorta*.

References: Ablasov & Chibichenko (1961, 1962); Akhumyan (1966); Babaev (1970); Bakke (1972); Bakke & Baruš (1976); Baruš (1964 c, 1974); Baruš, Ryšavý, Groschaft & Folk (1972*); Belogurov, Leonov & Zueva (1968); Bezubik (1956); Braun (1892); Cram (1936)*; Creplin (1839*, 1845—1846*); Creutz & Gottschalk (1969); Czaplinski (1962 b)*; Daiya 1967 a,b); Diesing (1851)*; Ellis & Williams (1973); Golikova (1959); Guildal (1964, 1966, 1968); Iksanov & Dikambaeva (1962); Jennings & Soulsby (1956, 1957, 1958); Kibakin (1965); Kosupko (1962, 1963); Kreis (1958); Krivonogova (1963); Krotov (1952); Kulachkova & Kochetova (1964); Kurochkin & Zablotsky (1961); Leonov (1958); Leonov & Belogurov (1963); Leonov & Shvetsova (1970); Linstow (1877 a,b, 1878, 1884, 1909); López-Neyra (1947 a,b*); Macko (1964 a,d); Madsen (1945)*; Mashtakov (1964); Michelson (1968); Muehling (1898); Okorokov & Tkachev (1969); Oshmarin (1963); Oshmarin & Paruhkin (1963); Parukhin & Truskova (1963); Pemberton (1963); Radulescu & Lustun (1967)*; Reimer (1973); Sergeeva (1969); Shakhtakhtinskaya (1953, 1959 a,b); Shigin (1961); Skrjabin (1923); Smetanina (1972); Smogorzhevskaya (1964); Smogorzhevskaya, Kornyushin, Iskova & Eminov (1965); Solonitsin (1928 a); Sonin & Larchenko (1974); Spasskaya (1949); Stossich (1890 c*, 1895); Threlfall (1965 b, 1967); Tsimbalyuk & Belogurov (1964); Vasilkova (1926, 1927, 1930); Vasilkova & Gushanskaya (1930)*; Volskis (1966, 1968); Yigis (1962).

Thominx spirale (Molin, 1858)

Fig. 7
Hosts: *Plegadis falcinellus* and *Platalea leucorodia*.
Localization: oesophagus.
Distribution: Europe (Italy, U.S.S.R. - Volga Delta, Ukraine), outside the Palaearctic Region, in Brazil.

Description: Body whitish in colour, cuticle with distinct transverse striations. Bacillary bands present and, according to Stossich (1890), the ventral band measures 3/4 of body diameter and the dorsal band only 1/3. Mouth slightly protruding, without ornamentation. Nerve ganglion situated 0.059 mm from anterior end of body.
Male: Body length 10.06 mm, body width 0.053—0.80 mm. Ratio of the anterior to posterior part of body 1 : 1.3. Oesophagus 4.54 mm long, including the muscular anterior part. Spicule weakly pseudochitinized, 0.41 mm long, average width 0.008 mm. Spicule sheath spinous, 0.208 mm long, 0.1 mm wide, spines numerous and more closely packed in the middle portion (measuring 0.048 mm) at a distance of 0.052 mm from posterior end of body, becoming widely dispersed proximally until they disappear. On terminal part of spicule sheath are 8 larger spines arranged in a crown. Posterior end of body bears two thick dorsolateral papillae interconnected by a cuticular membrane, and two small ventral papillae. Cloaca subterminal.
Female: Body length 11.50—17.75 mm, body width 0.04—0.128 mm. Ratio of anterior to posterior part of body 1 : 2.2—1 : 2.5. Oesophagus 4.17—4.23 mm long, including anterior muscular part 0.152 mm long. Vulva situated 0.136 mm from termination of oesophagus, its anterior margin protruding

slightly. Vagina very long, covered with small spines, which are more densely packed in the anterior part, as far as 0.072 mm from vulval opening. Eggs thick-shelled, 0.040—0.056 × 0.020—0.027 mm (according to Stossich 0.036 × 0.0198 mm). Posterior end of body blunt. Anus terminal.

This species was described from one female recovered from *Plegadis falcinellus* in Italy.

Notes: The incomplete original description was supplemented by Eberth

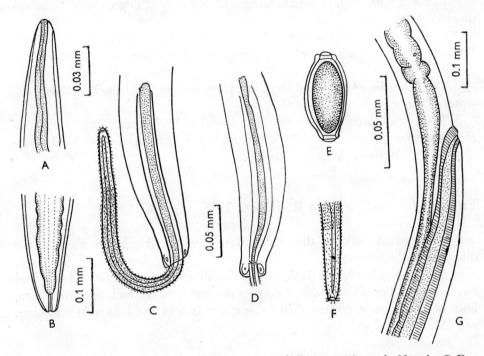

Fig. 7. *Thominx spirale* (Molin, 1858). A — anterior end; B — posterior end of female; C, D — posterior end of male (ventral view); E — egg; F — distal end of spicule sheath; G — region of vulva (lateral view). After Freitas & Almeida (1935).

(1863) and Stossich (1890). Later, this species was found in the same host in the U.S.S.R. by Dubinin (1938), who briefly described the females he recovered. López-Neyra (1947) concluded that the species *Capillaria venteli* Freitas & Almeida, 1935, found in Brazil in the hosts *Ajaja ajaia* and *Canchroma cochleare* was a synonymum of the species *T. spirale*.

References: Diesing (1861); Dubinin (1938)*; Dubinin & Dubinina (1940); Eberth (1863)*; López-Neyra (1947 b)*; Madsen (1945); Molin (1858 b*, 1861 b*); Parona (1894); Skrjabin, Shikhobalova & Orlov (1957)*; Smogorzhevskaya (1964); Stossich (1890 c)*.

SUBORDER *DIOCTOPHYMATA* SKRJABIN, 1927

Large nematodes with dense, transversely striated cuticle. Mouth opening surrounded by large papillae, arranged in two or three circles. Oesophagus simple, of more or less the same diameter throughout its length, usually very long. Posterior end of male ends in a muscular bell-shaped bursa. Spicule single, long, gubernaculum absent. Anus of female opens terminally. Vulva close to anus. Vagina very long. Genital apparatus of both sexes single. Eggs thick-shelled, with surface ornamented with knobs and depressions. Development indirect.

In the life-cycle of some species *(Eustrongylides)* two intermediate hosts (aquatic oligochaetes and fishes) are required. For other species fish serve as paratenic hosts.

The suborder includes two families, the parasites of birds, including fish-eaters, belonging to the *Dioctophymidae*.

The world fauna of birds has been found to be infected by 20 species of *Dioctophymata* (Karmanova 1968). Seven species have been recorded in fish-eating birds of the Palaearctic Region.

Family *Dioctophymidae* Railliet, 1915

The characteristics of the representatives of this family are given in the diagnosis of the suborder.

The family is divided into two subfamilies, namely, *Dioctophyminae* and *Eustrongylinae*. Parasites of birds belong to the latter, which includes two genera, both of which are represented in fish-eating birds in the Palaearctic Region.

KEY TO THE GENERA OF THE FAMILY *DIOCTOPHYMIDAE*

1 Anterior part of body armed with spines. Mouth surrounded by two circles of 12 papillae
. . . . *Hystrichis*
— Body smooth, without spines. Mouth surrounded by 12 to 18 papillae forming two to three circles, each of six papillae . . . . *Eustrongylides*

Genus *Eustrongylides* Jägerskiöld, 1909

Head end not swollen. Body wider in the middle region and attenuated towards extremities. Cuticle with transverse striations, thicker at both ends of body than in the mid-region. Mouth surrounded by 12 or 18 papillae arranged usually in two or three circles, each of six papillae. Oesophagus long, wider posteriorly. Bursa copulatrix of male muscular, thick-walled and without rib-like structures. Spicule long, needle-shaped. Posterior end of female blunt, vulva subterminal. Parasites of proventriculus of birds.

Type species *E. tubifex* (Nitzsch, 1819).

The genus includes 12 species of which 5 have been found in fish-eating birds of the Palaearctic Region.

KEY TO THE SPECIES OF THE GENUS *EUSTRONGYLIDES*

1 Bursa of male without deep notch on ventral margin 2
— Bursa of male with deep notch on ventral margin . . . . *E. excisus*
2 Papillae of outer circle very large, strongly salient, differing considerably in form and size from those of inner circle 3
— Papillae of outer circle smaller than those of inner circle or only slightly different in size 4
3 Papillae of inner circle with long slightly bent, spine-like process on top; papillae of outer circle large, digitiform; margin of male bursa without cuticular border . . . . *E. mergorum*
— Papillae of inner circle with rod-shaped short, straight spine on top; margins of male bursa with cuticular border . . . . *E. tubifex*
4 Papillae of inner and outer circle similar both in form and size, wider in middle and narrowed towards the base . . . . *E. africanus*
— Papillae of outer circle wider and flatter than those of inner circle . . . . . *E. sinicus*

Eustrongylides tubifex (Nitzsch, 1819)

Fig. 8

Hosts: *Gavia arctica, G. stellata, G. immer, Podiceps cristatus, P. ruficollis, Pelecanus crispus, Pel. onocrotalus, Phalacrocorax carbo, Ph. pygmaeus, Mergus albellus, M. merganser, M. serrator, Alca torda* and *Uria aalge*. Also recorded in birds of the family Anatidae even though not fish-eating.
Localization: wall of oesophagus and proventriculus.
Distribution: Europe (Italy, F.R.G., G.D.R., Bulgaria, Finland, U.S.S.R. — Ukraine, Volga Delta) and Asia (U.S.S.R.—Tajikistan, West and East Siberia, Far East.) Not reported outside the Palaearctic Region.

Description: Body fusiform, cuticle with transverse striations. Twelve cephalic papillae present. Height of papillae of inner circle 0.016 mm, diameter at base 0.011 mm.
Male: Body length 32—35 mm, maximum body width 1.56—2.0 mm. Vestibule 0.08—0.13 mm long. Oesophagus 9.3—11 mm long, maximum width 0.25—0.30 mm. Spicule 17.6—19.8 mm long. Bursa 0.272—0.35 mm long, well developed, bell-shaped with obliquely cut margins.
Female: Body length 35—44 mm, maximum body width 2.5—4.0 mm. Vestibule 0.08—0.14 mm long. Oesophagus 7.5—13.75 mm long, maximum width

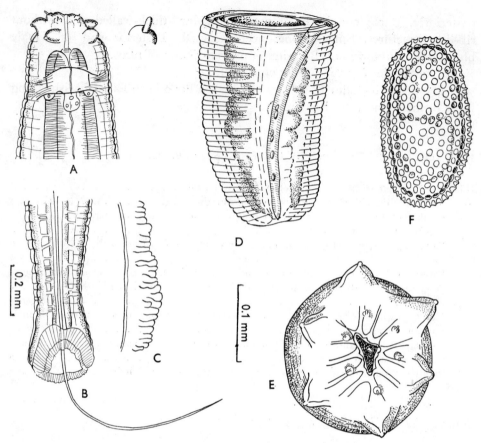

Fig. 8. *Eustrongylides tubifex* (Nitzsch, 1819). A — anterior end and papillae of inner circle; B — posterior end of male; C — cuticular ray of bursa margin; D — posterior end of female; E — anterior end (apical view); F — egg. After Karmanova (1968) — A, B, C, E and Jägerskiöld (1909). — D, F.

0.28—0.30 mm. Tail end blunt. Anus terminal, vulva close to anus. Eggs 0.06—0.08 × 0.03—0.44 mm, thick-shelled, with plugs at each pole, external surface of shell pitted.

The species was described under the name *Strongylus tubifex* from fish-eating birds and members of the *Anatidae* in Central Europe. It was transferred to the genus *Eustrongylides* by Jägerskiöld (1909).

Biology: The life-cycle of this species has not been studied. Larvae of *E. tubifex* have been found in fish *(Gobius* sp., *Rutilus rutilus)* from the Caspian Sea and in perch and true sturgeons from the Dnieper river.

Notes: Until Jägerskiöld's (1909) paper was published, the systematics of the nematodes which are now placed in the genus *Eustrongylides* was very confusing. It is therefore difficult to state which species *(E. tubifex, E. excisus* or

E. mergorum) are referred to in the earlier literature, as all three species may be found in the same species of definitive host.

References: Belogurov (1965); Belogurov & Smetanina (1965); Borgarenko (1970); Bremser (1824); Creplin (1845—1846); Daiya (1967 a*, b); Diesing (1851); Dubinin (1954); Dujardin (1845)*; Gubanov (1971); Gubanov & Daiya (1967); Gurlt (1845); Jägerskiöld (1909)*; Kambourov & Vasilev (1972); Karmanova (1968)*; Kontrimavichus & Bakhmeteva (1960)*; Krivonogova (1963); Leonov & Belogurov (1963); Parona (1894); Railliet (1895); Rudolphi (1809, 1819)*; Schneider (1866)*; Skrjabin (1923); Smogorzhevskaya (1962 a, 1964); Stossich (1899).

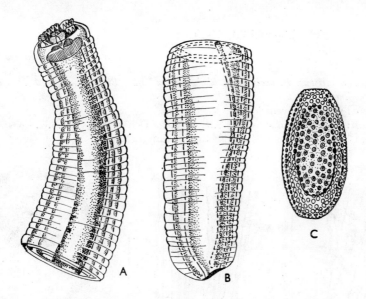

Fig. 9. *Eustrongylides africanus* Jägerskiöld, 1909. A — anterior end; B — posterior end of female; C — egg. After Jägerskiöld (1909).

Eustrongylides africanus Jägerskiöld, 1909

Fig. 9
Hosts: *Pelecanus onocrotalus, P. crispus* and *Platalea leucorodia*.
Localization: walls of proventriculus.
Distribution: Europe (U.S.S.R. — Volga Delta) and Asia (U.S.S.R. — Azerbaijan and Kazakhstan). Outside the Palaearctic Region in pelicans and *Ciconiiformes* from tropical Africa.

Description: Female only known. Body fusiform, of nearly the same thickness throughout length. Cuticle with annulations. 12 cephalic papillae, those of inner circle 0.056 mm tall and 0.056 mm wide at base, those of outer circle 0.048 mm tall and 0.096 mm wide at base. Body length 90—160 mm, maximum body width 1.5—2.5 mm. Width at head end 0.43—0.57 mm, at tail end

0.56 mm. Vestibule 0.08—0.10 mm deep. Nerve ring 0.16 mm from head end. Eggs oblong, oval, with plug at each pole and shell covered with shallow depressions.

The species was described from nematodes recovered from pelicans and *Ciconiiformes* from the Sudan.

References: Dubinin (1954); Dubinin & Dubinina (1940); Feyzullaev (1963 a, b); Jägerskiöld (1909)*; Karmanova (1968)*; Kasimov & Feyzullaev (1965); Panova (1927); Skrjabin (1915 a*, 1923).

Eustrongylides excisus Jägerskiöld, 1909

Fig. 10
Hosts: *Pelecanus crispus, P. onocrotalus, Phalacrocorax carbo, Ph. pygmeus, Ardeola bacchus* and *Nycticorax nycticorax*.
Localization: wall of proventriculus.
Distribution: Europe (Rumania, U.S.S.R. — Volga Delta); Asia (U.S.S.R. — Kazakhstan, Uzbekistan, China). Not reported outside the Palaearctic Region.

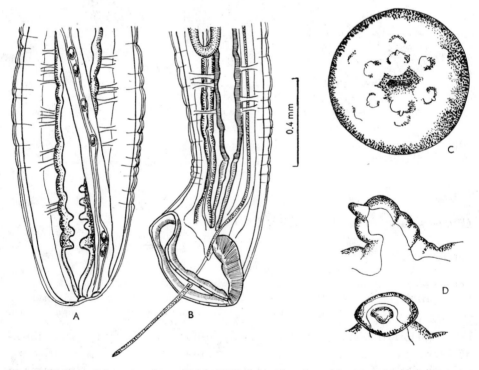

Fig. 10. *Eustrongylides excisus* Jägerskiöld, 1909. A — posterior end of female; B — posterior end of male; C — anterior end (apical view); D — papillae of inner circle. After Karmanova (1968) — A, B, D and Jägerskiöld (1909) — C.

Description: Body fusiform, widening gradually towards the middle. Cuticle transversely striated anteriorly and posteriorly, smooth in mid-body. Head end with 12 papillae arranged in two circles; 6 rounded papillae of inner circle each with finger-like projections at tip. Papillae of outer circle relatively flat, with wide base.

Male: Body length 50.0—56.0 mm long, maximum body width 1.5—1.6 mm. Width at level of outer circle of papillae 0.238 mm, at level of nerve ring 0.311—0.350 mm, at level of oesophagus 1.26 mm. Nerve ring 0.071—0.077 mm from anterior end. Oesophagus 6.0—6.68 mm long. Spicule very long (9.16—10.86 mm) and thin, with needle-like distal end. Bursa cup-shaped, with deep incision on ventral side. Inner surface of bursa covered with small papillae. Conical elevation (genital cone) present at base of bursa with cloaca at top. One row of papillae on margins of bursa which is 0.112—0.182 mm long, 0.54—0.63 mm wide.

Female: Body length 73.0—100.0 mm, maximum body width 2.10—2.25 mm, body width at level of outer circle of papillae 0.252 mm, at level of nerve ring 0.448 mm, at termination of oesophagus 1.54 mm. Nerve ring 0.196—0.252 mm from anterior end. Oesophagus 8.82—12.10 mm long. Eggs 0.067—0.072 × × 0.033—0.039 mm, oval, with operculum at each pole, shell thick, outer surface with shallow depressions.

The species was described by Jägerskiöld (1909) from specimens recovered from cormorants and obtained from the Vienna Museum. A more detailed redescription was published by Ciurea (1938) using specimens from cormorants from Rumania (Danube Delta).

Notes: According to Jägerskiöld (1909) the species *E. excisus* was known to Rudolphi (1819), who placed it in the species *Strongylus tubifex* Nitzsch, 1819. Stossich (1899), also incorrectly, regarded this species as *Hystrichis elegans* Olfers, 1816.

Biology: The life-cycle of this species was studied by Ciurea (1924, 1938), Dubinin (1949) and, in greater detail, by Karmanova (1968). According to Karmanova the development of the egg of *E. excisus* to the first-stage larva occurs in water and takes 21 to 30 days during summer in the Volga Delta. When eggs containing larvae are swallowed by the intermediate host, the larvae emerge and develop further in the oligochaetes *Lumbricullus*, *Tubifex* and *Limnodrilus*. The larvae penetrate first into the body-cavity and thence into the circulatory system of the intermediate host, where they moult twice during 18—49 days; development continues in the intermediate host for 60—70 days by which time the larvae are infective for the second intermediate hosts, usually benthic fish. In the second intermediate host the larvae are localized in the body-cavity where further growth and moulting occurs. The larvae become reddish in colour and capable of infecting the definitive host. At this stage of development paratenic hosts may also be involved. Among the latter are different species of predatory fish, frogs and reptiles. In the definitive host, the larvae become localized in the wall of the proventriculus and, after another moult (within 12 to 14 days) reach maturity.

References: Ciurea (1924, 1938); Chiriac (1965); Dubinin (1949, 1954); Dubinin & Dubinina (1940); Hoeppli, Hsü & Wu (1929)*; Jägerskiöld (1909)*; Karmanova (1958, 1968 a*b); Nikolskaya (1939); Panova (1927); Rudolphi (1819); Skrjabin (1915 a, 1923); Smogorzhevskaya (1962 a, 1964); Turemuratov (1962 a).

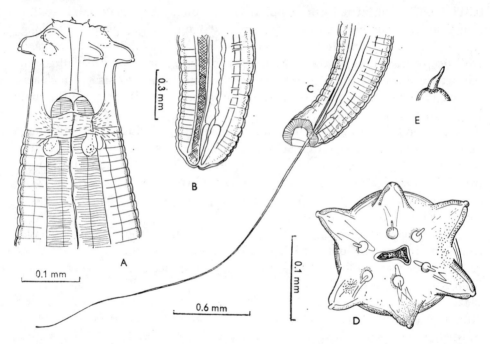

Fig. 11. *Eustrongylides mergorum* Rudolphi, 1809. A — anterior end; B — posterior end of female; C — posterior end of male; D — anterior end (apical view); E — papillae of inner circle. After Karmanova (1968).

Eustrongylides mergorum (Rudolphi, 1809)

Fig. 11
Hosts: *Gavia stellata, Podiceps cristatus, P. auritus, P. ruficollis, Ciconia nigra, Ardea cinerea, A. purpurea, Egretta garzetta, E. alba, Nycticorax nycticorax, Plegadis falcinellus, Mergus albellus, M. merganser, M. serrator, Larus ridibundus, Alca torda, Cepphus grylle* and *Uria lomvia*. Sometimes also in *Anatidae* and *Charadriiformes*, not fish-eating.

Localization: walls of proventriculus.
Distribution: Europe (England, Italy, Austria, G.D.R., Rumania, U.S.S.R. — Ukraine, North regions, the Volga region) and Asia (U.S.S.R.—Azerbaijan, Kazakhstan, Tajikistan, West and East Siberia, Far East; Japan). Outside the Palaearctic Region, in fish-eating birds of north-western regions of the Atlantic.

Description: Body fusiform, cuticle swollen and transversely striated at both extremities. Head end with 12 papillae; those of outer circle are longer and finger-like.
Male: Body length 18.0—35.5 mm, maximum body width 0.5—2.0 mm. Oesophagus 6.0—13.0 mm long, maximum width 0.24—0.40 mm. Nerve ring 0.14 mm from head end. Bursa 0.275—0.30 mm long with margins more or less dentate. Small cloacal cone present at base of bursa. Spicule 7.7—8 mm long.
Female: Body length 25—36 mm, maximum body width 2.0—3.2 mm. Nerve ring 0.15—0.20 mm from head end. Length of oesophagus 7.5—12 mm, maximum width 0.26—0.33 mm. Posterior end of body bluntly rounded. Vulva at level of anus. Eggs 0.06—0.08 ×0.03—0.05 mm, oval, with thick alveolate shell.

The species was described as *Strongylus mergorum* from specimens from *M. albellus* and *M. serrator* in Italy. It was transferred to the genus *Eustrongylides* by Cram (1927).
Biology: The life-cycle of *E. mergorum* has not been studied. Larvae of this species have been found in different species of fish (Schneider 1866; Linstow 1899; Dubinin 1949; and others).
Notes: Rudolphi (1809) described the species *Strongylus mergorum*, united it with the species *S. papillosus* Rudolphi, 1802 from nutcracker *(Nucifraga caryocatactes)* and in 1819 included *S. mergorum* in the species *S. tubifex*. *S. mergorum* has been referred to in the literature under the names *S. papillosus* and *S. tubifex*. In 1816 Olfers described the species *S. elegans*. Jägerskiöld (1909) recognized *S. elegans* as an independent species and transferred it to the genus *Eustrongylides*. Cram (1927) correctly stated that priority should be given to the name „*mergorum*". We agree with this view, as does Karmanova (1968).

References: Agapova & Zhatkanbaeva (1971); Alekseev (1970); Baylis (1928); Belogurov (1965); Belopolskaya (1959, 1963); Borgarenko (1972); Cram (1927)*; Daiya (1967 a*, b); Diesing (1851); Dubinin (1938, 1949, 1954); Dubinin & Dubinina (1940); Dubinina (1937); Golovin (1964); Gubanov & Daiya (1967); Karmanova (1958, 1968 a*, b); Kosupko (1963); Linstow (1877 a*, 1899 b); Maksimova (1966); Markov (1937, 1941); Nikolskaya (1939); Oshmarin (1963); Oshmarin & Parukhin (1963); Rudolphi (1809*, 1819); Ryzhikov (1963 a, b); Sailov (1966); Schneider (1866); Serkova (1948); Shigin (1957, 1959); Skrjabin (1915 a, 1923); Smetanina (1972); Smogorzhevskaya (1964); Sonin & Larchenko (1974); Tolkacheva (1967 a*, b); Vaidova (1965); Yamaguti (1935, 1941*).

Eustrongylides sinicus Wu & Liu, 1943

Hosts: *Ardea cinerea* and *Egretta garzetta*.
Localization: walls of proventiculus.
Distribution: Asia (U.S.S.R. — Far East; China).
Description: Body fusiform, cuticle with distinct transverse striations towards

extremities. Twelve cephalic papillae present; in the outer circle the lateral papillae are 0.012—0.014 mm tall and 0.04—0.044 mm in diameter and the submedian papillae are 0.01 × 0.038—0.04 mm; in the inner circle the lateral papillae are 0.019—0.26 mm tall and 0.03—0.036 mm in diameter and the submedian papillae are 0.017—0.024 × 0.072—0.032 mm. There is a row of small lateral papillae on both anterior and posterior ends.

Male: Body length 53.2—57.4 mm, maximum body width 1.0—1.2 mm. Vestibule 0.12 mm deep. Oesophagus 11.9—13.8 mm long. Nerve ring 0.19—0.20 mm from head end. Caudal bursa 11.38—11.73 mm long and 0.60—0.65 mm wide, with entire margin. Spicule simple, very long and thin. Female: Body length 41.9—65.4 mm, maximum body width 1.2—2.1 mm. Vestibule 0.12—0.18 mm deep. Oesophagus 13.8—15.5 mm long. Nerve ring 0.17—0.21 mm from head end. Anus terminal. Vulva at level of anus. Eggs 0.071—0.83 × 0.042—0.046 mm, with thick, striated shell.

The species was described from nematodes from *E. garzetta* in China. (No illustrations of this species have appeared in the literature.)

References: Karmanova (1968)*; Smetanina (1972); Wu & Liu (1943)*.

Genus *Hystrichis* Dujardin, 1845

Head end somewhat swollen, sometimes globular. Mouth surrounded by 12 salient papillae arranged in two circles. Anterior end of body armed with several rows of spines which, in 3rd- and 4th-stage larvae, may extend to the middle of the body length. Bursa copulatrix of male more or less marked. Spicule long and straight. Anus of female terminal, vulva at anal level. Eggs with plugs at both poles, shell with indentations. Parasitic in wall of proventriculus of aquatic birds.

Type species *H. tricolor* Dujardin, 1845.

The genus includes five species; two of them parasitize fish-eating birds of the Palaearctic Region.

KEY TO THE SPECIES OF THE GENUS *HYSTRICHIS*

1 Head end with more than 20 rows of large spines . . . . *H. tricolor*
— Head end with one row of large spines . . . . *H. coronatus*

Hystrichis tricolor Dujardin, 1845

Fig. 12 A-C
Hosts: *Egretta garzetta, Plegadis falcinellus, Mergus merganser* and *M. serrator*. Common parasite of domestic and wild ducks, geese and swans. Found also in *Charadriiformes* and *Ralliformes*.

Localization: connective tissue capsules in the wall of proventriculus.
Distribution: Europe (Italy, Bulgaria, G.D.R., U.S.S.R. — Ukraine), Asia (U.S.S.R. — Azerbaijan). In non-fish-eating birds of different Palaearctic and North American regions.

Description: Body large. Spines covering the swollen anterior end directed posteriorly and arranged in 28—34 alternating rows, 12—32 spines in each row. Largest spines are on swollen part of head end, number and size diminishing gradually posteriorly.
Male: Body length 27.3—30.4 mm, maximum body width 1.78—2.45 mm. Nerve ring 0.09—0.11 mm from head end. Length of oesophagus 7.95—8.72 mm, maximum width 0.215—0.248 mm. Spicule 1.06—1.18 mm long, with proximal end somewhat swollen.
Female: Body length 25—105 mm, maximum body width 3—5 mm. Nerve ring 0.12—0.16 mm from head end. Oesophagus 6.02—11.2 mm long, maximum width 0.49—0.62 mm. Eggs 0.08—0.088 × 0.042—0.048 mm.

The species was described from domestic and wild ducks in France.
Biology: The life-cycle of this species has been studied by Karmanova (1956, 1959, 1960 b, 1961, 1962, 1968). Under favourable conditions, larvae are formed in the eggs in 30 days. As soon as the eggs are swallowed by fresh-water oligochaetes *(Criodrillus lacuum, Allolobophora dubiosa* etc.), the larvae hatch, migrate in the body-cavity of the hosts and then move into the abdominal blood vessel where they moult twice. They reach the infective stage within 180 to 200 days of ingestion by the oligochaetes. In the definitive hosts (ducks) they reach the adult stage in 25 to 30 days.
Notes: According to Jägerskiöld (1909) a few nematodes identified by previous authors as *Strongylus tubifex* are, in fact, *H. tricolor*. The species *Strongylus anatis* Rudolphi, 1809, *S. tubifex* Nitzsch, 1819 and *Spiroptera tadorna* Bellingham, 1844 are synonyms of *H. tricolor*.

Karmanova (1959, 1960 a, 1968) studied the morphology of *H. tricolor* at different stages of development and, on the basis of these studies, synonymized with it the species *H. neglectus* Jägerskiöld, 1909, *H. orispinus* Molin, 1858, *H. varispinosus* Jägerskiöld, 1909 and *H. wedli* Linstow, 1879. We agree with this view.

References: Dujardin (1845)*; Feyzullaev (1963 a); Jägerskiöld (1909); Kambourov & Vasilev (1972); Karmanova (1956, 1965, 1958, 1959, 1960 a, b, 1961, 1962, 1968 a*); Molin (1861 a, b); Parona (1894); Skrjabin (1923); Smogorzhevskaya (1964).

Hystrichis coronatus Molin, 1861

Fig. 12 D, E
Host: *Mergus merganser*.
Localization: walls of proventriculus.
Distribution: Europe (Italy).

Description: Female only known. Body length 27.0 mm, maximum body width 3.0 mm. Circle of spines present on head end.

Notes: The species was described by Molin (1860) as *Hystrichis* sp. and named *H. coronatus* by him in 1861. In the same year, Diesing (1861) named it *H. mergimerganseris*. These nematodes have never been described adequately and the

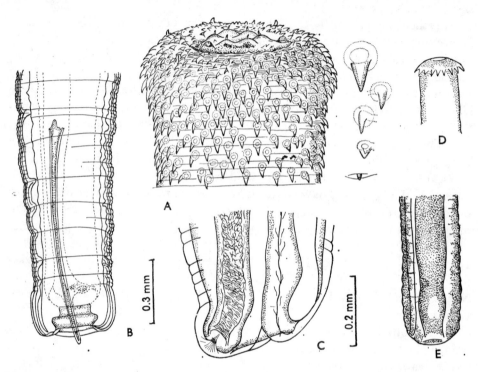

Fig. 12. *Hystrichis tricolor* Dujardin, 1845 — A, B, C, *Hystrichis coronatus* Molin, 1861 — D, E
A — anterior end with spines of upper, middle and lower rows; B — posterior end of male; C — posterior end of female. After Karmanova (1955). D — anterior end; E — posterior end. After Molin (1861).

ornamentation of the head end indicates that they do not belong in the genus *Hystrichis*. Nevertheless, we shall leave them, as did Karmanova (1968) in *Hystrichis* until further research enables us to resolve the systematic position of this species.

References: Diesing (1861); Jägerskiöld (1909)*; Karmanova (1968 a)*; Molin (1860 b*; 1861 a, b); Parona (1894); Skrjabin (1923); Stossich (1899).

Subclass *Secernentea* (Linstow, 1905) Dougherty, 1958

A characteristic feature of the representatives of this subclass is the presence of phasmids and cervical papillae but as, in many species, the cervical papillae are indistict, they are omitted from the diagnoses.

Of other characteristics common to all species of this subclass the following should be mentioned: absence of hypodermal glands and terminal tail glands, characteristic structure of amphids (pore-like or papilla-shaped) and cervical gland (bifurcated, composed of two cells).

Most of the parasitic nematodes belong to this subclass although the majority of the species of this subclass are free-living.

The subclass *Secernentea* is divided into three orders: *Rhabditida* (suborders *Rhabditata* and *Strongylata*), *Ascaridida* (suborders *Ascaridata* and *Oxyurata*) and *Spirurida* (suborders *Spirurata*, *Filariata* and *Camallanata*).

Fish-eating birds of the Palaearctic Region have been found to be parasitized by representatives of all three orders and of seven suborders.

SUBORDER *RHABDITATA* CHITWOOD, 1933

Heterogonic nematodes. Parasitic forms parthenogenetic, hermaphrodite or dioecious. Buccal capsule present or absent. Oesophagus with or without posterior bulb. Female tail short, vulva in middle or posterior third of body. Parasitic generation living in lungs or intestine of different vertebrates.

Free-living forms dioecious, rhabditiform. Stoma and lips usually developed. Oesophagus possesses corpus, isthmus and muscular bulb with valvular apparatus. One or two ovaries. Males with caudal alae supported by rib-like papillae (but typical bursa copulatrix not developed).

This suborder includes both free-living forms and parasites of plants, invertebrates and vertebrates. Only one member of the family *Strongyloididae* has been found in fish-eating birds.

Family *Strongyloididae* Chitwood & McIntosh, 1934

Only the parasitic forms of these nematodes have been encountered in vertebrates. They are parthenogenetic of dioecious, of small size and filiform. Stoma cup-shaped or greatly reduced. Oesophagus long, narrow, somewhat enlarged posteriorly but without bulb. Female tail short, vulva in posterior half of body. Two divergent uteri present, ovaries reflexed. Oviparous. Parasitic in gastrointestinal tract of different vertebrates.

Of the two genera included in this family (see Little 1966), only one member of the genus *Strongyloides* has been found in fish-eating birds.

Genus *Strongyloides* Grassi, 1879

Parasitic stage, a parthenogenetic female which lives embedded in mucosal epithelium of gastrointestinal tract, slender, up to 6 mm long by 0.075 mm wide (in most species 2—4 mm long). Cuticle finely striated. Tail short, abruptly tapered. Head with circumoral elevation which may or may not be lobed. Stoma shallow. Oesophagus long, filariform. Vulva about two-thirds of body length from anterior end, vagina very short. Oviduct short, with walls thicker than those of ovaries or uteri. Ovaries long, with reflexed loops spiralling around intestine or running parallel with it. Eggs in single row in uterus, elipsoidal with extremely thin walls, in early cleavage when deposited and may not undergo development before leaving the host.

To the present, more than 50 species of this genus parasitizing amphibians, reptiles, birds and mammals have been found. Birds have been found to harbour 8 species but only one, *S. turkmenicus*, has been recorded from fish-eating birds of the Palaearctic Region. The paper by Little (1966) is of great importance for an account of the more exact taxonomy of this genus.

The type species, *Strongyloides stercoralis* (Bavay, 1876) parazitizes man and some other warm-blooded vertebrates.

Strongyloides turkmenicus Kurtieva, 1953

Fig. 13
Host: *Larus canus*.
Localization: intestine.
Distribution: Norway; in the non-fish-eating bird *Himantopus candidus* in Turkmenistan. Not reported outside the Palaearctic Region.

Description: Body yellowish in colour, cuticle with indistinct transverse striations, inflated in some specimens. Body length 1.48—1.88 mm, width of head end 0.009—0.015 mm, at level of termination of oesophagus 0.023 to 0.030 mm, at level of vulva 0.026—0.030 mm, at level of anus 0.015—0.019 mm. Mouth terminal, hexagonal in shape. Circumoral elevation indistinctly divided into 6 small labial lobes. Stoma shallow, 0.0038 mm deep and 0.004 to 0.005 mm wide. Oesophagus filariform, 0.40—0.50 mm long, maximum width 0.018—0.020 mm. Vulva in form of transverse fissure with rounded, slightly protruding margins, situated 0.95—1.21 mm from anterior end. Vagina very short. Anterior ovary spiralled once around intestine, reaching level of termination of oesophagus or not quite as far. Posterior ovary extending to 0.062 mm from tail end. Eggs 0.038—0.042 × 0.019—0.023 mm, ellipsoidal, in single row in uterus. Tail short, blunt. Anus 0.034—0.038 mm from tail end, with margin of opening rounded and distinctly salient. Ratio of body length to oesophagus length from 3.5 : 1 to 4.2 : 1 and of body to tail 40.0 : 1

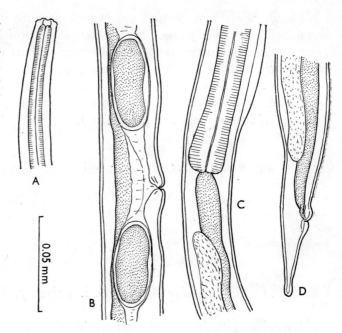

Fig. 13. *Strongyloides turkmenicus* Kurtieva, 1953. A — anterior end; B — region of vulva (lateral view); C — junction of oesophagus and intestine; D — posterior end (lateral view). Original.

to 55.5 : 1. Distance of vulva to anterior end of body 63.9—64.8 % of the total length of body.

The species was described from specimens recovered from *Himantopus candidus* in Turkmenistan (U.S.S.R.).

References: Bakke & Baruš (1976)*; Kurtieva (1953)*; Little (1966).

SUBORDER *STRONGYLATA* RAILLIET & HENRY, 1913

Nematodes of small size with thick cuticle. Distinct lips absent. Many species possess mouth capsule (vestibule) at the anterior end, others have head with complicated cuticular formation. Oesophagus club-shaped, slightly swollen at posterior end. Most characteristic feature of this suborder is the cuticular bursa at the end of the male tail. Bursa supported by a number of rays (three dorsal, two ventral and three lateral rays on each side). The size and configuration of the rays are characteristic of representatives of individual taxa of this suborder. Two spicules, equal and similar. Gubernaculum present in only some species. Tail end of female conoid. Vulva in posterior or anterior half of body.

Life-cycles of individual species are very different but many species develop without an intermediate host. Some *Syngamidae* utilize paratenic hosts.

Members of the *Strongylata* are usually parasites of mammals. Birds are parasitized by relatively few species, about 70 according to Yamaguti (1961). They occur most commonly in the digestive organs but some species *(Syngamidae)* are found in the respiratory organs.

Fish-eating birds of the Palaearctic Region have so far been found to be parasitized by 12 species belonging to five genera and three families *(Amidostomidae, Trichostrongylidae* and *Syngamidae)*.

KEY TO THE FAMILIES OF THE SUBORDER *STRONGYLATA*

```
1  Parasites of digestive tract (gizzard) . . . . . . . . . . . . . . . . . . 2
—  Parasites of respiratory tract . . . . . . . . . . . . . . . . . Syngamidae
2  Buccal capsule relatively large . . . . . . . . . . . . . . . . Amidostomidae
—  Buccal capsule rudimentary or absent . . . . . . . . . . . . Trichostrongylidae
```

Family *Amidostomidae* Baylis & Daubney, 1926

Buccal capsule shallow with relatively thin walls. One or three triangular teeth, each with sharp tip, situated at the bottom of the capsule. Caudal bursa of male wide, trilobate, with dorsal lobe feebly developed. Spicules relatively short, of complex structure with distal end divided into two or three branches. Gubernaculum present. Vulva in posterior half of body.

Parasites of water birds belonging mostly to the order *Anseriformes*.

The family contains two genera. The species reported from fish-eating birds belong to the type genus of the family.

Genus *Amidostomum* Raillet & Henry, 1909

The morphological characters of representatives of this genus are given above in the family diagnosis.

The species composition of the genus has recently been analysed by Kobuley & Ryzhikov (1968). According to the published data, 18 species of this genus have been described. The present authors have concluded that only 10 species can be recognized as valid. In order to confirm the independent status of four species further studies are necessary. Four species are synonymized with species described earlier.

Four species of this genus have been found in fish-eating birds of the Palaearctic Region but none of them is a characteristic parasite of fish-eating birds.

KEY TO THE SPECIES OF THE GENUS *AMIDOSTOMUM*

1 Buccal capsule with three teeth . . . . *A. anseris*
— Buccal capsule with one tooth 2
2 Spicules 0.17—0.21 mm long. Characteristic parasites of coot, found also in grebes and gulls . . . . *A. fulicae*
— Spicules shorter (0.13—0.14 mm long). Parasites of *Anseriformes* 3
3 Cephalic papillae large, projecting above the external margin of buccal capsule . . . .
  . . . . *A. orientale*
— Cephalic papillae small, not projecting above the margin of buccal capsule . . . . . .
  . . . . *A. acutum*

Amidostomum acutum (Lundahl, 1848)

Fig. 14
Hosts: *Mergus albellus* and *M. merganser* and occasionally in anatid birds.
Localization: under cuticle of gizzard.
Distribution: It has been found in *M. albellus* in many regions of the U.S.S.R. (Rybinsk water reservoir, Kazakhstan, Primorye) and has been observed to occur in *Anas* and *Aythya* in all parts of the Palaearctic Region.

Description: Buccal capsule with thin walls and one tooth at the bottom. Papillae indistinct, small, distributed on external margin of the capsule.
Male: Body length 9.3—12.8 mm, maximum body width 0.086—0.12 mm. Buccal capsule 0.009—0.012 × 0.012 mm. Oesophagus 0.642—0.707 mm long. Bursa well developed. Ventral rays close to one another, lateral rays running from a common trunk. Dorsal ray divided into two branches, each of which is further subdivided into two small branches. Spicules 0.125—0.140 mm long, distal part trifurcate. Gubernaculum rod-shaped, 0.070—0.076 mm long.
Female: Body length 9.1—14.2 mm, maximum body width 0.12—0.19 mm. Oesophagus 0.67—0.72 mm long. Vulva 1.9—2.3 mm from tail. Eggs 0.086 to 0.089 × 0.058—0.65 mm.

The species was described as *Strongylus acutus* from specimens recovered from *Anatidae* in Scandinavia. Seurat (1918) transferred it to the genus *Amidostomum*. In agreement with Czaplinski (1962) and Kobuley & Ryzhikov (1968) we regard the species *Amidostomum boschadis* Petrov & Feydushin, 1949 as a synonym of *A. acutum*. This species was mentioned under *A. boschadis* by all authors who found it in *Mergus albellus* (Shigin 1959, Oshmarin 1963, Maksimova 1967).

Biology: According to Zajíček (1964) the eggs of *A. acutum*, when released into the external environment are at the 16 or 32 blastomere stage. At temperatures of 18—24 °C the infective larva develops inside the egg within three days and

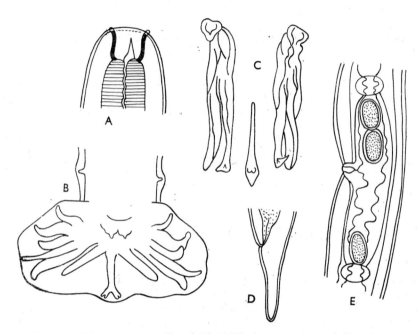

Fig. 14. *Amidostomum acutum* (Lundahl, 1848). A — anterior end; B — bursa copulatrix (ventral view); C — spicules and gubernaculum (ventral view); D — posterior end of female (lateral view); E — vulva (lateral view). After Ryzhikov (1967).

hatches immediately. Freshly hatched larvae were fed to 14-day-old ducklings and matured within 20 to 26 days.

References: Cram (1927)*; Czaplinski (1962 a)*; Kobuley & Ryzhikov (1968); Kurochkin (1964); Lundahl (1848)*; McDonald (1969); Maksimova (1967); Oshmarin (1963); Ryzhikov (1967 b); Seurat (1918); Shigin (1959); Sonin & Larchenko (1974); Zajíček (1964).

Amidostomum anseris (Zeder, 1800)

Fig. 15
Hosts: *Podiceps ruficollis* and *Ardea purpurea*. Both these birds are undoubtedly accidental hosts. The normal hosts are domestic goose and greylag goose *(Anser anser)*.
Localization: under cuticle of gizzard.

Distribution: In fish-eating birds in the U.S.S.R. (Ukraine, Uzbekistan). In geese in southern and middle areas of the Palaearctic Region. Outside the Palaearctic Region, in Asia, Africa and North America.

Description: Buccal capsule with three teeth. Three pairs of small, indistinct, papillae distributed along the margins of the capsule.

Male: Body length 9.6—14.0 mm, maximum body width 0.19—0.26 mm. Buccal capsule 0.044—0.052 × 0.020—0.024 mm. Oesophagus 1.01—1.29 mm long. Spicules 0.306—0.354 mm long, distal end of each trifurcate. Gubernaculum 0.120—0.158 mm long. Bursa well developed, its membrane with typical striations; each ray of lateral lobes arises independently, the shortest being the ventro-ventral ray and the longest the medio- and postero-lateral rays; dorsal ray with massive bifurcated trunk, each branch of which is further divided into two short branches.

Female: Body length 15.6—21.4 mm, maximum body width in region of vulva 0.27—0.39 mm. Oesophagus 1.13—1.53 mm long. Vulva situated 2.25—3.22 mm from tail end, dividing the body in ratio of 1 : 6. Eggs 0.088 to 0.100 × 0.52—0.064 mm.

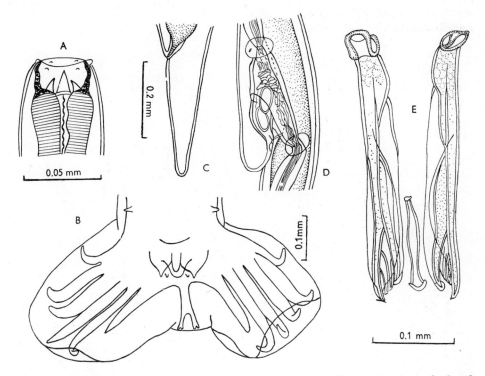

Fig. 15. *Amidostomum anseris* (Zeder, 1800). A — anterior end; B — posterior end of male (ventral view); C — posterior end of female (lateral view); D — region of vulva (lateral view); E — spicules and gubernaculum (ventral view). After Gorshkov (1937) in Ryzhikov (1967).

The species was described as *Strongylus anseris* from specimens recovered from *Anser anser f. domestica* in Central Europe. Railliet & Henry (1909) transferred it to the genus *Amidostomum*.

Biology: The life-cycle of this species has been studied by many authors. The eggs of the parasite are released to the exterior with the excrement of the birds. The larva develops within the egg and hatches. The birds are infected by swallowing the larvae in food or water but the infective larvae may also penetrate through the skin (Enigk & Dey Hazra 1968). The nematode matures within the goose in 17 to 22 days and survives in the host's body for 15 to 20 months.

References: Cram (1927)*, Enigk & Dey Hazra (1968); McDonald (1969); Railliet & Henry (1909); Ryzhikov (1967 b); Smogorzhevskaya (1964); Sultanov 1958); Zeder (1800)*.

Amidostomum fulicae (Rudolphi, 1819)

Fig. 16
Hosts: *Podiceps ruficollis, P. nigricollis* and *Larus ridibundus*. The coot *(Fulica atra)* is the characteristic host. Fish-eating birds are only occasional hosts.
Localization: under cuticle of gizzard.

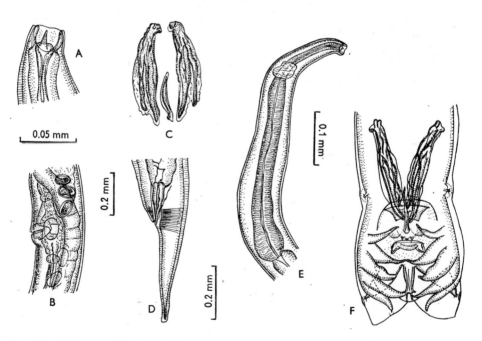

Fig. 16. *Amidostomum fulicae* (Rudolphi, 1819). A — anterior end; B — vulva (lateral view); C — spicules and gubernaculum (ventral view); D — posterior end of female (lateral view); E — anterior end (overall view); F — bursa copulatrix (ventral view). After Pavlov (1960).

Distribution: It has been found in fish-eating birds in two regions of the U.S.S.R., namely the Urals and Turkmenia and it has been encountered in coots in all parts of the Palaearctic Region and also in South Asia, Africa and North America.

Description: Buccal capsule with one large conoid tooth on the base of which are small tubercles.

Male: Body length 5.84—8.8 mm, maximum body width 0.21—0.27 mm. Oesophagus 0.6—1.0 mm long. Spicules equal, 0.16—0.19 mm long, complicated in form, with distal ends divided into two branches, of which the inner is shorter than the outer. Gubernaculum club-shaped, 0.06—0.108 mm long. Caudal bursa with two lateral lobes, their free margins bent towards ventral surface. All rays originating independently from base of lateral lobes; ventral rays narrow, rather long, separating from one another; latero-ventral rays longer than ventro-ventral rays, both pairs reaching margin of bursa; lateral rays shorter and somewhat thicker than ventral rays, the shortest of this group being the externo-lateral rays, which do not reach the margin of bursa; dorsal ray originates between bases of lateral lobes and is massive and trunk-shaped at its base, then bifurcates into a short lateral and a longer inner branch which are each divided by a shallow groove into two other small branches. One pair of large, sessile, contiguous papillae situated at posterior margin of cloaca.

Female: Body length 6.5—12.4 mm, maximum body width 0.16—0.23 mm. Buccal capsule 0.015—0.017 mm deep, 0.020—0.026 mm wide. Oesophagus 0.58—1.1 mm long. Tail 0.13—0.4 mm long. Vulva 1.3—2.5 mm from posterior end. Eggs 0.089—0.111 × 0.055—0.069 mm.

The species was described as *Spiroptera fulicae* from specimens recovered from the gizzard of *Fulica atra*. Seurat (1918) transferred it to the genus *Amidostomum*.

Notes: We regard the species *A. raillieti* Skrjabin, 1915 as a synonym of *A. fulicae*. These species were synonymized by Pavlov (1960) and subsequent workers studying the genus *Amidostomum* agreed with Pavlov's opinion. In grebes of the Urals the species was reported and described under the name *A. raillieti* (Okorokov & Tkachev 1969).

Biology: The exogenous phase of the life-cycle of this parasite was studied by Baruš (1964). The development of the egg to the infective larva takes 74—78 hours at 20 °C but only 40—44 hours at 27 °C. The infective larvae emerge from the eggs and survive in water for at most 85 days. The larvae are susceptible to desiccation.

References: Babaev (1970); Baruš (1964 b); Cram (1927); Kibakin & Babaev (1964); Kobuley & Ryzhikov (1968); Okorokov & Tkachev (1969); Pavlov (1960)*; Rudolphi (1819)*; Seurat (1918).

Amidostomum orientale Ryzhikov & Pavlov, 1959

Fig. 17
Hosts: *Mergus merganser*. Often occurs in ducks and pochards, accidental parasite of goosanders.
Localization: under cuticle of gizzard.
Distribution: In *M. merganser* in the U.S.S.R. — the lower reaches of the river Ob. Widely distributed in ducks and pochards in the north-eastern regions of Asia. Not reported outside the Palaearctic Region.

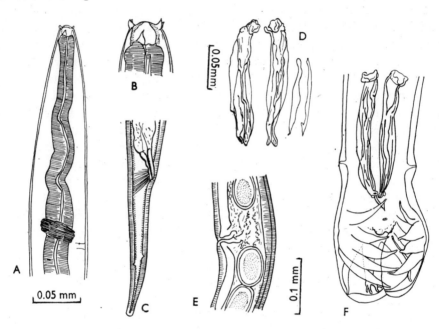

Fig. 17. *Amidostomum orientale* Ryzhikov & Pavlov, 1959. A — anterior end (overall view); B — anterior part of body (detail); C — posterior end of female (lateral view); D — spicules and gubernaculum (different views); E — vulva (lateral view); F — bursa copulatrix (ventral view). After Ryzhikov & Pavlov (1959).

Description: Buccal capsule with thin walls and one triangular tooth situated at the base. Four large papillae on external margin of capsule, projecting beyond it.
Male: Body length 8.9—12 mm, maximum body width 0.09—0.18 mm. Buccal capsule 0.010—0.014 mm in diameter and 0.007—0.010 mm deep. Oesophagus 0.61.—0.88 mm long. Spicules 0.12—0.14 mm long. Bursa with two large lateral lobes and a small ventral lobe. Bursal rays with following structure: ventral rays thinner, pointed at tips, bent ventrally and anteriorly with common base contiguous with bases of lateral rays. Lateral rays longer than ventral rays, also pointed at tips, not reaching margins of lobes of bursa.

Dorsal rays contiguous at their bases, then separate and join one or other of the latero-dorsal rays along the whole length. Medio-dorsal ray thick at base and thinner towards the tip which is subdivided into two branches, each further bifurcated. Externo-dorsal ray 0.065 mm long, medio-dorsal ray 0.060 mm long. Prebursal papillae present. Gubernaculum 0.06—0.09 mm long, in form of thin narrow plate, somewhat widened in its middle part, forming a hook at the distal end.

Female: Body length 12.3—18.3 mm, maximum body width 0.14—0.2 mm. Buccal capsule with external diameter 0.026 mm, internal diameter 0.013 mm, depth 0.011 mm. Oesophagus 0.71—1.02 mm long. Vulva in posterior part of body, 2.5—3.5 mm from tail end. Anus 0.27—0.34 mm from tip of tail. Eggs 0.042—0.049 × 0.077—0.084 mm.

The species was described from specimens recovered from *Clangula hyemalis* and *Somateria spectabilis* in northern Yakutia (U.S.S.R.).

References: Daiya (1967)*; Ryzhikov & Pavlov (1959)*.

Family *Trichostrongylidae* Leiper, 1912

Nematodes of medium size. Mouth capsule absent. Mouth cavity small, distinctly marked. In some species head with complicated epaulette-like formation. Bursa copulatrix well developed. Spicules of complicated structure. Gubernaculum present or absent. Vulva in posterior half of body. Tail end conical. Parasitic in digestive organs, mainly in mammals.

Most of the species parasitizing birds belong to the subfamily *Epomidiostominae*. Only one species has been reported in fish-eating birds and this belongs to the type genus of this subfamily.

Genus *Epomidiostomum* Skrjabin, 1915

Anterior end of body distinctly narrowed. Cuticle with marked transverse striations. Head with complex thickenings composed of four cuticular outgrowths directed laterally, epaulette-like formations and four protruding papillae surrounding mouth opening. Caudal bursa of male well developed, composed of two large lateral lobes and a small median lobe. Spicules equal, their posterior ends divided into three projections. Gubernaculum absent. Vulva in posterior half of body. Body narrows sharply behind anal opening forming a large finger-like projection which is bent ventrally.

Parasitic under cuticle of gizzard of aquatic birds, mostly *Anseriformes*. Only one species has been reported from fish-eating birds.

Epomidiostomum uncinatum (Lundahl, 1848)

Fig. 18
Hosts: *Podiceps cristatus*, *P. griseigena* and *P. ruficollis*, accidentally in grebes, usually parasitic in ducks and pochards.
Localization: under cuticle of gizzard.
Distribution: In grebes, reported from the U.S.S.R. (Ural). In ducks and pochards, it is cosmopolitan.

Description: Mouth opening surrounded by four projecting papillae. Epaulette-like formation of head with tridentate posterior margins.

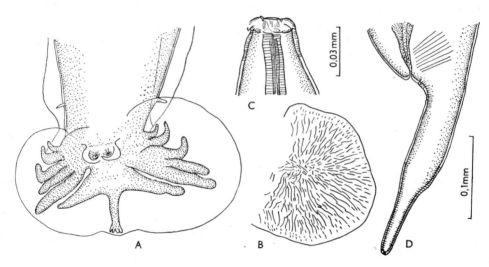

Fig. 18. *Epomidiostomum uncinatum* (Lundahl, 1848). A — bursa copulatrix (ventral view); B — cuticular structures on bursa copulatrix; C — anterior part of body; D — posterior end of female (lateral view). After Baruš & Lorenzo Hernández (1970).

Male: Body length 6.5—7.5 mm, maximum body width 0.15 mm. Oesophagus 0.8 mm long. Bursa copulatrix with irregular striations. Ventral rays separating slightly; lateral rays with common trunk; externo-dorsal rays thick, with bent distal ends; dorsal ray thick, short, distal end divided into four projections. One pair of papillae on genital cone. Prebursal papillae very small. Spicules 0.12—0.13 mm long, trifurcate posteriorly.
Female: Body length 10—11.5 mm, body width 0.23—0.24 mm. Vulva 2.5—3.2 mm from posterior end. Tail 0.16—0.19 mm long. Eggs 0.085 to 0.090 × 0.045—0.050 mm.

The species was described from specimens recovered from wild ducks in Scandinavia and originally assigned to the genus *Strongylus*. Seurat (1918) transferred it to the genus *Epomidiostomum* which was erected by Skrjabin (1915) for *E. anatinum*, described from the domestic duck. Seurat considered *E. anatinum* a synonymum of *E. uncinatum* but not all authors of subsequent papers agreed with his opinion and the synonymy of these two species remains questionable. Skrjabin et al. (1954), in validation of *E. uncinatum* showed that it was originally very insufficiently described and proposed, therefore, to regard it as species inquirenda. However, many authors agreed with Seurat, as, for example, did M. M. Ali (1971), who published a revision of the species composition of the genus *Epomidiostomum* and we also agree with those authors who consider both species to be identical. The specimens recovered from grebes were described as *E. anatinum* (Okorokov & Tkachev 1969; Tkachev 1971).

Biology: The life-cycle of this species was studied by Kurochkin (1954). The eggs are expelled from the bird host at the beginning of cleavage, development of the larva within the egg takes three days at 23 °C, and the larvae are infective when hatched. On dissection of birds infected with 4- to 5-day-old larvae young parasites were found after 18 days.

References: Cram (1927)*; Kurochkin (1954); Lundahl (1848)*; Okorokov & Tkachev (1969); Seurat (1918); Skrjabin (1915 b); Skrjabin, Shikhobalova & Shults (1954)*; Sonin & Larchenko (1974); Tkachev (1971).

Family *Syngamidae* Leiper, 1912

Nematodes with markedly developed buccal capsule. Margin of buccal capsule thick, in form of six festoons. Teeth radially distributed at base of buccal capsule. Males distinctly smaller than females. In some species male is permanently attached by bursa copulatrix to vulva of female. Spicules equal or subequal. Gubernaculum present or absent. Posterior end of body of females conical. Vulva in anterior or middle third of body, uteri parallel. Oviparous, eggs with small polar plugs or without plugs. Parasites of respiratory tract of birds and mammals.

KEY TO THE GENERA OF THE FAMILY *SYNGAMIDAE*

1 Externo-dorsal rays of the same length as, or longer than, medio-dorsal ray . . . . *Syngamus*
— Externo-dorsal rays markedly shorter than the medio-dorsal ray, not exceeding the level of its distal division 2
2 Medio-dorsal ray with medial projection extending beyond margin of bursa copulatrix . . . . *Cyathostoma*
— Medio-dorsal ray without medial projection, not extending beyond the margin of bursa copulatrix . . . . *Hovorkonema*

Genus *Syngamus* Siebold, 1836

Cuticular buccal collar always present, more or less developed. Inner wall of buccal capsule smooth, without longitudinal ribs. Males and females permanently in copula. Bursa copulatrix markedly wider than long, medial lobe indistinct. Externo-dorsal rays as long as medio-dorsal rays, not extending past margin of bursa copulatrix. Spicules short (not longer than 0.10 mm). Eggs with two opercula. Parasites of respiratory organs of birds.

Type species *S. trachea* (Montagu, 1811).

Of the 8 species so far described in this genus, two have been found in fish-eating birds in the Palaearctic Region.

KEY TO THE SPECIES OF THE GENUS *SYNGAMUS*

1 Size of eggs 0.091—0.12 × 0.038—0.049 mm . . . . *S. trachea*
— Size of eggs 0.077—0.086 × 0.043—0.049 mm . . . . *S. arcticus*

Syngamus trachea (Montagu, 1811)

Fig. 19
Hosts: *Pelecanus onocrotalus, Phalacrocorax carbo, Ciconia ciconia, C. nigra, Larus argentatus, L. fuscus, Sterna hirundo, S. paradisea* and, most frequently, birds of the orders *Passeriformes* and *Galliformes*.
Localization: trachea.
Distribution: cosmopolitan, in fish-eating birds found in Europe (England, Norway, U.S.S.R.—Latvia), Asia (U.S.S.R.—Azerbaijan, Far East).

Description: Body usually bright red in colour. Cuticle with fine transverse striation. Anterior end of body of both sexes bluntly truncate. Mouth circular, surrounded by cuticular collar. Three pairs of cephalic papillae. Mouth opening leading into large buccal capsule with six to eleven pseudochitinous teeth at base. Anterior border forming festoon divided by six incisions into dorsal, ventral and two double lateral parts.
Male: Body length 2.15—7.85 mm, maximum body width 0.19—0.56 mm. Width of buccal collar 0.32—0.69 mm. Internal diameter of buccal capsule 0.10—0.35 mm, depth 0.10—0.27 mm. Oesophagus 0.53—0.99 mm long, maximum width 0.080—0.18 mm. Posterior end of body formed by bell-shaped bursa copulatrix which is divided on the ventral side by a slit which reaches nearly to its base. Ventral rays lying close to each other for nearly all their length, with rounded tips not reaching edge of bursa copulatrix. Group of three lateral rays distinctly divided from dorsal and ventral rays. Lateral rays either apart from or close to each other with rounded tips. Groups of dorsal rays arising from common base, with an externo-dorsal ray extending laterally

from each side, none reaching margin of bursa copulatrix. Medio-dorsal ray divided into two or more branches on which there may be additional protuberances. Spicules equal or subequal, 0.057—0.096 mm long. Proximal end of spicules 0.003—0.005 mm wide.

Female: Body length 8.14—36.27 mm, maximum body width 0.039—1.23 mm. Width of buccal collar 0.46—1.07 mm. Internal diameter of buccal capsule 0.21—0.81 mm, depth 0.21—0.78 mm. Oesophagus 0.53—1.37 mm long, maximum width 0.12—0.32 mm. Vulva at anterior end of body, forming rounded prominent protuberance. Relation of prevulval to postvulval part of body ranges from 1 : 2.1 to 1 : 8.1. Posterior portion of female body conical, with sharp point. Anus 0.17—0.45 mm from tail tip. Eggs 0.091—0.12 × 0.038 to 0.049 mm, oval, with operculum at each pole, shell light brown, containing 8 blastomeres when laid.

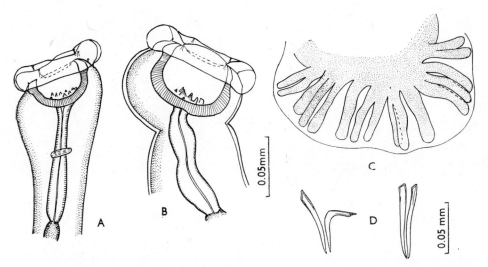

Fig. 19. *Syngamus trachea* (Montagu, 1811). A — anterior end of male; B — anterior end of female; C — bursa copulatrix (ventral view); D — spicules. After Baruš (1964).

The species was described as *Fasciola trachea* from specimens recovered from *Phasianus colchicus* in England. It was transferred to the genus *Syngamus* by Siebold (1836).

Biology: Life-cycle is direct, with paratenic hosts (Mégnin 1881, 1882, Ortlepp 1923, Ryzhikov 1949, Baruš & Blažek 1965 and others).

Notes: This species is of great importance, because it is known to be very pathogenic, causing serious poultry disease. Systematic problems are dealt with in the papers by Chapin (1925), Madsen (1950) and Baruš (1964). A survey of the systematics, morphology and biology of this parasite is given in the monograph by Ryzhikov (1949).

References: Bakke (1973); Baruš (1964 a)*; Baruš & Blažek (1965)*; Belogurov, Leonov & Zueva (1968); Chapin (1925)*; Creplin (1849); Ellis & Williams (1973); Krivonogova (1963); Madsen (1950 a); Molin (1861 b); Montagu (1811)*; Ortlepp (1923)*; Oshmarin & Parukhin (1963); Pemberton (1963); Ryzhikov (1949)*; Sailov (1962, 1965 b); Siebold (1836); Skrjabin (1923); Threlfall (1965 a, b, 1967).

Syngamus arcticus Ryzhikov, 1952

Fig. 20
Hosts: *Gavia stellata*.
Localization: trachea.
Distribution: Europe (U.S.S.R. — White Sea).

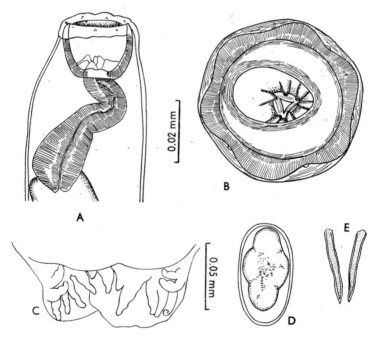

Fig. 20. *Syngamus arcticus* Ryzhikov, 1952. A — anterior end; B — anterior end (apical view); C — bursa copulatrix; D — egg; E — spicules. After Ryzhikov (1952).

Description: Body with smooth cuticle. Buccal capsule small with six rounded protuberances on its outer margin and six sessile cephalic papillae in the incisions between them, and six teeth of approximately equal size at its base. Cuticular collar around mouth not fully developed.
Male: Body length 2.0 mm, maximum body width 0.51 mm. Internal dia-

meter of buccal capsule 0.19 mm, depth of buccal capsule 0.14 mm; oesophagus with bulbous widening posteriorly, length 0.58 mm, maximum width 0.18 mm. Bursa copulatrix relatively small and narrow. Bursal rays massive, with bluntly rounded ends. Ventral rays with common base, their inner margins closely adjoining. Lateral rays joined along nearly whole length and longer than ventral rays. Group of dorsal rays consists of two externo-dorsal rays, which may or may not have relatively large inner protuberance, and a medio-dorsal ray which is strong, may or may not be bifurcated, and is shorter than externo-dorsal rays. Spicules 0.051 mm, with distal end pointed, proximal end blunt. Female: Body length 9.1 mm, maximum body width 0.8 mm. Internal diameter of buccal capsule 0.36 mm, depth 0.27 mm, maximum width of its wall 0.04 mm. Oesophagus 0.94 mm long, maximum width 0.27 mm. Vulva 1.3 mm from anterior end. Eggs regularly oval, 0.077—0.086 × 0.043—0.049 mm, shell without opercula. Posterior end of body conical.

The description was based on specimens recovered from *G. stellata* in the U.S.S.R.

Reference: Ryzhikov (1952)*.

Genus *Cyathostoma* Blanchard, 1849

Buccal collar absent. Inner wall of buccal capsule smooth, without longitudinal ridges. Males and females usually separated, permanent copulation exceptional. Bursa copulatrix of strongyliform type, medial lobe distinct. Externo-dorsal rays distinctly shorter than medio-dorsal ray, which has thorn-like outgrowth which overlaps the margin of bursa copulatrix. Spicules 0.08—0.4 mm long, gubernaculum present. Eggs with one or two opercula. Parasites of respiratory organs of birds in the Old and New World.

Type species *C. lari* Blanchard, 1849.

Four species of this genus are parasites of fish-eating birds in the Palaearctic Region and one in the Neotropical Region.

KEY TO THE SPECIES OF THE GENUS *CYATHOSTOMA*

```
1   Spicules longer than 0.2 mm . . . . . . . . . . . . . . . . . . . . 2
—   Spicules shorter than 0.2 mm (0.08—0.15 mm) . . . . . . . . . . C. microspiculum
2   Cuticle in circumoral region covered with small bosses . . . . . . .  C. verrucosum
—   Cuticle in circumoral region without bosses . . . . . . . . . . . . . . . . . 3
3   Spicules 0.30—0.55 mm long . . . . . . . . . . . . . . . . . . . . . . C. lari
—   Spicules 0.24—0.29 mm long . . . . . . . . . . . . . . . . . . . . C. trifurcatum
```

Cyathostoma lari Blanchard, 1849

Fig. 21

Hosts: *Ardea cinerea, Larus canus, L. argentatus, L. fuscus, L. ridibundus* and birds of the order *Passeriformes* and *Charadriiformes;* experimentally also *Galliformes*.
Localization: nasal and orbital cavities.
Distribution: Europe (England, Ireland, Italy, G.D.R., Netherlands, Norway, Czechoslovakia, U.S.S.R.—central regions); outside the Palaearctic Region, in *L. argentatus* in Canada.
Description: Living worm moderately robust and bright red in colour. Cuticle with fine transverse striations. Anterior end blunt, posterior end running into conical tail. Buccal capsule well pseudochitinized with nine to twelve teeth at base.

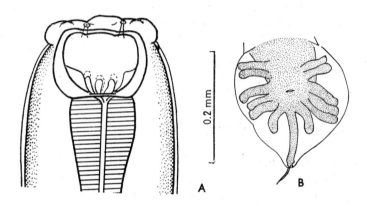

Fig. 21. *Cyathostoma lari* Blanchard, 1849. A — anterior end of female; B — bursa copulatrix (ventral view). After Baruš & Tenora (1972).

Male: Body length 6.0—10.0 mm, maximum body width 0.26—0.38 mm; external diameter of buccal capsule 0.067—0.097 mm, depth 0.051—0.088 mm. Oesophagus 0.50—0.61 mm long, muscular, clavate, widest towards posterior end. Caudal bursa with entire margin, projecting posteriorly and with digitiform ray. Dorsal rays with accessory branches, variable in form and arrangement, and medial protuberance which usually overlaps the bursal margin. Externo-dorsal and antero-lateral rays do not reach the rim but are bent outward and applied to outer face of bursa. Postero-lateral and medio-lateral rays apposed throughout their lengths, as also are the two ventral rays. Spicules equal or subequal, brown, slender, with thin, fimbriated flange along inner margin except at tips where they unite, 0.30—0.55 mm long. Gubernaculum 0.064—0.077 mm long.
Female: Body length 9.58—25.80 mm, maximum body width 0.46—1.50 mm. Oesophagus 0.64—1.01 mm long. Oesophago-intestinal valve lies in the

widening, thin-walled anterior end of intestine. Neither ventriculus nor intestinal caecum present. Vulva posterior to middle of body. Uteri amphidelphic, opening into ovejector. Eggs in the form of prolate spheroids, 0.072—0.088 × 0.039—0.048 mm, with two opercula; segmented before oviposition. Posterior end of body tapering gradually to the tail tip.

The description original was based on specimens recovered from *L. ridibundus* in Italy.

Biology: The life-cycle of this species is direct or with paratenic hosts (Pemberton 1959; Zavadil 1961; Threlfall 1965, 1966; Baruš 1970).

Notes: The description given in previous papers (Blanchard 1849; Siebold 1837) were completed and revised by Burt & Eadie (1958) who also re-defined the diagnosis of the genus.

References: Bakke (1969); Bakke & Baruš (1976); Baruš (1970 c); Baruš & Tenora (1972); Blanchard (1848—1949)*; Broek & Jansen (1964); Burt & Eadie (1958)*; Creplin (1829); Creuts & Gottschalk (1969); Diesing (1851)*; Molin (1861b); Muehling (1898); Parona (1894); Parukhin (1964 c); Parukhin & Truskova (1963); Pemberton (1959, 1963); Shigin (1961); Siebold (1837)*; Skrjabin (1923); Threlfall (1965 b, c, 1966 a, b, d, 1967); Zavadil (1961)*.

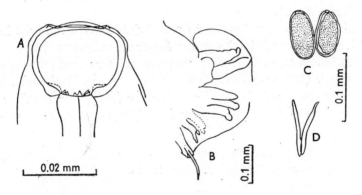

Fig. 22. *Cyathostoma microspiculum* (Skrjabin, 1915). A — anterior end of male (dorsal view); B — bursa copulatrix (lateral view); C — eggs; D — spicules. After Lengy (1969).

Cyathostoma microspiculum (Skrjabin, 1915)

Fig. 22
Hosts: *Phalacrocorax carbo*, *P. pygmaeus*, *Phalacrocorax* sp. and *Pelecanus onocrotalus*.
Localization: trachea.
Distribution: Europe (U.S.S.R. — Dagestan), Asia (U.S.S.R. — Azerbaijan, Kazakhstan, Tajikistan, Uzbekistan; Israel).

Description: Buccal capsule without cuticular collar. Anterior part of buccal capsule with six projections, of which the dorsal and ventral are longer than the lateral ones, and between which are sessile papillae. Six teeth arranged radially at base of buccal capsule.

Male: Body length 5.91 mm, maximum body width 0.43 mm. Diameter of mouth opening 0.204 mm. Diameter of buccal capsule 0.142 mm, depth 0.160 mm. Oesophagus 0.693 mm long, maximum width 0.204 mm. Ventral rays of bursa copulatrix divergent throughout their length. Lateral rays of equal length, also divergent along their length. Externo-dorsal rays shorter than and separated from medio-dorsal, arising as long, thin filament and ending in sharp thorn-like point overlapping bursal margin. Medio-dorsal ray with two small projections on each side in lower third of its length. Spicules 0.15 mm long.

Female: Body length 13.73 mm, width in region of buccal capsule 0.520 mm, at level of anus 0.317 mm. Diameter of mouth opening 0.215 mm. Internal diameter of buccal capsule 0.327 mm, depth 0.300 mm. Oesophagus 0.927 mm long, maximum width 0.289 mm. Vulva 3.80 mm from anterior end. Anus 0.215 mm from posterior end. Tail conical. Eggs 0.070—0.075 × 0.046 mm, with operculum at one pole.

The description was based on specimens recovered from *Ph. carbo* in the U.S.S.R. (Uzbekistan).

Notes: This species was redescribed from Skrjabin's type material by Ryzhikov (1949), who placed it in the genus *Syngamus*, subgenus *Ornithogamus* Ryzhikov, 1948. The morphology of this species was later described in more detail by Turemuratov (1963) who transferred it to the genus *Cyathostoma*. We agree with the systematic conclusions of Turemuratov (1963) and consider the new name *(C. turemuratovi)* used by Lengy (1969) to be a synonym of *C. microspiculum*.

References: Dubinina & Serkova (1951); Lengy (1969); Ryzhikov (1949)*; Skrjabin (1915 a)*; Turemuratov (1963 a, b)*; Vaidova (1963).

Cyathostoma trifurcatum (Hovorka & Macko, 1959)

Fig. 23
Host: *Ciconia nigra*.
Localization: trachea.
Distribution: Europe (Czechoslovakia, Rumania, Poland, U.S.S.R. — Latvia).

Description: Body brown-red in colour, cylindrical, tapering to both ends. Cuticle transparent, with transverse striations anteriorly and posteriorly. Buccal capsule very strongly developed, with stout chitinized wall and seven teeth of different sizes at base. Oesophagus muscular, widening at the posterior end. Nerve ring mid-way along oesophagus.

Male: Body length of two males 7.46 and 8.2 mm respectively, maximum body width 0.198—0.205 mm. Diameter of head portion with circumoral collar 0.109—0.127 mm. Border of circumoral collar smooth, buccal capsule longer than wide, measuring 0.09—0.112 by 0.081—0.096 mm, barrel-shaped. Oesophagus 0.631—0.722 mm long. Cloaca and well developed bursa copulatrix at posterior end of body. Bursal walls supported by dorsal, lateral and ventral rays, of which the independent dorsal ray is most characteristic; this has 3/4 of its length within the bursal wall and then emerges with its free end terminating in a spiniform projection (0.048—0.054 mm long), turned at a right angle to the main axis of the ray; just below the emergence of the spiniform

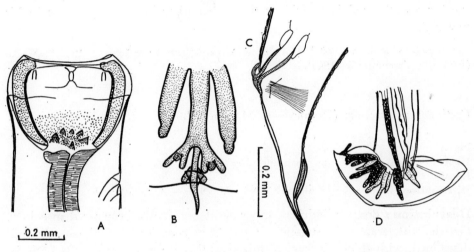

Fig. 23. *Cyathostoma trifurcatum* (Hovorka & Macko, 1959). A — anterior end; B — dorsal rays; C — posterior end of female (lateral view); D — bursa copulatrix (lateral view). After Hovorka & Macko (1959).

projection from the bursal wall the dorsal ray gives off dorsally a pair of latero--medial branches, which are fixed to the upper edge of the bursal membrane; ventrally to these and at 2/3 the length of the dorsal ray, it gives off a pair of latero-medial saddle-shaped trifurcate branches. The externo-dorsal ray originates independently and its tip is tapering and spherical. The tips of the lateral and ventral rays are similar. The three lateral rays originate from a common base, with the postero-lateral the longest and the antero-lateral the shortest. Spicules subequal, thread-like, slightly curved, 0.240—0.243 mm and 0.288 mm long. Gubernaculum present.

Female: Body length 26.0—26.7 mm. Head diameter with circumoral collar 0.504—0.654 mm. Buccal capsule cup-shaped, wider than long, 0.286 to 0.408 × 0.441—0.572 mm. Mouth opening rounded, diameter 0.327 to 0.384 mm. Oesophagus 1.354—1.380 mm long, 0.294—0.504 mm wide.

Anus 0.543 mm from posterior end. Body ending in sharp process, 0.299 mm long. Tail of female is characteristic. At its beginning there are a pair of papillae and lateral cuticular alae lying on a slightly widened portion in front of the sharp posterior end. Vulva rimose, 1/3 body length from anterior end. Eggs oval, 0.081 × 0.042 mm, with one operculum.

The species was described as *Calcaronema trifurcatum* from specimens recovered from *C. nigra* in Czechoslovakia. It was transferred to the genus *Cyathostoma* by Turemuratov (1963).

Notes: This species was recorded by Iordăchescu (1962) from the territory of Rumania under the name *Cyathostomum variegatum*. From his description and the figures attached it is evident that the author had *C. trifurcatum* in his material.

References: Gundlach (1969)*; Hovorka & Macko (1959)*; Iordăchescu (1962)*; Macko (1964 c)*; Turemuratov (1963).

Cyathostoma verrucosum (Hovorka & Macko, 1959)

Fig. 24

Hosts: *Pelecanus crispus, P. onocrotalus* and *Ciconia ciconia*.
Localization: trachea.
Distribution: Europe (Czechoslovakia), Asia (U.S.S.R. — Uzbekistan; Israel).

Description: Body cylindrical, tapering at both ends, brown-red in colour. Cuticle with transverse striations towards extremities. Buccal capsule well developed, with thick, darker chitinized wall and seven teeth at its base, its upper edge with circumoral cuticular collar densely covered with small bosses. Mouth opening rounded. Oesophagus strong, muscular, with a bulbous widening. Nerve ring at half the length of oesophagus.
Male: Body length 15.9 mm, maximum body width 0.447 mm. Buccal capsule wider than long, cup-shaped, inner dimensions 0.226 wide × 0.301 mm deep. Oesophagus 1.022 mm long, 0.218 mm wide. Cloaca and well-developed bursa copulatrix at posterior end of body. Dorsal ray most characteristic, overlapping bursal margin by a thorn-like median projection almost at right angles to the main axis and giving off two pairs of lateral branches at different levels; of these, the upper pair is longer, the lower shorter and fixed to the upper edge of the bursal membrane. Externo-dorsal rays extending as far as mid-point of medio-dorsal ray, with rounded ends. Group of lateral rays with common base, with postero-lateral longest and antero-lateral shortest with a spherical widening at its end; antero-ventral ray extending as far as mid point of postero-lateral ray. Spicules threadlike, subequal, 0.339—0.345 mm long, slightly inflected. Gubernaculum short and broad.
Female: Body length 21.8—25.5 mm, width of anterior part of head region

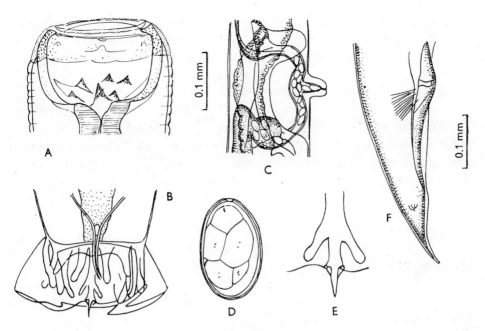

Fig. 24. *Cyathostoma verrucosum* (Hovorka & Macko, 1959). A — anterior end; B — bursa copulatrix (dorsal view); C — vulva (lateral view); D — egg; E — distal end of mediodorsal ray; F — posterior end of female (lateral view). After Hovorka & Macko (1959).

0.315—0.681 mm, maximum width in first third of body length 0.490—0.769 mm. Buccal capsule cup-shaped, 0.205—0.274 mm wide, 0.171—0.185 mm deep. Walls of buccal capsule with maximum thickness at the inferior fold (0.041 mm). Oesophagus 0.831—1.322 long by 0.204—0.313 mm wide. Anus protrudes above body surface and lies 0.312—0.349 mm from the posterior end of body, which passes into a sharp tail 0.204—0.274 mm long. On each side is a papilla directed laterally. Vulva in anterior half of body, 8.033—15.637 mm from the head end, salient in some females. Eggs oval, 0.075—0.084 × × 0.049—0.051 mm, with operculum at one pole.

The description was based on specimens recovered from *C. ciconia* in Czechoslovakia. It was transferred to the genus *Cyathostoma* by Turemuratov (1963).

Notes: Hovorka & Macko (1959) created for this species (and for *C. trifurcatum*) a new genus *Calcaronema*. Later, Turemuratov (1963) placed the genus *Calcaronema* in synonymy with *Cyathostoma* which he divided into two subgenera — *Cyathostoma* and *Hovorkonema*. Lengy (1969) recognized in the family *Syngamidae* only the genus *Syngamus* with four subgenera. We consider that *Cyathostoma* is a valid genus as defined by Chapin (1925), Burt & Eadie (1958) and Baruš & Tenora (1972).

References: Baruš & Tenora (1972); Burt & Eadie (1958); Chapin (1925 a); Hovorka & Macko (1959)*; Lengy (1969)*; Macko (1963 b)*; Turemuratov (1963 b)*.

Genus *Hovorkonema* Ryzhikov, 1967

Inner wall of buccal capsule smooth, without longitudinal ribs. Buccal collar absent. Males and females usually separate. Bursa copulatrix of strongyliform type, median lobe distinct. Externo-dorsal ray divided by a shallow incision into branches which can be further divided. Spicules 0.45—0.80 mm long. Eggs with one or two opercula. Parasites of respiratory organs of birds of the Old and New Worlds.

Type species *H. tadornae* (Chatin, 1847).

Only one species of this genus is known to parasitize fish-eating birds in the Palaearctic Region.

Hovorkonema variegatum (Creplin, 1849)

Fig. 25

Hosts: *Ciconia ciconia*, *C. nigra* and birds of the other orders; found once in mammals—*Ursus beringiana (Carnivora)*.
Localization: bronchi and trachea.
Distribution: Europe (G.D.R., Czechoslovakia), Asia (U.S.S.R. — Azerbaijan, Far East), outside the Palaearctic Region in North America and India.

Description: Cuticle smooth, with fine transverse striations at the anterior end. Buccal capsule cup-shaped, thick-walled, cylindrical in males, tapering towards base in females. Cuticular collar narrow. At the base of buccal capsule six teeth arranged in circle, the seventh situated excentrically.

Male: Body length 9.5—10.7 mm, maximum body width in the middle of body 0.32—0.33 mm. External diameter of buccal capsule 0.145—0.150 mm, internal diameter 0.099 mm, depth of buccal capsule 0.109—0.132 mm. Length of oesophagus 0.65—0.78 mm, maximum width 0.14—0.16 mm. Nerve ring midway along length of oesophagus. Posterior portion of body slightly widened, forming a bursa. Its maximum width when open is 0.512 mm, depth 0.208 mm. Ventral rays shortest, 0.140—0.165 mm long, with usual structure, contiguous with one another along their lengths, end slightly bent ventrally; lateral rays of different lengths, medio-lateral ray 0.21—0.26 mm long; externo-lateral ray short, thick, with globular thickening at end, arising from the same base as medio-dorsal ray which bifurcates at half its length with each branch further subdivided into two to three accessory branches of irregular shape. Length of externo-dorsal ray 0.144—0.170 mm, of medio-dorsal 0.330—0.363 mm.

Spicules thin, 0.580—0.720 mm long, brown in colour, with proximal ends somewhat widened and distal pointed and slightly bent ventrally.

Female: Body length 25—40 mm, maximum body width 0.62—0.83 mm. Mouth opening rounded, 0.192 mm in diameter. External diameter of buccal capsule 0.41—0.56 mm, internal diameter 0.352—0.384 mm, depth 0.304—0.368 mm. Oesophagus 1.12—1.15 mm long, maximum width 0.272—0.304

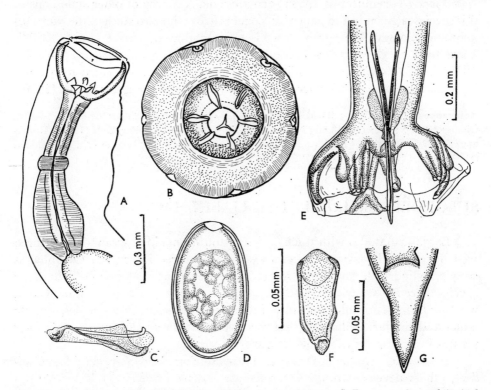

Fig. 25. *Hovorkonema variegatum* (Creplin, 1849). A — anterior end; B — anterior end (apical view); C — gubernaculum (lateral view); D — egg; E — bursa copulatrix (ventral view); F — gubernaculum (ventral view); G — posterior end of female (ventral view). After Ryzhikov & Zavadil (1958).

mm. Vulva without salient appendages situated at the border of the first and second third of body. In females 37 and 35 mm long, the vulvar opening was 12 mm from anterior end of parasite, and in females 40 mm long, 14.5 mm from anterior end. Vulval loops at anterior portion extend up to oesophagus, sometimes nearer to anterior end, at posterior portion up to anal opening. Posterior part of body conical, blunt, anal opening 0.272—0.336 mm from posterior end. Eggs 0.073—0.083 × 0.052—0.056 mm, oval, with one operculum.

The species was described by Nathusius (1837) as *Strongylus trachealis* from specimens recovered from *Ciconia nigra*. Creplin (1849) described this species from *Corvus cornix* in Europe. It was transferred to the genus *Hovorkonema* by Ryzhikov (1967a).

Biology: The life-cycle of this nematode is direct, with a paratenic host (Vogel 1928, Romanova 1948, Zavadil 1957, Ryzhikov & Zavadil 1958).

Notes: Ryzhikov (1967 a) and Baruš & Tenora (1972) raised the subgenus *Hovorkonema* Turemuratov, 1963 to generic rank. A survey of older synonyms of this species is given in the paper by Vogel (1928) who also studied the morphology and biology. We agree with his opinion that *C. brantae* Cram, 1928 is a synonym of *C. variegatum* (Creplin, 1849). The species *C. boularti* (Mégnin, 1884) and *C. bronchialis* (Muehling, 1884) have also been relegated to synonymy with this taxon, the former by Zavadil (1958) and the latter by Ali (1970).

References: Ali (1970);*; Baruš & Tenora (1972); Chapin (1925 a); Creplin (1849)*; Diesing (1851)*; Linstow (1890); Nathusius (1837); Oshmarin & Parukhin (1963); Romanova (1948)*; Ryzhikov (1967 a); Ryzhikov & Zavadil (1958)*; Skrjabin (1923); Vaidova (1964); Vogel (1928)*; Zavadil (1957 a, b)*.

SUBORDER *ASCARIDATA* SKRJABIN, 1915

Large nematodes with thick cuticle. Mouth opening surrounded by three well developed lips. Some species also possess three interlabia. Oesophagus simple, cylindrical, sometimes somewhat widened at posterior end. In some species there is a ventriculus situated between oesophagus and intestine. Caeca running from ventriculus and intestine may be present. Two spicules, usually equal and similar. Gubernaculum rarely present. Preanal papillae sometimes present in males. Vulva in middle part of body.

Some species utilize intermediate hosts during their life-cycle *(Anisakidoidea)*, others develop without intermediate hosts *(Ascaridoidea)*. Reservoir parasitism is characteristic for many species.

Parasites of all classes of vertebrates. The world fauna of *Ascaridata* parasitizing birds includes 152 species (Yamaguti 1961). Fish-eating birds of the Palaearctic Region are parasitized by 24 species which belong to one family and two genera.

Family *Anisakidae* Skrjabin & Karokhin, 1945

Cuticle without spines and additional rib-like or ridge-like formations. Oesophagus with or without posterior ventriculus. Ventriculus with or without posterior hollow or solid appendix. Intestinal diverticula present or absent. Parasites of vertebrates.

KEY TO THE GENERA OF THE FAMILY *ANISAKIDAE*

1. Oesophagus with posterior ventriculus, with solid posterior appendix, intestinal caecum present . . . . *Contracaecum*
— Oesophagus with posterior ventriculus, without ventricular appendix, intestinal caecum present . . . . *Porrocaecum*

Genus *Contracaecum* Railliet & Henry, 1912

Lips without dentigerous ridges; interlabia present, usually well developed. Ventriculus reduced, with solid posterior appendix. Intestinal caecum present. Male without definite caudal alae. Postanal papillae number up to eight pairs, partly subventral and partly lateral. Praecloacal papillae numerous. Spicules long, alate, equal or subequal. Female with vulva in anterior region of body. Oviparous. Parasites of fishes, birds and piscivorous mammals.

Type species *C. microcephalum* (Rudolphi, 1809).

Fish-eating birds of the Palaearctic Region are parasitized by 15 species of this genus (more than 50 species are known).

KEY TO THE SPECIES OF THE GENUS *CONTRACAECUM*

1. Gubernaculum absent 2
— Gubernaculum present . . . . *C. praestriatum*
2. Tail with numerous cuticular spines near the tip . . . . *C. matwejewi*
— Tail smooth, without cuticular spines near the tip 3
3. Interlabia shorter than midlength of lips . . . . *C. ovale*
— Interlabia longer than midlength of lips 4
4. Tip of interlabia not bifurcated (rounded) in adult specimens 5
— Tip of interlabia bifurcated in adult specimens 8
5. Seven pairs of postcloacal papillae, all of them simple; length of spicules 1.4—3.7 mm . . . . *C. microcephalum*
— Five to six pairs of postcloacal papillae 6
6. Six pairs of postcloacal papillae, all of them simple; length of spicules 12 mm or more . . . . *C. septentrionale*
— Five to six pairs of postcloacal papillae, first pair doubled 7
7. Length of spicules 3—6 mm . . . . *C. yamaguti*
— Length of spicules 1.0 mm . . . . *C. andersoni*
8. Eight pairs of postcloacal papillae . . . . *C. pandioni*
— Seven or fewer pairs of postcloacal papillae 9
9. All postcloacal papillae simple 11
— One or more pairs of postcloacal papillae doubled 10
10. First pair of postcloacal papillae doubled, other pairs simple . . . . . *C. travassosi*
— Three pairs of postcloacal papillae doubled, only one pair simple . . . . *C. micropapillatum*
11. Distal end of spicules pointed, with thorn-like projection . . . . *C. variegatum*

— Distal end of spicules blunt or rounded 12
12 Length of spicules 4—10 mm, lip pulp divided into two lobes anteriorly. . . . . . .
. . . . C. rudolphii
— Length of spicules 10 mm or more, lip pulp with two anterior lobes and additional oval
lobe present near anterior margin of lips . . . . C. milviensis

Note: *C. bodenheimeri* and *C. haliaeti*, which were described on the basis of females only, are omitted from the Key but descriptions are given on p. 73.

Contracaecum microcephalum (Rudolphi, 1809)

Fig. 26
Hosts: *Podiceps cristatus, P. auritus, P. griseigena, Pelecanus onocrotalus, Pel. crispus, Phalacrocorax carbo, Ph. pygmaeus, Ph. pelagicus, Ardea cinerea, A. purpurea, Egretta alba, E. garzetta, Ardeola bacchus, Ar. ralloides, Botaurus stellaris, Bubulcus ibis, Nycticorax nycticorax, Butorides striatus, Ixobrychus*

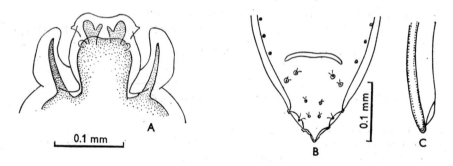

Fig. 26. *Contracaecum microcephalum* (Rudolphi, 1809). A — dorsal lip.; B — posterior end of male (ventral view); C — distal end of spicule. After Hartwich (1964).

minutus, Platalea leucorodia, Plegadis falcinellus, Ciconia nigra, Larus crassirostris, L. ichthyaetus and *Sterna hirundo*. It occurs only rarely in birds which do not eat fish.
Localization: stomach, intestine.
Distribution: Europe (England, Albania, Yugoslavia, Italy, F.R.G., G.D.R., Czechoslovakia, Bulgaria, Rumania, Poland, U.S.S.R.—Ukraine, the Volga Region), Asia (U.S.S.R.—republics of the Transcaucasus Region and Central Asia, Kazakhstan, West and East Siberia, Far East; China; Japan), Africa (A.R.E.). Outside the Palaearctic Region in South-East Asia, Africa, North and South America.

Description: Lips hexagonal, rounded, distinctly longer than wide. Lip pulp forms two lobes anteriorly, each further divided into markedly rounded lateral lobe and finger-like medial lobe. Interlabia reach the length of four-fifths of lips, with tips not bifurcated but distinctly rounded.
Male: Body length 13.10—36.92 mm, maximum body width 0.27—0.70 mm. Oesophagus 1.69—4.10 mm long, appendix and ventriculus 0.66—1.18 mm long, intestinal caecum 1.34—3.22 mm long. Posterior end of body conical,

with pointed tip. Cloaca 0.17—0.30 mm from tip of tail; 22—30 pairs of precloacal papillae arranged in two longitudinal rows: seven pairs of postcloacal papillae, all of which are simple; first two pairs of postcloacal papillae located a small distance behind cloaca, the other five pairs in lower half of tail (three pairs situated more laterally, two pairs more medially). Spicules with narrow longitudinal alae, similar, subequal (left 1.41—3.65 mm, right 1.40—3.50 mm long).

Female: Body length 12.37—37.20 mm, maximum body width 0.27—0.85 mm. Oesophagus 1.85—4.33 mm long, ventriculus with appendix 0.70 to 1.24 mm long, intestinal caecum 1.40—3.29 mm long. Posterior end of body conical, tip rounded. Anus 0.19—0.46 mm, phasmid 0.074—0.194 mm, respectively, from tip of tail. Vulva 5.26—12.10 mm from anterior end. Eggs 0.057—0.068 × 0.041—0.050 mm, spherical to oval, with granulated to wrinkled surface.

The species was described as *Ascaris microcephala* from specimens from *Ardeola ralloides* from Central Europe. It was transferred to the genus *Contracaecum* by Baylis (1920).

Biology: The life-cycle of this species was studied in detail by Mozgovoy, Semenova & Shakhmatova (1965, 1968). They found that the eggs are released to the outside before cleavage begins. First-stage larvae are formed within six to seven days at 25—28 °C. These larvae moult relatively speedily and leave the eggs. They move actively in water and can survive for 35 days at temperatures of 16—18 °C. Different species of cyclops were found to serve as intermediate hosts and the authors showed that the stay in cyclops was necessary for further development of the larvae. As second intermediate hosts dragonflies, chironomids and fry of various carp species are mentioned and in these the larvae moult for the second time and become infective for the definitive hosts — fish-eating birds.

References: Ablasov, Iksanov & Chibichenko (1960); Babaev (1970); Babič (1936); Baird (1853); Bashkirova (1960); Baylis (1920, 1928, 1936); Belopolskaya (1959, 1963); Belous (1971); Bezubik (1956); Borgarenko (1972); Chiriac (1965); Creplin (1845 — 1846); Daiya (1967 a); Diesing (1851); Dubinin & Dubinina (1940); Dubinina (1937); Dubinina & Serkova (1951); Feyzullaev (1963 a, b); Ginetsinskaya (1952); Gubanov & Daiya (1967); Gubsky (1960); Gurlt (1845); Gvozdev & Kasymzhanova (1965); Hartwich (1964)*; Hoeppli & Hsü (1929); Hsü (1957); Iordăchescu (1957); Iordăchescu-Lăzărescu (1963); Ivanitsky (1940); Kasimov & Feyzullaev (1965); Kibakin (1965); Kosupko (1963); Kurashvili (1953, 1957); Kurochkin & Zablotsky (1961); Layman & Andronova (1926)*; Leonov (1956, 1960 a); Linstow (1883, 1896, 1906*); Macko (1961 a, 1964 b); Molin (1861 b); Mozgovoy (1953)*; Mozgovoy, Semenova & Shakhmatova (1965, 1968); Myers, Kuntz & Wells (1962); Okorokov & Tkachev (1969); Oshmarin (1963); Oshmarin & Parukhin (1963); Panova (1927); Parona (1887 a, b, 1894); Rudolphi (1809*, 1819); Sailov (1962, 1965 b); Schneider (1866)*; Serkova (1948); Shakhtakhtinskaya (1959 a, b); Shakhtakhtinskaya & Sadykov (1967); Shigin (1957); Skrjabin (1915a, 1923); Smetanina (1972); Smogorzhevskaya (1961, 1962 a, 1964); Smogorzhevskaya, Kornyushin, Iskova & Eminov (1965);

Sonin & Larchenko (1974); Sprehn (1962); Stossich (1890 b, 1892 b, 1893, 1901); Sultanov (1959 a, b, 1963); Sultanov, Ryzhikov & Kozlov (1960); Tsacheva-Petrova (1971); Turemuratov (1962 a, b, 1963 a, 1964, 1966); Vaidova (1965, 1969); Vojtěchovská-Mayerová (1952); Yamaguti (1935*, 1941); Zhatkanbaeva (1964).

Contracaecum andersoni Vevers, 1923

Fig. 27
Hosts: *Podiceps ruficollis* and *Phalacrocorax pygmaeus*.
Localization: small intestine.
Distribution: Asia (U.S.S.R. — Tajikistan). Outside the Palaearctic Region, in South America.

Description: Lips relatively large, hexagonal, 0.16 mm long, each bearing two double papillae. Interlabia long, triangular, their tips not bifurcated. Two blind processes present at level of junction of oesophagus with intestine, one directed backwards, 0.5 mm long and the other forwards, 2.75 mm long. Nerve ring 0.325 mm from anterior end of body.

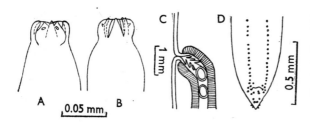

Fig. 27. *Contracaecum andersoni* Vevers, 1923. A — anterior end (dorsal view); B — anterior end (ventral view); C — region of vulva (lateral view); D — posterior end of male (ventral view). After Vevers (1923).

Male: Body length 15.0—17.0 mm, body width 0.35 mm. Tail short, conical. Cloaca 0.085 mm from tail end. Large number of caudal papillae; five pairs of postcloacal papillae, the first one being doubled, arranged in two rows on both sides of tail; five pairs of paracloacal papillae situated on sides of cloaca. Precloacal papillae (20—25 pairs) arranged in two longitudinal rows. Spicules equal, 1.0 mm long and 0.01 mm wide with sharp distal ends.
Female: Body length 19—22 mm, body width 0.75 mm. Tail conical, distincly pointed. Anus 0.13 mm from tail end. Vulva some distance in front of middle of body length, 9.5 mm from anterior end (in specimens 20 mm long.). Vagina muscular, 0.15 mm long. Eggs oval, 0.055 × 0.045 mm, with thick mosaic shell.

This species was described from specimens recovered from herons, *Florida caerulea*, in South America (British Guiana).

References: Borgarenko (1972); Vevers (1923)*.

Contracaecum bodenheimeri Witenberg, 1929

Fig. 28
Host: *Procellaria diomedea*.
Localization: intestine.
Distribution: Africa (A.R.E.); no records since Witenberg's original description.

Description (female): Body length 19.0 mm, maximum body width 0.64 to 0.68 mm. Body tapering towards both extremities. Cuticle with fine transverse striations, which at a distance of 0.03—0.04 mm from head are very deep and form peculiar collar consisting of narrow cuticular foldings. Mouth with three lips and three interlabia. Lips measure 0.09 mm, interlabia 0.03 to 0.05 mm. Dorsal lip bears two oval papillae, latero-ventral lips each bear one

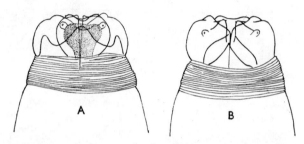

Fig. 28. *Contracaecum bodenheimeri* Witenberg, 1929. A — anterior end (dorsal view); B — anterior end (ventral view). After Witenberg (1929).

similar papilla. Otherwise, all three lips are similar, rounded, with evagination on anterior margin and two conical teeth on the inner surface of each lip, each of which bears a small oblique thorn directed outwards. Pulp of dorsal lip heart-shaped. All lips hollow on inner surface and form a dome-like mouth cavity. Interlabia conical, with wide base, of different lengths and sometimes bear small longitudinal appendage on their tips. Both cervical papillae flat, situated 0.38—0.46 mm from anterior end of body. Oesophagus 3.7 mm long, with a terminal appendage measuring 0.04—0.10 mm. Intestinal caecum 2.72 mm long. Vulva situated between anterior and middle third of body. Anus 0.26—0.31 mm from conical posterior end of body, which bears a small chitinized thorn 0.014 mm long.

The species was described on the basis of female specimens (only) recovered from *P. diomedae* in A.R.E. (Suez).

References: Witenberg (1929)*.

Contracaecum haliaeti Baylis & Daubney, 1923

Fig. 29
Hosts: *Haliaetus leucoryphus*.
Localization: intestine.

Distribution: Asia (Mongolia). Outside the Palaearctic Region in North America and South-East Asia.

Description (female): Lips hexagonal, somewhat longer than wide, with anterior margin distinctly saddle-shaped in the middle, extending laterally over the two not very prominent ear-like projections, which are directed slightly backwards. Lobes of pulp consist of a short-tipped median and a larger, rounded, lateral lobe. Interlabia not bifurcated distally, measuring more than 3/4 of length of lips. Cervical alae absent. Body of female 50.0 mm long,

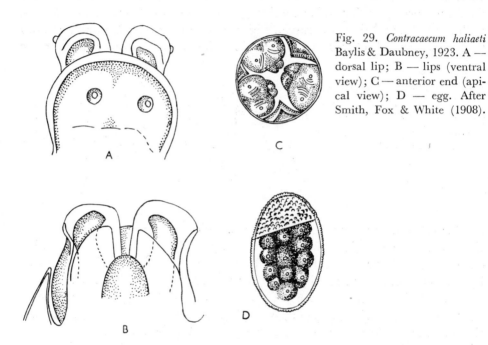

Fig. 29. *Contracaecum haliaeti* Baylis & Daubney, 1923. A — dorsal lip; B — lips (ventral view); C — anterior end (apical view); D — egg. After Smith, Fox & White (1908).

maximum width 1.5 mm. Oesophagus 5.44 mm long, intestinal caecum 4.14 mm long. Ventriculus with appendix 1.37 mm long. Vulva 18.7 mm from anterior end of body. Posterior end of body pointed, conical, tail 0.49 mm long. Eggs 0.05—0.06 × 0.071—0.090 mm, with shell covered with bosses. (Only females known.)

This species was described from the host *Haliaetus leucocephalus* in the U.S.A. as *Ascaris aquillae* Smith, Fox & White, 1908. On the basis of their study of type material, Baylis & Daubney (1923) transferred it to the genus *Contracaecum* and renamed it *C. haliaeti* as the original name was preoccupied by *Ascaris aquillae* Gmelin, 1790. Baylis & Daubney (1923) also assigned *Ascaris zeylanica* Linstow, 1907 and *A. fissicollis* Linstow, 1906 to this species as synonyms.

References: Baylis & Daubney (1923)*; Cram (1927)*; Hartwich (1966)*; Mozgovoy (1953)*; Smith, Fox & White (1908)*.

Contracaecum matwejewi Layman & Mudretsova, 1926

Fig. 30
Host: *Sterna paradisea*.
Localization: stomach, intestine.
Distribution: U.S.S.R. — the Barents Sea.

Description: Mouth with three lips and three interlabia. Dorsal lip rounded without deepening and dentigerous ridges, with two papillae symmetrically

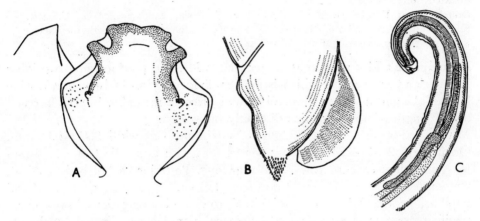

Fig. 30. *Contracaecum matwejewi* Layman & Mudretsova, 1926. A — dorsal lip and interlabium; B — posterior end of male (lateral view); C — anterior end (overall view). After Layman & Mudretsova (1926).

arranged on its surface. Lip pulp forming four processes anteriorly which reach up to anterior margin. Ventriculus with appendix directed caudally. Intestinal caecum directed anteriorly. Length of appendix and intestinal caecum nearly the same (0.950—0.969 mm).
Male: Body length 30.0—39.0 mm, body width 0.49—0.57 mm. Oesophagus 2.295 mm long. Cuticular ridge, characteristic of this species, situated on dorsal side of tail end. Cloaca 0.115 mm from tail end. Tip of tail conical, covered with minute cuticular spines. Spicules equal (2.311 mm), with blunt proximal end, cuticular spicular sheaths present. Caudal papillae have not been studied.
Female: Body length 35.9 mm, body width 0.66—0.69 mm. Oesophagus 2.95 mm long. Anus 0.23 mm from tail end. Tip of tail armed with spines (as

in male). Vulva 0.328 mm from anterior end of body. Eggs spherical, 0.059 mm in diameter.

This species was found only once and was described from the type host *(S. paradisea)* in the U.S.S.R. (Barents Sea).

References: Layman & Mudretsova (1926)*; Mozgovoy (1953)*.

Contracaecum micropapillatum (Stossich, 1890)

Fig. 31

Hosts: *Podiceps ruficollis, Pelecanus crispus, P. onocrotalus, Pelecanus* sp., *Phalacrocorax carbo, Ph. pygmaeus, Ardea purpurea, Egretta alba, Mergus squamatus* and *Stercorarius pomarinus*. Not reported from other orders of birds.
Localization: stomach, intestine.
Distribution: Europe (England, Yugoslavia, Bulgaria, Rumania), Asia (U.S.S.R. — Kazakhstan and republics of Middle Asia, Far East; Afghanistan). Outside the Palaearctic Region in North America, Africa, Australia.

Description: Lips somewhat longer than wide. Lip pulp forming two lobes in anterior part, each of which is further divided by an incision into a wide lateral lobe and a finger-like small lobulus directed medially. Interlabia reach 3/4 of length of lips, with tips distinctly bifurcated.
Male: Body length 10.30—25.20 mm, maximum body width 0.41—0.88 mm. Oesophagus 2.18—3.74 mm long, ventriculus with appendix 0.50—1.16 mm long, intestinal caecum 1.58—2.69 mm long. Posterior end of body conical, tip pointed. Cloaca 0.21—0.32 mm from tail end. 29—45 pairs of precloacal papillae, arranged in two longitudinal rows and four pairs of postcloacal papillae present: three pairs of postcloacal papillae are formed by double papillae, one pair by simple papillae; the first pair of double papillae being situated at the end of first third of tail and the other pairs in lower half; the pair of single papillae lies laterally to the last pair of double papillae. Spicules similar, subequal (left 1.21—3.46 mm, right 1.29—3.53 mm long), with longitudinal alae.
Female: Body length 11.80—30.80 mm, maximum body width 0.40—0.77 mm. Oesophagus 2.62—4.12 mm long, ventriculus with appendix 0.67—1.54 mm long, intestinal caecum 2.00—3.47 mm long. Vulva 6.0—10.3 mm from anterior end of body. Posterior end of body conical, with pointed tip. Phasmids 0.083—0.118 mm and anus 0.28—0.41 mm, respectively, from tip of tail. Eggs 0.056—0.062 × 0.047—0.049 mm, with characteristic surface structure.

The species was originally described as *Ascaris micropapillata* from specimens recovered from *Pelecanus* sp. in Yugoslavia. It was transferred to the genus *Contracaecum* by Baylis (1920).
Note: Hartwich (1964) considers *Contracaecum bancrofti* Johnston & Mawson, 1941 to be a synonym of this species.

Biology: According to Semenova (1971), in *C. micropapillatum* cleavage does not begin until the eggs have been expelled from the host. Depending on the temperature of the environment the eggs become segmented and the larvae form within 4 to 25 days. The first moult occurs within the eggs, five to six days after formation of the first-stage larva. The 2nd-stage larvae are free-living and after actively hatching from the egg may live in the environment at temperatures of 19—25 °C for 18 to 25 days. After infection of the intermediate hosts, which are different species of *Copedoda* and fry, the larvae shed the cuticular sheath in the intestine of the copepode and penetrate into haemocoel. After further four to six days they moult again and become infective to the definitive

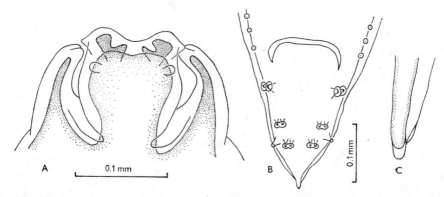

Fig. 31. *Contracaecum micropapillatum* (Stossich, 1890). A — dorsal lip; B — posterior end of male (ventral view); C — distal end of spicule. After Hartwich (1964).

hosts. It was found that infective larvae may also concentrate in paratenic hosts, which are dragon-flies, tadpoles and different fish species in which the larvae encyst and enlarge but do not change in their morphology.

References: Ablasov, Iksanov & Chibichenko (1960); Babaev (1970); Babič (1936); Baruš, Kullmann & Tenora (1972); Baylis (1920); Chiriac (1965); Hartwich (1964)*; Jančev (1958); Mozgovoy (1953)*; Semenova (1971, 1972); Smetanina (1972); Smogorzhevskaya, Kornyushin, Iskova & Eminov (1965); Stossich (1890 a)*; Turemuratov (1962 a, b, 1963 a, 1964, 1966).

Contracaecum milviensis Karokhin, 1937

Fig. 32
Hosts: *Haliaetus albicilla*. Reported also from *Falconiformes* which do not eat fish.
Localization: stomach and small intestine.
Distribution: Asia (U.S.S.R. — Azerbaijan). Not recorded outside the Palaearctic Region.
Description: Lips hexagonal, with rounded tips, narrowed at their base. Dorsal lip with two large subdorsal papillae, latero-ventral lips with a subventral papilla and a small lateral papilla. Lip pulps with deep incision in

medial part. Anterior lobes divided by a short incision into small lateral and medial projections. Interlabia only a little shorter than lips, with bifurcated tips. Intestinal caecum approximately four times as long as ventricular appendix.
Male: Body length 23.0—25.0 mm long, maximum body width 0.61—0.75 mm. Oesophagus 3.39—3.63 mm long, ventriculus 0.153 mm, ventricular appendix 0.684 mm and intestinal caecum 2.479—2.922 mm long, respectively. Tail end tapering, with 38 pairs of precloacal papillae and seven pairs of post-

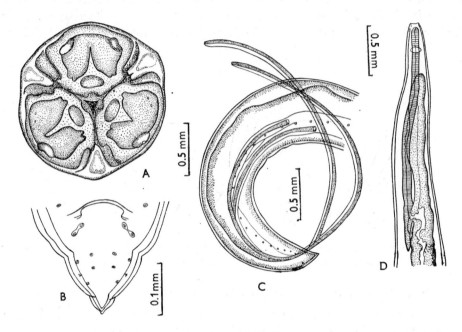

Fig. 32. *Contracaecum milviensis* Karokhin, 1937. A — anterior end (apical view); B — caudal end of male (ventral view); C — posterior end of male (lateral view); D — anterior end of female (lateral view). After Karokhin (1937).

cloacal papillae. Two first pairs of postcloacal papillae situated close to one another, a short distance from cloaca, remaining papillae occupying lower half of tail (three pairs medial, two pairs lateral). Precloacal papillae arranged in two longitudinal rows. Cloaca 0.222 mm from tail end. Spicules 10.20 to 10.84 mm long, each with two longitudinal alae. Width of spicules with alae 0.056—0.058 mm.
Female: Body length 21.0—23.0 mm, maximum body width 0.72 mm. Interlabia 0.068—0.084 mm long. Oesophagus 3.42—3.52 mm long, ventriculus 0.136 mm, ventricular appendix 0.632—1.036 mm and intestinal caecum 2.48—2.67 mm long respectively. Vulva 8.0—9.0 mm from anterior end. Posterior end of body conical, anus 0.290 mm from tail end. Phasmids 0.085

mm from tail end. Eggs 0.066 × 0.044 mm, shell with longitudinal shallow depressions on surface, giving corrugated appearance.

The species was described from specimens recovered from *Milvus korschun* in the U.S.S.R. (West Siberia).

References: Karokhin (1937)*; Mozgovoy (1953)*; Samedov (1967 a,b, 1969).

Contracaecum ovale (Linstow, 1907)

Fig. 33
Hosts: *Podiceps cristatus, P. auritus, P. nigricollis, P. griseigena, P. ruficollis, Platalea leucorodia, Egretta alba* and *Ardea purpurea*. Recorded also in *Ralliformes* and *Anseriformes*.
Localization: stomach, intestine.
Distribution: Europe (England, G.D.R., Rumania, Czechoslovakia, U.S.S.R. — Ukraine, Lithuania, North and Central Regions), Asia (U.S.S.R.—republics of Middle Asia and Kazakhstan, West Siberia, Far East). Outside the Palaearctic Region in Central and South America, South Africa and Australia.

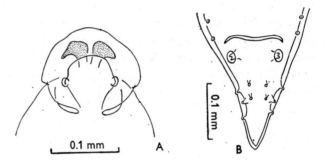

Fig. 33. *Contracaecum ovale* (Linstow, 1907). A — dorsal lip; B — posterior end of male (ventral view). After Hartwich (1964).

Description: Mouth with three lips, the upper margins of which are markedly rounded. Base of lips distinctly narrowed. Lip pulp with two rounded lobes protruding forwards and widening in anterior part (sometimes forming a finger-like lateral projection). Interlabia small, reaching hardly one third of length of lips.
Male: Body length 16.10—32.80 mm, maximum body width 0.34—0.67 mm. Oesophagus 2.46—5.07 mm long, ventriculus with appendix 1.33—2.59 mm, and intestinal caecum 1.70—4.10 mm long. Cloaca 0.18—0.25 mm from tail end. There are 27 to 38 pairs of precloacal papillae arranged in two longitudinal rows. Of the six pairs of postcloacal papillae, the first is double and situated near cloaca, the remaining five pairs are in the lower half of tail, two of them medial and three lateral in position. Posterior end of body conical with rounded tip. Spicules subequal, each with two longitudinal alae, right spicule 1.55—2.86 mm long and left spicule 1.59—3.01 mm long.

Female: Body length 17.35—50.20 mm, maximum body width 0.32—0.96 mm. Oesophagus 2.59—5.51 mm long, ventriculus with appendix 1.20 to 2.72 mm, and intestinal caecum 1.78—3.97 mm long. Vulva 6.50—16.50 mm from anterior end of body. Anus and phasmids 0.35—0.69 mm and 0.125 to 0.235 mm, respectively, from tail end. Eggs oval, 0.064—0.069 × 0.050 to 0.060 mm.

The species was described as *Ascaris ovalis* from specimens recovered from *Podiceps cristatus* in Central Europe. Baylis (1920) transferred it to the genus *Contracaecum*.

Notes: Macko (1961) and Hartwich (1964) regarded *Contracaecum nehli* Karokhin, 1949, *C. spasskii* Mozgovoy, 1950 and *C. ruficolle* Vuylsteke, 1953 as synonyms of *C. ovale* and Baruš & Zajíček (1967) relegated *C. podicipitis* Johnston & Mawson, 1949 to its synonymy also.

Biology: The life-cycle of *C. ovale* has been studied only by Mozgovoy, Shakhmatova & Semenova (1965 a). These authors found that the eggs are expelled into the external environment before the beginning of cleavage. First-stage larvae form within two to three days at temperatures of 27—29 °C. Two to three days later the larvae moult for the first time and some days later leave the eggs. The free larvae survive in water for 15 to 21 days. *Copepoda* (of the genus *Macrocyclops*) were found to serve as intermediate hosts. Other intermediate hosts are dragon-fly larvae (of the genus *Agrion* and *Coenagrion*) and fry of different species of fresh-water fishes. The infectivity of larvae of *C. ovale* from different intermediate hosts was verified by experimental infection of definitive hosts *(Podiceps cristatus* and *P. ruficollis)*.

References: Ablasov (1957); Ablasov & Chibichenko (1961, 1962); Agapova & Zhatkanbaeva (1971); Alekseev & Smetanina (1968); Babaev (1970); Baruš & Zajíček (1967)*; Bashkirova (1960); Baylis (1920, 1939); Chiriac (1965); Daiya (1971); Golovin (1964); Gvozdev & Kasymzhanova (1965); Hartwich (1964)*; Iksanov & Dikambaeva (1962); Karokhin (1949)*; Linstow (1907)*; Macko (1961 b)*; Michelson (1968); Mozgovoy (1950, 1953)*; Mozgovoy, Shakhmatova & Semenova (1965 a); Oshmarin (1950, 1963); Ryzhikov & Kozlov (1959); Sailov (1966); Shigin (1957); Skrjabin (1923)*; Smetanina (1972); Smetanina & Alekseev (1967); Smogorzhevskaya (1960, 1962 a, 1964); Sonin & Larchenko (1974); Sultanov (1963); Turemuratov (1964, 1965 a, 1966); Vaidova (1965); Yigis (1962); Zhatkanbaeva (1964).

Contracaecum pandioni Sobolev & Sudarikov, 1939

Fig. 34
Host: *Pandion haliaetus*.
Localization: stomach, intestine.
Distribution: Europe (U.S.S.R. — the Volga Region), Asia (U.S.S.R. — Azerbaijan, West Siberia, Far East). Not reported outside the U.S.S.R.
Description: Dorsal lip with straight anterior margin and slightly convex

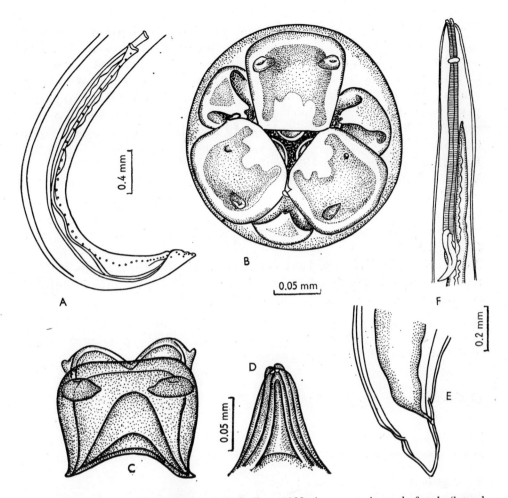

Fig. 34. *Contracaecum pandioni* Sobolev & Sudarikov, 1939. A — posterior end of male (lateral view); B — anterior end (apical view); C — dorsal lip; D — interlabium; E — posterior end of female (lateral view); F — anterior end of female (lateral view). After Sobolev & Sudarikov (1939).

lateral margins, two rounded projections present on inner side of lips; dorsal lip bears one pair of large papillae, latero-ventral lips each with only one large papilla. Lip pulp divided anteriorly by shallow incision into two long lobes directed forwards, each lobe bearing a short finger-like projection directed medially. Interlabia 0.117 mm long, with divided tips. Cervical papillae 0.5 mm from anterior end of body. Oesophagus measures 1/9 of total length of body, intestinal caecum reaches half the length of oesophagus and appendix is nearly as long as intestinal caecum (1/18 of the total body length).

Male: Body length 18.0—34.0 mm, maximum body width 0.46—0.78 mm.

Posterior end of body pointed. Cloaca 0.25 mm from tail end. There are 41 pairs of precloacal papillae arranged in two longitudinal rows, one pair of paracloacal papillae and eight pairs of postcloacal papillae. First pair of postcloacal papillae situated close to cloaca, two other pairs more distant and more ventral, three further pairs still more distant and ventral. The last two pairs of papillae are on their sides. Spicules subequal, 3.49—5.55 mm long. Proximal end of spicules wide, measuring 0.11—0.19 mm, distal end measuring 0.03 to 0.08 mm.
Female: Body length 73—75 mm, maximum body width 1.14—1.15 mm. Vulva in anterior part of body, 11.12—13.42 mm from cephalic end. Tail conical, pointed, 0.34—0.46 mm long. Eggs 0.06 × 0.09 mm, shell covered with small bosses.

The species was described from *Pandion halieatus* in the Volga Region (U.S.S.R.).

Notes: *C. pandioni* is believed to be conspecific with *C. haliaeti* Baylis & Daubney, 1923 but as only females of the latter are known, the question of their synonymy cannot be resolved with certainty.

References: Daiya (1967 a); Mozgovoy (1953)*; Oshmarin (1963); Oshmarin & Parukhin (1960, 1963); Sailov (1962, 1965 b); Shigin (1954, 1959); Sobolev & Sudarikov (1939)*; Vaidova (1965); Zablotsky (1962).

Contracaecum praestriatum Mönnig, 1923

Fig. 35

Hosts: *Podiceps nigricollis*. Outside the Palaearctic Region, found in *P. capensis* (Republic of South Africa).
Localization: not given in original description.
Distribution: Europe (Poland), Asia (U.S.S.R. — West Siberia).

Description: Lips thick, wide, of unequal size. Dorsal lip with pair of double papillae, latero-ventral lips each with one double papilla. Dentigerous ridges absent. Margins of lips with thick cuticular border. Interlabia short, triangular, with very wide base. Tip of tail pointed in both males and females.
Male: Body length 19—22 mm, maximum body width 0.65—0.80 mm. Oesophagus 3.30—3.94 mm long, intestinal caecum 2.00—2.06 mm, and ventricular appendix 1.94—2.00 mm long. Tail 0.16—0.30 mm in length, bearing five pairs of postcloacal papillae of which first pair is double, situated close to cloaca; other four pairs in middle third of tail, second and fourth pair ventral, third and fifth pair lateral. More than 17 pairs of precloacal papillae arranged in two longitudinal rows on sides of body, of which first pair is situated 3.6 mm in front of cloaca. Spicules equal, 1.96—2.00 mm long, with longitudinal alae which taper markedly at the end of central section and then form

with it a small blunt protrusion. Smal gubernaculum measures 0.12—0.14 mm in length.

Female: Body length 30.0—33.0 mm, body width 0.8 mm. Oesophagus with ventriculus together 4.1 mm in length. Appendix 2.0 mm and intestinal caecum 2.3 mm long. Nerve ring 0.6 mm from anterior end of body. Anus 0.36 mm from tail end. Vulva situated at level of transition from thick transverse cuticular striations to finer ones and thus divides body in the ratio of 5 : 12. Ovejector runs first caudally, then turns and divides into two uterine branches. Eggs thin-shelled, 0.040—0.056 × 0.032—0.050 mm.

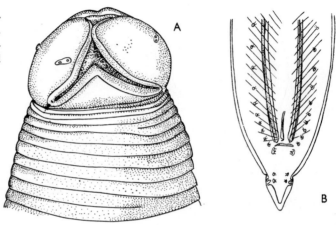

Fig. 35. *Contracaecum praestriatum* Mönnig, 1923. A—anterior end; B — posterior end of male (ventral view). After Mönnig (1923).

The species was described from specimens recovered from *Podiceps capensis* in Africa.

Notes: In the Palaearctic Region this species has been recorded by Serkova (1948) and Bezubik (1956) only. However, these authors gave no data on the morphology of their material. Considering that the species characteristic for the hosts in this region is *C. ovale* (see the survey of findings), these records of *C. praestriatum* are doubtful.

References: Bezubik (1956); Mönnig (1923)*; Mozgovoy (1953)*; Serkova (1948).

Contracaecum rudolphii Hartwich, 1964

Fig. 36

Hosts: *Gavia adamsii, G. arctica, G. stellata, Podiceps cristatus, P. auritus, P. nigricollis, P. griseigena, P. ruficollis, Podiceps* sp., *Fulmarus glacialis, Pelecanus crispus, P. onocrotalus, Phalacrocorax carbo, Ph. aristotelis, Ph. pelagicus, Ph. pygmaeus, Ph. urile, Ph. capillatus, Phalacrocorax* sp., *Ardea cinerea, A. purpurea, Botaurus stellaris, Nycticorax nycticorax, Egretta alba, Mergus merganser, M. albellus, M. serrator, M. squamatus, Stercorarius parasiticus, S. longicaudatus, Rissa tridactyla, R. brevirostris,*

Rhodosthethia rosea, Larus canus, L. argentatus, L. crassirostris, L. hyperboreus, L. fuscus, L. genei L. ichthyaetus, L. marinus, L. minutus, L. ridibundus, Chlidonias hybrida, Ch. leucoptera, Hydroprogne tschegrava, Gelochelidon nilotica, Sterna hirundo, S. alaeutica, S. paradisea, S. sandvicensis, Fratercula corniculata, Lunda cirrhata, Synthliboramphus antiquus, Cepphus carbo, C. grylle, Uria aalge, U. lomvia and *Alca torda*.
Localization: intestine, stomach.
Distribution: Europe (England, Iceland, Netherlands, Yugoslavia, Italy, Switzerland, Rumania, F.R.G., G.D.R., Bulgaria, Czechoslovakia, Denmark, Norway, Poland, U.S.S.R.— Ukraine, Lithuania, North and Central Regions, Volga Region), Asia (U.S.S.R.—republic of Transcaucasus Region and North Asia, Kazakhstan, Turkmenistan, West and East Siberia, Far East), Japan, Africa (A.R.E.), Tunis. Cosmopolitan.

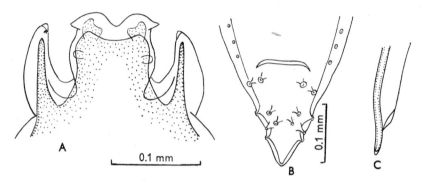

Fig. 36. *Contracaecum rudolphii* Hartwich, 1964. A — dorsal lip; B — posterior end of male (ventral view); C — distal end of spicule. After Hartwich (1964).

Description: Lips slightly wider than long, with marked medial deepening on upper margin. Lip pulp in anterior part forming two lobes, each divided into a rounded lateral lobulus and a rounded medial lobulus. Interlabia reach 4/5 of length of lips, with tip distinctly bifurcated.
Male: Body length 12.10—33.90 mm, maximum body width 0.24—0.95 mm. Oesophagus 2.03—4.26 mm long, ventriculus with appendix 0.58—1.37 mm, and intestinal caecum 1.53—3.68 mm long. Posterior end of body conical, with tip rounded. 27—43 pairs of precloacal papillae arranged irregularly in two longitudinal rows. Cloaca 0.14—0.24 mm from tail end. Seven pairs of simple postcloacal papillae; first two pairs a small distance behind cloaca; remaining five pairs in middle third of tail length and of these three pairs are more lateral, and two pairs more medial. Spicules similar, subequal (left 4.05—9.98 mm, right 4.46—9.19 mm long), with longitudinal alae.
Female: Body length 10.10—57.60 mm, maximum body width 0.29 to 1.51 mm. Oesophagus 1.62—5.48 mm long, ventriculus with appendix 0.62 to 1.58 mm, and intestinal caecum 1.28—4.12 mm long. Vulva 5.12—17.70 mm from anterior end of body. Posterior end of body conical, with rounded tip. Phasmids open 0.068—0.165 mm from tail end, anus 0.19—0.63 mm

from tail end. Eggs spherical to oval, 0.059—0.073 × 0.041—0.059 mm, surface of shell with shallow depressions.

This species was described from specimens from the type host *Phalacrocorax carbo* from G.D.R.

Biology: The life-cycle of *C. rudolphii* (=*C. spiculigerum*) was studied by Huizinga (1966) and Mozgovoy, Shakhmatova & Semenova (1965, 1968). The eggs are expelled to the outside environment before the start of cleavage; at temperatures of 21—29 °C larvae are formed within three to seven days. According to Huizinga (1966) and Mozgovoy et al. (1968) the first moult occurs within the egg. The first intermediate hosts or, if not true intermediate hosts, the transport hosts, of the larvae of this nematode are different species of *Copepoda*. The larvae are located in the haemocoel of their hosts. The second (or first) intermediate hosts are fishes and *Odonata*, in which the larvae moult for the second time and become infective for the definitive hosts — fish-eating birds. Mozgovoy et al. (1968) fed infective larvae from dragon-flies to fishes (*Tinca tinca*) where they encysted in the body cavity. The authors indicated this fish species as a paratenic host of the infective larvae.

Notes: Following Mozgovoy (1953) we consider *C. himeu* Yamaguti, 1941 and *C. umiu* Yamaguti, 1941 to be synonyms of *C. rudolphii*.

References: Ablasov (1957); Ablasov & Chibichenko (1961, 1962); Ablasov, Iksanov & Chibichenko (1960); Agapova & Zhatkanbaeva (1971); Alekseev (1970); Alekseev & Smetanina (1968); Babaev (1970); Babič (1936); Baird (1853); Bakke (1972); Bakke & Baruš (1975); Bashkirova (1960); Baylis (1923, 1928, 1939); Belogurov (1965); Belogurov, Leonov & Zueva (1968); Belopolskaya (1952, 1959, 1963); Bezubik (1956); Braun (1892); Broek & Jansen (1964); Chiriac (1963, 1965); Chiriac, Udrescu & Theodorescu (1970); Cram (1927); Creplin (1829, 1845—1846); Daiya (1967 a); Danzan (1964); Diesing (1851); Dubinin (1954); Dubinin & Dubinina (1940); Dubinina & Serkova (1951); Ellis & Williams (1973); Feyzullaev (1963 a, b); Golikova (1959); Gubanov (1971); Gubanov & Daiya (1967); Gubanov & Sergeeva (1971); Gurlt (1845); Hartwich (1964)*; Huizinga (1966)*; Iksanov & Dikambaeva (1962); Iordăchescu (1957); Jägerskiöld (1894); Kasimov & Feyzullaev (1965); Kibakin (1965); Kontrimavichus & Bakhmeteva (1960); Korpaczewska & Sulgostowska (1967); Kosupko (1963); Kowalewski (1904); Kreis (1958); Krivonogova (1963); Kurashvili (1957); Kurochkin & Zablotsky (1961); Leonov (1956, 1958); Leonov & Belogurov (1963); Lewis (1926, 1927); Linstow (1883, 1886, 1901 a, 1903); Machida (1966); Markov (1941); Mozgovoy (1950*, 1953*); Mozgovoy & Romanova (1956); Mozgovoy, Shakhmatova & Semenova (1965 b, 1968); Muehling (1898); Nicoll (1927); Nikolskaya (1939); Okorokov (1957, 1964); Okorokov & Tkachev (1969); Oshmarin (1963, 1965); Oshmarin & Parukhin (1963); Panova (1927); Parona (1887 a, b, 1894); Pemberton (1963); Petrov & Chertkova (1950); Radulescu (1966)*; Reimer (1969, 1973); Rudolphi (1809*, 1819); Ryšavý (1958); Ryzhikov (1963 a, b); Ryzhikov & Daiya (1967); Ryzhikov & Kozlov (1959); Sailov (1962, 1965 a, b, 1970); Sergeeva (1969); Sergienko (1963, 1972); Serkova (1948); Seurat (1918); Shakhtakhtinskaya (1959 a, b); Shigin (1959); Škarda (1964); Skrjabin (1915 a, 1923, 1924, 1926); Smetanina (1972); Smetanina & Alekseev (1967); Smogorzhevskaya (1962 a, b, 1964, 1967); Smogorzhevskaya, Kornyushin, Iskova & Eminov (1965); Solonitsin (1928 a, b); Sonin & Larchenko (1974); Spasskaya (1949); Sprehn (1962); Stossich (1889, 1892 b, 1893, 1895, 1896 b, 1902); Sultanov (1959 a, b, 1963); Sultanov,

Ryzhikov & Kozlov (1960); Threlfall (1965 a, b, 1966 c, 1967); Tkachev (1971); Tolkacheva (1967); Tsimbalyuk (1965); Tsimbalyuk & Belogurov (1964); Turemuratov (1962 a, b, 1963 a, 1964, 1965 a); Vaidova (1963, 1965, 1969); Vevers (1920); Volskis (1966, 1968); Williams (1961); Yamaguti (1941); Zajíček & Páv (1961); Zhatkanbaeva (1964, 1965, 1966 a, 1971); Zhelyazkova-Paspaleva (1962 a, b).

Contracaecum septentrionale Kreis, 1955

Fig. 37
Host: *Phalacrocorax aristotelis*.
Localization: stomach.
Distribution: Iceland.

Description: Length of lips 0.103—0.147 mm. Both sublateral lips conical, with very wide base, each bearing a papilla and pulp divided into wide basal part and triangular anterior part. Dorsal lips distinctly divided into two parts, of which the basal is widely rounded anteriorly and the anterior wing-like

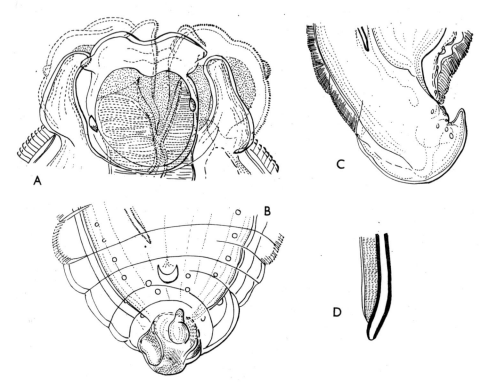

Fig. 37. *Contracaecum septentrionale* Kreis, 1955. A — anterior end (dorsal view); B — posterior end of male (ventral view); C — posterior end of male (lateral view); D — distal end of spicule. After Kreis (1955).

part is invaginated in the middle, with conical processes on sides of wing. Labial pulp consists of small narrow basal part, wide middle part and flat triangular anterior part. Interlabia conical, in males 0.064—0.086 mm long, in females 0.076—0.118 mm long.

Male: Body length 31.6—38.0 mm, maximum body width 0.88—1.24 mm. Form of oesophagus typical of genus. Length of ventriculus 0.168—0.210 mm, appendix 0.756—1.029 mm, and intestinal caecum 2.05—2.14 mm. Spicules 12.57—15.16 mm long. Along the whole length of spicules runs a finely striated lateral membrane, not reaching the terminal point. 30—35 pairs of precloacal and six pairs of postcloacal papillae present, two of latter being situated immediately behind cloaca nearly in ventral line. Behind it there are two lateroventral pairs of papillae and, close to tip of tail, two other pairs of very small papillae. Tail 1.21—1.66 mm long.

Female: Body length 39—45 mm, maximum body width 1.30—1.89 mm. Length of ventriculus 0.189—0.378 mm, appendix 1.00—1.13 mm, and intestinal caecum 1.51—2.08 mm. Vulva in first third of body; relatively long vagina bifurcating to two uterine branches. Tail conical, with blunt tip. Tail 0.9—0.95 mm long.

This species was described from specimens recovered from *Ph. aristotelis* from Iceland. Not reported since first described.

References: Kreis (1955*, 1958).

Contracaecum travassosi Gutierrez, 1943

Fig. 38
Hosts: *Phalacrocorax carbo*. Outside the Palaearctic Region, reported from cormorants and ospreys.
Localization: stomach, intestine.
Distribution: Europe (U.S.S.R. — Ukraine), also North and South America.

Description:
Male: Body length 34.0—58.0 mm; body width 0.576—0.847 mm and at level of lips 0.256—0.320 mm. Oesophagus 3.5—7.0 mm long. Nerve ring 0.487—0.705 mm from anterior end of body. Lips 0.153—0.192 mm wide. Interlabia present. Intestinal caecum 2.1—4.4 mm long. Cloaca 0.303 mm from tail end. Spicules subequal, left 7.5—12.9 mm, right 7.2—11.4 mm long. 30 pairs of precloacal papillae and six pairs of postcloacal papillae present, the first of them (close to cloaca) formed of double papillae.
Female: Body length 53.0—55.0 mm, body width 1.1—1.3 mm, at level of lips 0.375—0.384 mm. Oesophagus 5.0—5.1 mm long. Nerve ring 0.75—0.76 mm from anterior end. Lips 0.210—0.217 mm wide. Intestinal caecum 4.0 to 4.3 mm long. Anus 0.43—0.448 mm from tail end. Vulva 24.0 mm from an

terior end of body. Eggs 0.062 × 0.043 mm. Two branches of uterus present.

The species was described from specimens recovered from *Phalacrocorax albiventris* in Argentine; later it was found by Morgan, Schiller & Rausch (1949) in *Pandion haliaetus* in the U.S.A. It has only once been recorded in the

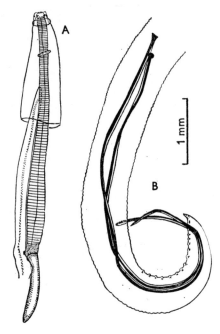

Fig. 38. *Contracaecum travassosi* Gutierrez, 1943. A — anterior end (overall view); B — posterior end of male (lateral view). After Lent & Freitas (1948).

Palaearctic Region, in the U.S.S.R. (Ukraine — Sergienko 1963) but the nematode was not described. The occurrence of this species in the Palaearctic Region seems, therefore, questionable.

References: Gutierrez (1943)*; Morgan, Schiller & Rausch (1949)*; Sergienko (1963).

Contracaecum variegatum (Rudolphi, 1809).

Fig. 39
Hosts: *Gavia immer, G. stellata, Rissa tridactyla, Larus canus, L. ichthyaetus, Alca torda, Uria aalge* and *U. lomvia*.
Localization: stomach, intestine.
Distribution: Europe (F.R.G., G.D.R.), Asia (U.S.S.R.—Far East, Mongolia, Japan). Outside the Palaearctic Region—Central America and Australia.

Description: Mouth with rounded hexagonal lips. Base of lips of same width as opposite margin. Lip pulp protruding forward by two small lobes only, each of which bears a small medial finger-like process. Interlabia reach two thirds of length of lip with bifurcate tips.

Male: Body length 8.60—21.84 mm, maximum body width 0.25—0.65 mm. Oesophagus 1.82—3.08 mm long, ventriculus with appendix 0.77—1.17 mm and intestinal caecum 1.25—2.43 mm long. Posterior end of body conical, with rounded tip. Cloaca 0.12—0.22 mm from tail end. 29—46 pairs of precloacal papillae arranged in two longitudinal rows, and seven pairs of postcloacal papillae present. Two pairs of latter are located a small distance behind cloaca, others in middle third of tail length (three pairs more lateral, two pairs more medial). Spicules subequal (right 2.59—5.56 mm, left 2.59—5.60 mm long), similar, with longitudinal alae.

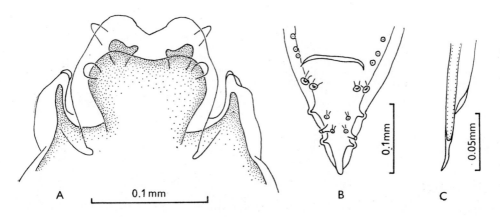

Fig. 39. *Contracaecum variegatum* (Rudolphi, 1809). A — dorsal lip; B — posterior end of male (ventral view); C — distal end of spicule. After Hartwich (1964).

Female: Body length 8.50—36.60 mm, maximum body width 0.25 to 0.79 mm. Oesophagus 1.36—3.70 mm long, ventriculus with appendix 0.52—1.07 mm and intestinal caecum 0.90—2.88 mm long. Vulva 4.10 to 10.20 mm from anterior end of body. Anus 0.18—0.41 mm from tail end. Posterior end of body conical, with pointed tip (thorn-like). Eggs 0.061—0.106 × 0.049—0.071 mm, shell surface is finely granulated.

This species was described as *Ascaris variegata* from specimens from *Gavia stellata* in G.D.R. Hartwich (1964) transferred it to *Contracaecum* as a valid species.

Note: Hartwich (1964) regards *C. magnipapillatum* Chapin, 1925, *C. torquatum* Yamaguti, 1935, *C. oschmarini* Mozgovoy, 1950 and *C. magnicollare* Johnston & Mawson, 1941 as synonyms of *C. variegatum*.

Referents: Bakke (1972); Chapin (1925 b)*; Hartwich (1964*, 1966*); Krotov & Delyamure (1952); Mozgovoy (1950*, 1953*); Rudolphi (1809*, 1819*); Yamaguti (1935)*.

Contracaecum yamaguti Mawson, 1956

Fig. 40
Host: *Mergus merganser.*
Localization: intestine.
Distribution: Asia (Japan), outside the Palaearctic Region in Canada.

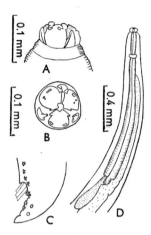

Fig. 40. *Contracaecum yamaguti* Mawson, 1956. A — anterior end (lateral view); B — anterior end (apical view); C — caudal end of male (lateral view); D — anterior end of female (lateral view). After Mawson (1956).

Description:
Male: Body length 15.0—34.0 mm, maximum body width 0.45—1.0 mm. Cuticle forms deep circular folds behind lips. Lips 0.06—0.11 mm high. Pulp of each lip with strong medial lobe and two small antero-medial lobes. Interlabia 0.042—0.1 mm long, in young specimens their tip is bifurcated. Nerve ring and cervical papillae 0.38—0.63 mm and 0.42—0.65 mm, respectively, from anterior end of body. Oesophagus 2.0—4.0 mm long, ventriculus 0.15 to 0.2 mm, appendix 0.5—1.1 mm, and intestinal caecum 1.5—2.9 mm long. Posterior end of body conical, tail 0.13—0.25 mm long. Spicules equal, similar, 3.4—5.8 mm long, with longitudinal alae. 28—34 pairs of precloacal papillae present, first pair 3.0—7.1 mm from cloaca. Five to six pairs of postcloacal papillae present, three pairs of subventral and two to three lateral pairs on each side of tail. First pair of caudal papillae formed by large double papillae, middle pair of middle group sometimes absent or rudimentary.
Female: Body length 32.0 mm, maximum body width 0.66 mm. Lips 0.088 mm long, interlabia 0.07 mm long. Nerve ring and cervical papillae 0.56 to 0.6 mm from anterior end of body. Oesophagus 3.1 mm long, ventriculus 0.15 mm, appendix 0.85 mm and intestinal caecum 2.38 mm long. Tail conical, 0.4 mm long, with two small lateral papillae near the tip. Vulva 10.0 mm from anterior end of body. Eggs nearly spherical or oval, 0.06—0.072 × 0.038 to 0.054 mm.

This species was described as *Contracaecum* sp. from *Mergus merganser* in Japan. Mawson (1956) proposed the name *C. yamaguti* for this species.

References: Mawson (1956)*; Mozgovoy (1953)*; Yamaguti (1941)*.

Genus *Porrocaecum* Railliet & Henry, 1912

Mouth with three distinct lips, interlabia usually small. Posterior to oesophagus the ventriculus and intestinal caecum are directed anteriorly. Oesophageal process absent. Lips armed with small denticles near their inner margin. Gubernaculum absent in most species. Vulva in middle part of body. Eggs spherical to oval, their surface with distinct net-like or other structure. Adults parasites of digestive tract of birds, rarely also of reptiles and fishes.

Type species *Porrocaecum crassum* (Deslongchamps, 1824), common parasite of hosts of the family *Anatīdae* in the Palaearctic Region, rarely encountered also in fish-eating birds.

About 40 species are known to belong to this genus. Only nine of them have been recovered from fish-eating birds in the Palaearctic Region.

KEY TO THE SPECIES OF THE GENUS *PORROCAECUM*

1   Gubernaculum absent 2
—   Gubernaculum present (about 0.28 mm long) . . . . *P. reticulatum*
2   Intestinal caecum at most twice as long as ventriculus or shorter 3
—   Intestinal caecum three or more times as long as ventriculus 4
3   Cervical alae absent . . . . *P. ensicaudatum*
—   Cervical alae present . . . . , *P. semiteres*
4   Cervical alae absent 5
—   Cervical alae present 7
5   Lobes of lip pulp each with anterior projection . . . . *P. phalacrocoracis*
—   Lobes of lip pulp each with two or three finger-like anterior projections 6
6   Lobes of lip pulp each with three short finger-like anterior projections; denticles reach only to level of papillae . . . . *P. crassum*
—   Lobes of lip pulp each with two finger-like anterior projections; denticles reach nearly to base of lips . . . . *P. depressum*
7   Lobes of lip pulp with short finger-like process directed latero-caudally; denticles reach only to level of papillae . . . . *P. ardeae*
—   Lobes of lip pulp rounded on lateral side (without lateral projection); denticles reac nearly to base of lip . . . . *P. angusticolle*

Note: The species *P. praelongum* (Dujardin, 1845) was not included in the Key, because the morphological data available are insufficient for the unequivocal differentiation from other species of this genus. A description is given on p. 100.

Porrocaecum crassum (Deslongchamps, 1824)

Fig. 41
Hosts: *Podiceps griseigena;* common parasite of birds of the family *Anatidae;* very rare in fish-eating birds.
Localization: stomach and small intestine.
Distribution: Asia (U.S.S.R. — West Siberia, Azerbaijan). In birds not eating fish in different areas of the Palaearctic Region and India.

Description: Mouth with three lips and three interlabia. Lips hexagonal, their base distinctly narrower than width of anterior margin. Labial pulp with deep grove dividing it into two lobes, each lobe bearing 3 short finger-like projections. Inner margin of lips armed with a row of denticles. Interlabia triangular, relatively long. Ventriculus oval, intestinal caecum approximately three times as long as ventriculus. Cervical alae absent.
Male: Body length 11.6—52.0 mm, maximum body width 0.5—1.8 mm.

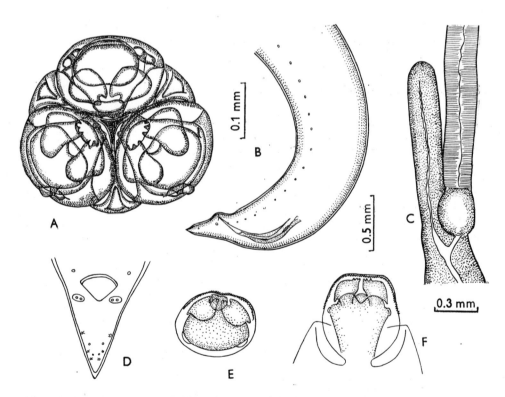

Fig. 41. *Porrocaecum crassum* (Deslongchamps, 1824). A — anterior end (apical view); B — posterior end of male (lateral view); C — region of oesophagus; D — posterior end of male (ventral view); E, F — dorsal lip. A, B, C — After Mozgovoy (1953); D, E, F — After Hartwich (1959).

Length of oesophagus 3.1—3.34 mm, ventriculus 0.35—0.49 mm, and intestinal caecum 1.5—1.63 mm. Spicules equal, 0.338—1.14 mm long, with narrow alae. Gubernaculum absent. Posterior end of body conical, tail with rounded tip. 12—14 pairs of precloacal papillae present, arranged in two longitudinal rows. Double postcloacal pair of papillae situated near cloaca. Five pairs of simple postcloacal papillae situated on posterior half of tail, of which the first, third and fifth pairs lie more laterally and the second and fourth pairs more ventrally to median line.

Female: Body length 43.0—66.0 mm, maximum body width 2.0—2.2 mm. Length of oesophagus 3.4—4.65 mm, ventriculus 0.4—0.46 mm, and intestinal caecum 1.41—1.9 mm. Vulva 19.0—32.0 mm from anterior end of body. Anus 0.39—0.85 mm from tail end. Eggs 0.091—0.118 × 0.068—0.091 mm.

The species was described as *Ascaris crassa* from *Anas platyrhynchos* f. *domestica* in France. Railliet & Henry (1912) transferred it to the genus *Porrocaecum*.

Biology: The life-cycle of this species was studied by Mozgovoy (1952 a, b). The eggs are expelled into the external environment at the blastomere stage. The larvae develop in the eggs for five to six days and the first moult occurs. These larvae are already infective to the earthworm intermediate hosts which, according to Karmanova (1960), are *Allolobophora dubiosa* and *Criodrilus lacuum*. The second moult occurs in the intermediate hosts within two months and the larvae become infective to the definitive hosts. The birds infect themselves by eating the earthworms containing infective larvae. The nematodes reach maturity in the definitive hosts within three weeks.

References: Deslongchamps (1824)*; Dujardin (1845)*; Hartwich (1959)*; Karmanova 1960 b); Mozgovoy (1952 a, b, 1953)*; Paskalskaya (1968); Shakhtakhtinskaya & Sadykov 1967); Skrjabin (1923*; 1926*).

Porrocaecum angusticolle (Molin, 1860)

Fig. 42

Hosts: *Egretta intermedia*, *Pandion haliaetus* and *Haliaetus albicilla*. Also in many species of *Falconiformes*.
Localization: stomach, intestine.
Distribution: Europe (Ukraine, central regions of Russia), Asia (U.S.S.R. — Georgia, East Siberia; Japan). In birds not eating fish in different areas of the Palaearctic Region and also in South-East Asia and North America.

Description: Lips of irregular hexagonal shape, with rounded angles. Base of lips narrower than upper margin. Labial pulp forming two lobes anteriorly, the tips of which are directed towards median line of lips. Row of small denticles on inner margin of lips, reaching nearly to base of lips. Interlabia triangular, reaching half the length of lip. Distinct cervical alae present. Intestinal caecum five to seven times as long as ventriculus.

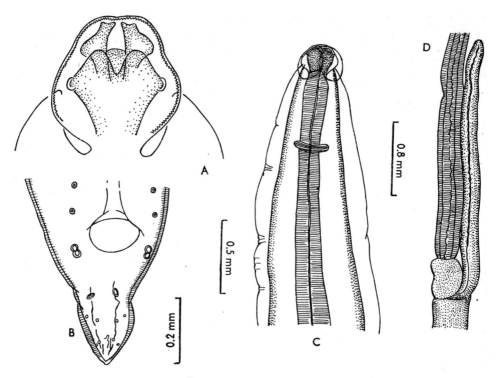

Fig. 42. *Porrocaecum angusticolle* (Molin, 1860). A — dorsal lip; B — posterior end of male (ventral view); C — anterior end (ventral view); D — region of oesophagus. A, B — After Hartwich (1959); C, D — After Mozgovoy (1953).

Male: Body length 59.0—144.0 mm, maximum body width 0.82—1.4 mm. Length of oesophagus 2.28—5.72 mm, ventriculus 0.31—0.83 mm, and intestinal caecum 1.54—4.47 mm. Posterior end of body conical. Cloaca 0.2—0.4 mm from posterior end of body. Tail distinctly tapered in middle. 18 to 22 pairs of precloacal papillae present, arranged in two longitudinal rows, of which the first pair is situated 2.9 mm in front of cloaca. Of the six pairs of postcloacal papillae, the first is double and is situated near cloaca, the other five pairs are on ventral side of lower half of tail (third and fifth pair more laterally). Spicules equal, 0.577—1.21 mm long, with small alae.

Female: Body length 75.0—196.0 mm, maximum body width 0.74—1.62 mm. Length of oesophagus 4.14—6.43 mm, ventriculus 0.45—0.75 mm, and intestinal caecum 2.94—5.36 mm. Vulva 26.0—51.0 mm from anterior end of body. Anus 0.45—0.79 mm from tail end. Posterior end of body conical, with one pair of lateral papillae situated 0.114—0.149 mm from tip of tail. Eggs 0.091—0.118 × 0.068—0.091 mm.

This species was described as *Ascaris angusticollis* from *Buteo buteo* in Austria and transferred to the genus *Porrocaecum* by Baylis & Daubney (1922).

Biology: Osche (1955) found the larvae of this species in cysts on the intestinal wall and mesentery of insectivores and verified their systematic position by experimental infection of the definitive host. In Spassky's opinion small mammals are the reservoir (or paratenic) hosts of infective *Porrocaecum* larvae.

References: Baylis & Daubney (1922); Daiya (1968); Golovin (1964); Gubanov & Daiya (1967); Hartwich (1959)*; Kurashvili (1961); Molin (1860 c)*; Mozgovoy (1953)*; Osche (1955); Oshmarin & Parukhin (1960); Skrjabin (1923); Smogorzhevskaya (1964); Spassky (1952).

Porrocaecum ardeae (Frölich, 1802)

Fig. 43
Hosts: *Ardea cinerea, A. purpurea, Egretta alba, E. garzetta, Nycticorax nycticorax, Butorides striatus* and *Sterna paradisea*. Common parasite of herons. Parasitic also in cranes.

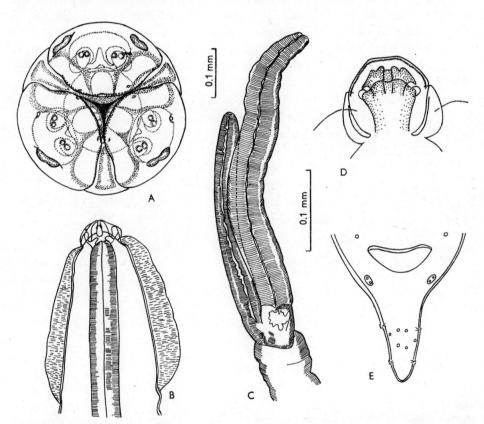

Fig. 43. *Porrocaecum ardeae* (Frölich, 1802). A — anterior end (apical view); B — anterior end overall view); C — region of oesophagus; D — dorsal lip; E — posterior end of male (ventral view). A, B, C — After Mozgovoy (1953); D, E — After Hartwich (1959).

Localization: intestine.
Distribution: Europe (England, Italy, G.F.R., G.D.R., Rumania, U.S.S.R. — Ukraine, Moldavia, Leningrad Region, Volga Region, Ural), Asia (U.S.S.R.—Azerbaijan, Kazakhstan, republics of Middle Asia, Far East). Outside the Palaearctic Region, South-East Asia, North and South America, tropical Africa.

Description: Lip approximately hexagonal, its base nearly as wide as opposite upper margin. Shallow incision divides lip pulp into two lobes. Lateral finger-like projections of pulp short. Median lobe extending far beyond anterior margin of main pulp. Row of denticles near inner margin of lip, reaching nearly the height of subdorsal papillae. Length of interlabia 3/5 of lip length.
Male: Body length 32.0—70.0 mm, maximum body width 0.75—2.5 mm. Oesophagus 2.94—4.24 mm long. Cervical alae distinct and of the same width throughout. Length of ventriculus 0.4—0.54 mm, and intestinal caecum 2.1—3.3 mm. Cloaca 0.31—0.456 mm from tail end. Posterior end of body markedly tapering at the end of its first third, terminating in round tip. 14—22 pairs of precloacal papillae present, arranged in two longitudinal rows. Six pairs of postcloacal papillae present, of which the first pair is double, other five pairs situated on posterior half of tail (first lateral, second ventral, third subventral, fourth ventral and fifth lateral). Spicules equal, 0.994—1.35 mm, with longitudinal alae.
Female: Body length 67.0—160.0 mm, maximum body width 1.45—4.0 mm. Length of oesophagus 3.12—5.47 mm, ventriculus 0.49—0.72 mm and intestinal caecum 2.38—5.4 mm. Vulva 18.7—27.5 mm from anterior end of body. Anus 0.78—2.1 mm from tail end. Eggs 0.099—0.115 × 0.077—0.091 mm.

The species was described as *Ascaris ardeae* from *Ardea cinerea* in Central Europe. Baylis (1936) transferred it to the genus *Porrocaecum*.
Notes: Baylis (1936) considered the species *P. serpentulus* (Rudolphi, 1809) to be a synonym of *P. ardeae* and we fully agree with this opinion.

References: Babaev (1970); Baird (1853); Baylis (1920, 1936); Chiriac (1965); Diesing (1851)*; Dubinin & Dubinina (1940); Dubinina (1937); Dubinina & Serkova (1951); Frölich (1802)*; Gurlt (1845); Hartwich (1959)*; Mozgovoy (1953)*; Oshmarin (1963); Parona (1894); Rudolphi (1809, 1819); Sailov (1962, 1965); Shakhtakhtinskaya (1959 b); Shigin (1957); Skrjabin (1923*, 1926); Smetanina (1972); Smogorzhevskaya (1955); Sonin & Larchenko (1974); Stossich (1896 a)*; Sultanov (1959 a, b, 1963); Zhatkanbaeva (1964).

Porrocaecum depressum (Zeder, 1800)

Fig. 44
Hosts: *Haliaetus albicilla*, *H. leucoryphus* and *H. pelagicus*. Reported also from wide range of hosts of the order *Falconiformes* and *Strigiformes*, more rarely from birds of other orders.
Localization: intestine.
Distribution: Europe (F.R.G., G.D.R., Italy, U.S.S.R. — central regions), Asia (U.S.S.R. —

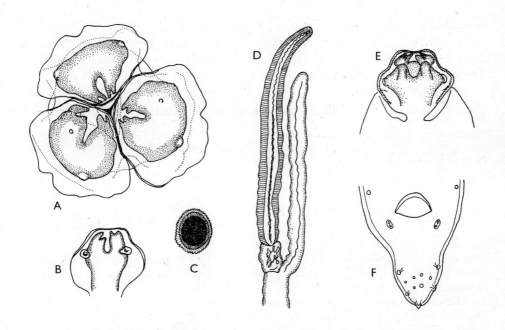

Fig. 44. *Porrocaecum depressum* (Zeder, 1800). A — anterior end (apical view); B, E — dorsa lip; C — egg; D — region of oesophagus; F — posterior end of male (ventral view). A, B, C, D — After Mozgovoy (1953); E, F — After Hartwich (1959).

Kirghizia, Far East). Outside the Palaearctic Region—South-East Asia, North and South America and Australia.

Description: Lips hexagonal, very narrow at base. Row of denticles present on inner side of lips, reaching to base on sides behind papillae. Labial pulp bilobed anteriorly, each lobe with two finger-like processes on anterior margin. Interlabia short, approximately half the length of lip. Intestinal caecum five to seven times as long as ventriculus.

Male: Body length 27.0—100.0 mm, maximum body width 0.46—1.5 mm. Length of oesophagus 1.85—4.48 mm, ventriculus 0.27—0.58 mm, and intestinal caecum 1.56—3.78 mm. Tail somewhat tapering in its middle part. Rounded tip of tail with a short terminal process. 15—33 pairs of precloacal papillae present, arranged in two longitudinal rows. One pair of double postcloacal papillae situated close to cloaca, five other pairs of postcloacal papillae on posterior half of tail (first, third and fifth pairs more medial). Spicules equal, 0.723—1.504 mm, with longitudinal alae. Cloaca 0.15—0.41 mm from tail end.

Female: Body length 30.0—125.0 mm, maximum body width 0.68—2.2 mm.

Length of oesophagus 2.4—6.62 mm, ventriculus 0.23—0.78 mm and intestinal caecum 1.78—4.22 mm. Vulva 20.0—44.0 mm from anterior end of body. Posterior end of body conical. Anus 0.33—0.87 mm from tail end. Eggs 0.091—0.115 × 0.074—0.087 mm.

The species was described as *Fusaria depressa* from *Haliaetus albicilla* in Central Europe. Baylis (1920) transferred it to the genus *Porrocaecum*.

References: Ablasov & Chibichenko (1961, 1962); Baylis (1920); Belogurov (1965); Diesing (1851); Golovin (1958, 1964); Gurlt (1845); Hartwich (1959)*; Molin (1861 a, b); Parona (1894); Rudolphi (1809); Skrjabin (1923); Zeder (1800)*.

Porrocaecum ensicaudatum (Zeder, 1800)

Fig. 45

Hosts: *Stercorarius longicaudatus, Larus canus, L. argentatus, L. ichthyaetus, L. ridibundus, Larus* sp., *Chlidonias nigra, Sterna sandvicensis, S. hirundo* and *S. paradisea*. Common parasite of *Passeriformes*, reported also from *Charadriiformes*.
Localization: intestine, stomach and, more rarely, oesophagus.
Distribution: Europe (England, F.R.G., U.S.S.R. — Ukraine, Lower Kuban and Lower Volga), Asia (U.S.S.R.—West Siberia). Also in North and South America, outside the Palaearctic Region, in birds which are not fish-eating.

Description: Lips hexagonal, their base narrower than upper margin. Labial pulp divided by a deep fissure into two lobes, each of which runs into a long lateral projection. Inner margin of lips bears row of denticles reaching posteriorly to level of subdorsal papillae. Interlabia only $1/2$ to $3/5$ of lip length. Cervical alae absent. Intestinal caecum short, reaching $1/2$ length of ventriculus, usually smaller and often hardly visible.
Male: Body length 21.5—44.5 mm, maximum body width 0.5—1.02 mm. Length of oesophagus 1.66—3.04 mm, ventriculus 0.3—0.66 mm, and intestinal

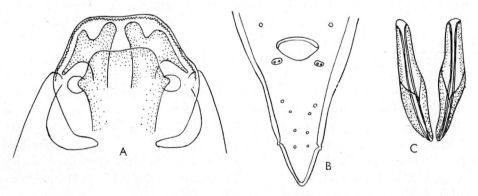

Fig. 45. *Porrocaecum ensicaudatum* (Zeder, 1800). A — dorsal lip; B — posterior end of male (ventral view); C — spicules. After Hartwich (1959).

caecum 0.07—0.32 mm. Cloaca 0.22—0.36 mm from tail end. Tip of tail rounded. 13—20 pairs of precloacal papillae present, arranged in two longitudinal rows and six pairs of postcloacal papillae, the first of which is double, situated near cloaca. Five other pairs of postcloacal papillae lie on ventral side of tail, the fifth being more lateral in position. Spicules equal, 0.56 to 0.735 mm, with longitudinal alae.

Female: Body length 22.5—68.0 mm, maximum body width 0.6—1.4 mm. Length of oesophagus 2.15—3.4 mm, ventriculus 0.35—0.7 mm, and intestinal caecum 0.075—0.175 mm. Vulva 12.5—29.5 mm from anterior end of body. Posterior end of body conical. Anus 0.46—1.02 mm from tail end. Eggs 0.085—0.107 × 0.064—0.079 mm.

This species was described as *Fusaria ensicaudata* from *Turdus merula* in Central Europe. Baylis (1920) transferred it to the genus *Porrocaecum*.

Biology: An experimental study of the life-cycle of this species was carried out by Ryšavý (1959). The eggs are deposited before the beginning of cleavage which occurs at temperatures of 20—22 °C. The first-stage larva is formed and is infective to the intermediate host within 12 to 14 days. *Lumbricus terrestris* was found to serve as intermediary. The larvae are localized in the dorsal and ventral circulatory system where further development occurs. They reach the infective stage within six weeks. After experimental infection of the definitive host, *T. merula*, they develop first under the cuticle of the gizzard and then in the small intestine.

References: Baylis (1920); Hartwich (1959)*; Kurochkin & Zablotsky (1961); Mozgovoy (1953)*; Osche (1959); Pemberton (1960); Ryšavý (1959); Sergeeva (1969); Skrjabin (1926); Smogorzhevskaya (1962 a, 1964); Sonin & Larchenko (1974); Threlfall (1965 b); Zeder (1800)*.

Porrocaecum phalacrocoracis Yamaguti, 1941

Fig. 46
Host: *Phalacrocorax capillatus*.
Localization: stomach.
Distribution: Asia (Japan).

Description:
Male: Body length 53.0—67.0 mm, maximum body width 0.77—0.95 mm, body width at level of lip 0.25 mm. Cuticle transversely striated. Lips 0.18 to 0.20 mm long, dorsal lip with two double papillae, latero-ventral lips each with a double and a single papilla. Pulp of each lip projects forwards in two lobe-like projections. Dentigerous ridges of lips fine. Interlabia 0.07—0.1 mm long, triangular. Nerve ganglion 0.6—0.65 mm from anterior end of body. Oesophagus 2.9—3.2 mm long, 0.16—0.2 mm wide. Ventriculus 0.3—0.35 mm

long, 0.2—0.22 mm wide. Cloaca 0.24—0.33 mm from tail end. Tail tapering distinctly in middle, terminating in small thorn-like projection. Spicules equal, 0.75—0.83 mm long. 21—23 pairs of precloacal and five pairs of postcloacal papillae present. First pair of precloacal papillae situated 3.1—4.5 mm in front of cloaca. First pair of postcloacal papillae double, situated near cloaca.
Female: Body length 65.0 mm, maximum body width 1.0 mm. Body width at level of lips 0.31 mm. Lips 0.3 mm and interlabia 0.11 mm long. Oesophagus 2.6 mm long and 0.225 mm wide, ventriculus 0.35 mm long and 0.25 mm

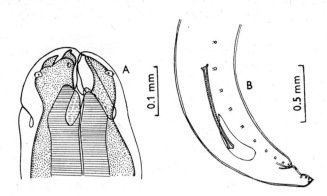

Fig. 46. *Porrocaecum phalacrocoracis* Yamaguti, 1941. A — anterior end; B — posterior end of male (lateral view). After Yamaguti (1941).

wide. Length of intestinal caecum 2.15 mm. Tail with rounded tip, 0.33 mm long. Vulva divides body in ratio of 1 : 2. Eggs nearly spherical, 0.087 to 0.105 × 0.070—0.090 mm.

This species was described by Yamaguti (1941) and has not been reported since.

References: Yamaguti (1941)*.

Porrocaecum praelongum (Dujardin, 1845)

Host: *Podiceps auritus*.
Localization: intestine.
Distribution: Europe (most probably Central Europe).

Description: Body white, coiled, filiform, ten times as long as wide, head relatively large, 0.4 mm wide. Oesophagus 4.3 mm long, nearly cylindrical, 0.4 mm wide. Ventriculus 0.4 mm long, intestinal caecum 3 mm long, 0.6 mm wide. Cuticle with transverse striations.
Male: Body length 90 mm, body width 0.9 mm. Tail attenuated, terminating in short conical tip. Anus 0.33 mm from tail end. Caudal alae membranous, long and narrow, each supported by row of 15 papillae (not 5 as stated by

Cram 1927) on either side of posterior part in front of anus. Spicules equal, 0.9 mm long, 0.058 mm wide, spicular sheath $1/3$ longer than spicules.

Female: Body length 154 mm, body width 1.3 mm. Tail straight, conical, pointed. Anus 0.6 mm from tip of tail. Vulva situated at $1/4$ of body length, 44 m from head. Uterus thin, nearly filiform and sinuous up to 45 mm from vulva, then swollen and divided into two parallel branches 35.5 mm long and 0.75 mm wide, sinuous at first, then attenuated, running into oviducts and filiform ovaries. Eggs spherical, 0.11—0.12 mm in diameter, with membranous shell, which is irregularly folded or net-like.

This species was described as *Ascaris praelonga* from *Colymbus auritus*, probably from Central Europe (the material described by Dujardin was obtained from the Vienna Museum but the locality was not given). It has not been reported since the original description and has not been illustrated in the literature.

References: Dujardin (1845)*; Mozgovoy (1953)*.

Porrocaecum reticulatum (Linstow, 1899)

Fig. 47

Hosts: *Ardea cinerea, A. purpurea, Egretta alba, E. intermedia, E. garzetta, Ardeola ralloides, A. bacchus, Nycticorax nycticorax* and *Bubulcus ibis*. Obligate parasite of birds of the family Ardeidae.
Localization: intestine.
Distribution: Europe (Rumania, U.S.S.R. — Ukraine, Middle and Lower Volga Region),

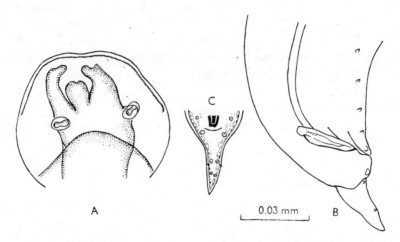

Fig. 47. *Porrocaecum reticulatum* (Linstow, 1899). A — dorsal lip; B — posterior end of male (lateral view); C — posterior end of male (ventral view). A — After Mozgovoy (1953). B — After Baylis & Daubney (1922); C — After Hsü (1933).

Asia (U.S.S.R.—republics of Transcaucasus region and Middle Asia, Far East, China, Japan), Africa (A.R.E.). Outside the Palaearctic Region in South-East Asia, tropical Africa and North America.

Description: Body with marked transverse striations. Lips with denticles on inner margin arranged in one row. Labial pulp projects anteriorly as two long lobes and a short medial lobe. Dorsal lip with two double papillae, latero-ventral lips each with one double and one single papilla.
Male: Body length 54.9—80.0 mm, body width 1.25—1.6 mm, body width at level of lips 0.3—0.4 mm. Lips 0.28—0.35 mm long, interlabia 0.15—0.24 mm long. Length of oesophagus 3.75—5.75 mm, ventriculus 0.45—0.50 mm, and intestinal caecum 2.85—4.74 mm. Tail conical, tapering in middle. Cloaca situated 0.44—0.45 mm from tail end. Large number of precloacal papillae and six to seven pairs of postcloacal papillae present. First pair of precloacal papillae 1.0—1.4 mm in front of cloaca. Spicules equal, 0.4—0.61 mm long. Gubernaculum pointed at distal end, oblique at proximal end, 0.28 mm long and 0.075 mm wide.
Female: Body length 46.0—100.0 mm, body width 0.9—1.8 mm. Length of lips 0.2—0.37 mm, and of interlabia 0.13—0.25 mm. Length of oesophagus 3.2—6.4 mm, ventriculus 0.25—0.50 mm, and intestinal caecum 2.4—5.0 mm. Tail conical, with one pair of caudal papillae situated 0.338 mm from its tip. Anus 0.5—1.25 mm from tail end. Vulva situated at beginning of middle third of body. Eggs nearly spherical or slightly oval, 0.095—0.11 × 0.08—0.105 mm.

This species was described as *Ascaris reticulata* from *Ardea cacoi* from tropical Africa. Baylis & Daubney (1922) transferred it to the genus *Porrocaecum*.
Biology: Dubinin (1949) fed some *Porrocaecum* larvae from the body-cavities of *Pelecus cultratus*, *Cottus* sp. and *Cyprinus carpio* to two young, helminth-free *Ardea cinerea*. After 22 days mature nematodes of *Porrocaecum reticulatum* were recovered at autopsy. Spassky (1952) regards fishes as possible paratenic (or reservoir) hosts of larvae of this species.

References: Babaev (1970); Bashkirova (1960); Baylis & Daubney (1922)*; Chiriac (1965); Dubinin (1949); Dubinin & Dubinina (1940); Dubinina (1937); Feyzullaev (1963 a, b); Hoeppli & Hsü (1929); Hsü, H. F. (1933); Hsü, W. N. (1957); Kasimov & Feyzullaev (1965); Leonov (1956, 1960 a); Linstow (1899 a)*; Mozgovoy (1953)*; Myers, Kuntz & Wells (1962); Oshmarin & Parukhin (1963); Sailov (1962, 1965 b); Shigin (1957); Smogorzhevskaya, Kornyushin, Iskova & Eminov (1965); Sultanov (1959 a, b, 1963); Turemuratov (1962 a); Vaidova (1965, 1969); Yamaguti (1935*, 1941*).

Porrocaecum semiteres (Zeder, 1800)

Fig. 48
Hosts: *Stercorarius longicaudatus*, *S. pomarinus*, *Larus argentatus*, *L. fuscus* and *Sterna paradisea*. Common parasite of birds of the order *Charadriiformes*, rarely of *Passeriformes* and *Lariformes*.

Localization: intestine.
Distribution: Europe (England, U.S.S.R. — Karelian A.S.S.R.), Asia (Low Yenisei). Outside the Palaearctic Region in *L. argentatus* in Canada. In non-fish-eating birds in different areas of the Palaearctic Region.

Description (Data in parentheses are according to Sergeeva 1969 who described specimens from *Sterna paradisea*): Nematodes yellowish in colour. Lateral cuticular alae present. Mouth with three lips and interlabia. Labial pulp divided anteriorly by a narrow medial incision into two lobes, each of which projects laterally. Median lobe extends beyond margin of main pulp. Denticles on inner side near upper margin do not reach the level of mediodorsal papillae. Interlabia reach 2/3 of lip length. Intestinal caecum as long as ventriculus, or may be longer or shorter. Length of ventriculus 0.096 to 0.138 mm, and intestinal caecum 0.08—0.12 mm.

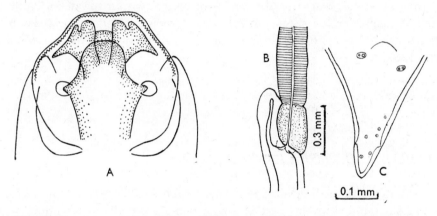

Fig. 48. *Porrocaecum semiteres* (Zeder, 1800). A — dorsal lip; B — region of oesophagus; C — posterior end of male (ventral view). A — After Hartwich (1959); B, C — After Baylis (1920).

Male: Body length 22.0—40.0 (9.3—15.0) mm, width 0.56—1.0 mm (at level of lips 0.07—0.1 mm, at level of ventriculus 0.14—0.16 mm). Length of oesophagus 1.76—2.05 (0.8—0.9) mm, ventriculus 0.22—0.37 (0.09—0.12) mm, and intestinal caecum 0.24—0.65 (0.1—0.12) mm. Posterior end strongly tapered. 15—17 (13) pairs of precloacal and six pairs of postcloacal papillae present. Spicules equal or subequal, similar, 0.60—1.24 (0.09—0.18) mm long.
Female: Body length 21.0—58.0 (6.8—8.02) mm, body width 0.62—1.45 mm (at level of lips 0.06—0.068 mm, at level of ventriculus 0.14—0.16 mm). Length of oesophagus 1.5—2.98 (0.58—0.8) mm, ventriculus 0.2—0.47 (0.096 to 0.138) mm, and intestinal caecum 0.22—0.62 (0.08—0.1) mm. Posterior end of body conical, with rounded tip. Anus 0.46—0.75 mm from tail end.

Vulva 9.95—24.7 mm from anterior end of body. Eggs 0.091—0.093 × 0.07 to 0.084 mm. (Fully mature females with eggs were not present in the material described by Sergeeva in 1969).

The original description of this species was based on specimens recovered from *Pluvialis apricarius* from Central Europe. Baylis (1920) transferred it to the genus *Porrocaecum*.

Biology: The life-cycle of this nematode was studied by Mozgovoy & Bishaeva (1959) and by Yigis (1967). The eggs are expelled into the external environment before the onset of cleavage. Cleavage and development of larvae occur outside and within 28 to 30 days (at about 20 °C) the first moult may be observed (Yigis 1967). Second-stage larvae are infective for the intermediate hosts. Within two to four months the larvae moult for the second time in the earthworms *(Lumbricus rubellus, Eisenia foetida)* and become infective for the definitive hosts. It was shown by infection of a facultative host *(Sturnus vulgaris)*, that the nematodes reach maturity in two months within the definitive host.

Note: We consider the species *P. heteroura* (Creplin, 1829) to be a synonym of *P. semiteres*. The identity of these species was proved by Hartwich (1959) and Yigis (1967).

References: Hartwich (1959)*; Kulachkova & Kochetova (1964); Mozgovoy (1953)*; Mozgovoy & Bishaeva (1959); Sergeeva (1969)*; Threlfall (1965 a, b); Yigis (1967); Zeder (1800)*.

SUBORDER *OXYURATA* SKRJABIN, 1923

Nematodes of medium size. Males somewhat smaller than females. Mouth opening surrounded by three or six lips, which are not always well developed. Oesophagus relatively short, with characteristic swelling (bulb) at posterior end. Inner surface of oesophagus covered with cuticle forming folds and thickenings. Spicules two in number (some species parasitic in mammals have only one spicule or none). Gubernaculum present in some species. Males usually with preanal sucker. Genital apparatus single in male and double in female. Tail end of male pointed, often with lateral alae. Posterior end of females tapering to form a fine pointed tail. Vulva in anterior part of body.

There are different types of life-cycles of *Oxyurata:* some forms develop directly, others require an intermediate host, usually an insect.

All classes of vertebrates are definitive hosts of this suborder of nematodes and the adults of many species parasitize arthropods.

According to Yamaguti (1961), 119 species of the world fauna of *Oxyurata* are parasites of birds.

Fish-eating birds of the Palaearctic Region are parasitized by three species belonging to two genera and two families.

KEY TO THE FAMILIES OF THE SUBORDER *OXYURATA*

1 Mouth capsule indistinct or absent; tail end of male with rounded precloacal sucker with chitinized border . . . . *Heterakidae*
— Chitinized mouth capsule very distinct, with projecting chitinized teeth at the bottom (usually three); tail end of male with oval precloacal sucker without chitinized border
. . . . *Subuluridae*

Family *Heterakidae* Railliet & Henry, 1914

Mouth with three distinct lips which are rounded. Oesophagus widening posteriorly to bulb. Cervical papillae absent. Caudal alae of males long, narrow. Pedunculate caudal papillae short and thick or long and thin. Precloacal pseudosucker with marked sclerotized support. Spicules equal or unequal, similar or dissimilar. Gubernaculum usually absent. Vulva situated in middle part of body or slightly anterior to it. Oviparous. Parasites of birds and mammals.

Type genus *Heterakis* Railliet & Henry, 1912. The family is divided into two subfamilies — *Heterakinae* Railliet & Henry, 1912 and *Meteterakinae* Inglis, 1957 but is, however, not typical of the fauna of fish-eating birds. Three species which occur in fish-eating birds belong to the genus *Heterakis*.

Genus *Heterakis* Dujardin, 1845

Lateral cuticular alae usually present. Oesophagus formed of a short pharynx anteriorly, a cylindrical middle part and a posterior bulb with valvular apparatus. Precloacal sucker with sclerotized support. Vulva in mid-body or slightly in front of it. Uteri amphidelphic. Eggs thick-shelled. Parasites of intestines of birds and mammals.

Type species *H. gallinarum* (Schrank, 1788), cosmopolitan parasite of domestic and some wild gallinaceous birds.

A large number of species belonging to the genus parasitize terrestrial birds.

KEY TO THE SPECIES OF THE GENUS *HETERAKIS*

1 12 pairs of caudal papillae 2
— 10 pairs of caudal papillae . . . . *H. kurilensis*
2 Distal end of short spicule in form of double hook . . . . *H. pavonis*
— Distal end of spicule in form of simple hook . . . . *H. gallinarum*

Heterakis gallinarum (Schrank, 1788)

Fig. 49
Hosts: *Podiceps* sp. and *Ciconia ciconia*. Common parasite of birds of the order *Galliformes*.
Localization: caecum and large intestine.
Distribution: Cosmopolitan; in fish-eating birds only in Asia (U.S.S.R. — Uzbekistan, Turkmenistan).

Description: Cuticle with transverse striations. Mouth with three lips, each bearing two small papillae. Narrow lateral cuticular alae along nearly whole length of body.
Male: Body length 5.84—11.14 mm, maximum body width 0.27—0.39 mm. Pharynx 0.058 mm long and 0.039 mm wide. Oesophagus widening posteriorly, ending in bulb. Oesophagus 0.81—1.10 mm long, bulb 0.26—0.31 mm long

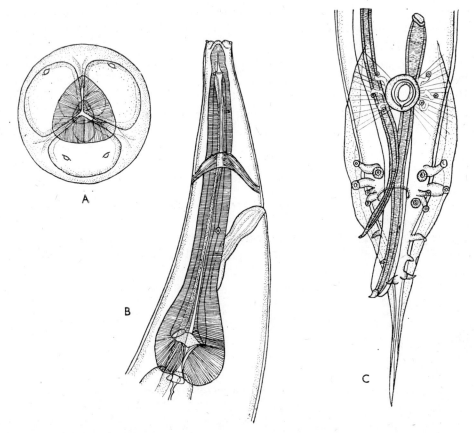

Fig. 49. *Heterakis gallinarum* (Schrank, 1788). A — anterior end (apical view); B — anterior end of female (lateral view); C — posterior end of male (ventral view). After Skrjabin & Shikhobalova (1949).

and 0.14—0.24 mm wide. Cloaca 0.36—0.65 mm from posterior end of body. Width of body at level of cloaca 0.13—0.23 mm. Rounded pseudosucker, 0.070—0.082 mm in diameter, situated 0.14—0.15 mm in front of cloaca. One unpaired papilla present near lower margin of pseudosucker, plus 12 pairs of caudal papillae, of which two pairs are situated on sides of pseudosucker, four pedunculate and two sessile pairs of papillae are near the cloaca and four other pairs are postcloacal in position. Relatively large caudal alae present. Spicules unequal and dissimilar, left 1.62—2.10 mm, right 0.54—0.72 mm long. Longer spicule tapers distally and ends in a sharp point, the shorter one bears two longitudinal alae and has hook-like distal end.

Female: Body length 7.92—11.43 mm, maximum body width 0.27—0.45 mm. Oesophagus 0.97—1.17 mm long, bulb 0.27—0.33 mm long and 0.18—0.23 mm wide. Anus 0.90—1.24 mm from tail end. Body width at level of anus 0.070—0.12 mm. Vulva 4.38—6.44 mm from anterior end of body. Eggs thick-shelled, 0.050—0.070 × 0.030—0.039 mm.

This species was described by Schrank (1788) as *Ascaris gallinarum* from specimens recovered from *Gallus gallus* f. *domestica*. An historical analysis was published in the monograph by Skrjabin, Shikhobalova & Lagodovskaya (1961).

Biology: The life-cycle of this species has been studied by many authors. The development of embryos into infective larvae occurs within the eggs at temperatures ranging from 10 to 42 °C. The eggs are very resistant to environmental factors and remain infective for a long time. The definitive hosts are infected through the mouth when swallowing food contaminated with eggs containing infective larvae. The larvae hatch in the digestive tract and develop further in the caeca. They reach sexual maturity after 27 days. (For more detailed data see the monograph by Skrjabin et al. 1961.)

References: Babaev (1970); Schrank (1788)*; Skrjabin, Shikhobalova & Lagodovskaya (1961)*; Sultanov (1958, 1959 a, b, 1963).

Heterakis kurilensis Oshmarin, 1950

Fig. 50 A, B
Host: *Aethia cristatella*.
Localization: intestine.
Distribution: Asia (U.S.S.R. — Kuril Islands). This species has not been found outside the Palaearctic Region.

Description: Anterior end of body with three prominent lips. Oesophagus slender anteriorly, posteriorly widening to form bulb which contains valvular apparatus. Excretory pore situated approximately at mid-length of oesophagus. Posterior end of body long and thin.

Male: Body length 7.4 mm, maximum body width 0.145 mm. Length of oesophagus with bulb 0.935 mm. Tail end with narrow caudal alae. Cloaca situated 0.47 mm from posterior end. Precloacal sucker, 0.062 mm in diameter, 0.13 mm from cloaca. Two small papillae present a short distance in front of sucker. Three pairs of long caudal papillae anterior to cloaca, and two pairs of papillae, of which the first pair is longer than the second, a short distance posterior to cloaca. Another pair of papillae situated nearer to tail end and three pairs of short papillae arranged in tandem at end of caudal alae. Right spicule straight, with slightly swollen proximal part, tapering gradually distally, 2.015 mm long. Left spicule 0.65 mm long, slightly swollen at proximal end.

Female: Body length 7.28 mm, body width 0.165 mm. Oesophagus with bulb 0.666 mm long. Length of tail 1.01 mm. Vulva situated 3.25 mm from anterior end of body. Oval eggs, 0.060 × 0.036 mm.

The description was based on specimens recovered from *A. cristatella* in the U.S.S.R. and no other finding has hitherto been reported.

References: Oshmarin (1950)*; Skrjabin, Shikhobalova & Lagodovskaya (1961)*.

Heterakis pavonis Maplestone, 1931

Fig. 50 C—E
Host: *Nycticorax nycticorax.*
Localization: caecum.
Distribution: Asia (Japan); outside the Palaearctic Region, in South-East Asia in *Phasianidae*.

Description:
Male: Anterior end of body bent dorsally; body length 8.3—10.9 mm, body width 0.26—0.35 mm. Lateral alae begin posterior to cephalic region. Pharynx measures 0.045—0.06 × 0.036—0.048 mm. Oesophagus 1.0—1.15 mm long, bulb 0.16—0.225 mm wide, containing valvular apparatus. Nerve ring 0.33—0.35 mm and excretory pore 0.42—0.53 mm from anterior end. Outer diameter of caudal pseudosucker 0.084—0.12 mm, inner diameter 0.06—0.076 mm. Centre of pseudosucker 0.18—0.25 mm from cloaca. Caudal alae well developed, 12 pairs of caudal papillae present, two pairs on both sides of pseudosucker, four pairs of pedunculate papillae and two pairs of sessile inner papillae on both sides of cloaca and four pairs of postcloacal papillae. Tail thorn-like, 0.21—0.33 mm long. Long spicule, 1.2—1.9 mm long, with sharp distal end, short spicule, 0.7—0.98 mm long, with distal end forming a double hook. Lateral alae present along the whole length of short spicule.

Female: Anterior end of body bent dorsally, lateral alae present. Body length 11.8—13.7 mm, body width 0.35—0.38 mm. Pharynx 0.045—0.07 × 0.038

to 0.051 mm, oesophagus 1.16—1.3 mm long, bulb 0.18—0.24 mm wide. Posterior end of body tapering to slender tail, 1.0—1.55 mm long. Vulva near middle of body. In some specimens cuticular ridge-like formations occur near vulva, one anterior and two or three posterior to it. Eggs thick-shelled, 0.066 to 0.078 × 0.036—0.042 mm.

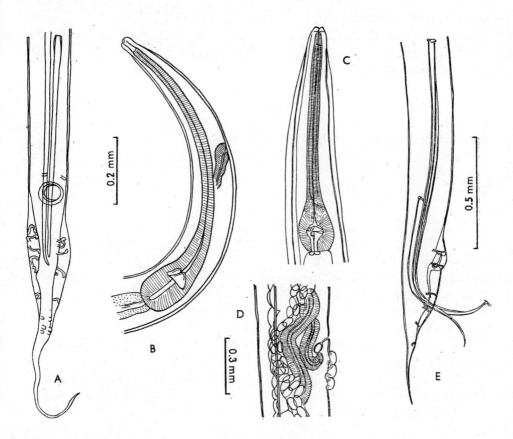

Fig. 50. *Heterakis kurilensis* Oshmarin, 1950 — A, B and *H. pavonis* Maplestone, 1931 — C, D, E. A — posterior end of male (ventral view); B — anterior end of female (lateral view); C — anterior end (dorso-ventral view); D — region of vulva (lateral view); E — posterior end of male (lateral view). A, B — After Oshmarin (1950); C, D, E — After Yamaguti (1941).

Yamaguti (1941) described the species *H. yamadori* from *Galliformes* and *N. nycticorax* in Japan. Like Madsen (1950 b) and Inglis (1958), we consider *H. yamadori* to be synonymous with *H. pavonis*.

References: Inglis (1958); Madsen (1950 b); Maplestone (1931)*; Skrjabin, Shikhobalova & Lagodovskaya (1961)*; Yamaguti (1941)*.

Family *Subuluridae* Yorke & Maplestone, 1926

Nematodes of small or medium size. Mouth usually without lips, sclerotized and with well developed vestibule from the bottom of which grow sclerotized pharyngeal teeth, usually three in number. Oesophagus ending in bulb, intestine straight. Characteristic muscular pseudosucker, without pseudochitinous support, present in front of cloaca. Two spicules and gubernaculum usually present. Precloacal, paracloacal and postcloacal caudal papillae present. Female tail long and sharp, vulva in mid-body. Oviparous. Parasites of digestive tract of birds and mammals.

The numerous species of this family were divided by Skrjabin, Shikhobalova and Lagodovskaya (1964) into two subfamilies *(Subulurinae* and *Leipoanematinae)*. Only one species of the type genus *Subulura* has been found in fish-eating birds from the Palaearctic Region.

Genus *Subulura* Molin, 1960

Mouth without lips, rounded, oval or hexagonal. Mouth capsule simple or divided. From the bottom of vestibule grow three sclerotized pharyngeal toothlike elements. Lateral cuticular alae usually present. Oesophagus with bulb containing valvular apparatus. Spicules equal or unequal. Gubernaculum present. Precloacal muscular pseudosucker fusiform. Vulva of female near middle of body. Eggs thin-shelled, globular to oval, containing a coiled larva. Parasites of digestive tract of birds and mammals.

Type species *S. acutissima* Molin, 1860, parasite of owls and cuckoos in South America.

The genus comprises a large number of species (about 75), of which most are parasites of birds, and only a few of mammals. Only one species has been found in fish-eating birds.

Subulura suctoria (Molin, 1860)

Fig. 51

Host: *Plegadis falcinellus*. Only once reported from fish-eating birds. This species often parasitizes domestic and wild gallinaceous birds (in subtropical and tropical zones), birds of the order *Caprimulgiformes*, rarely *Gruiformes, Cuculiformes, Strigiformes* and *Anseriformes*.
Localization: caecum.
Distribution: Asia (U.S.S.R. — Azerbaijan).

Description: Anterior part of body rounded, bent dorsally. Mouth hexagonal, without lips. Vestibule strongly sclerotized, with three sclerotized, rounded pharyngeal elements at base. Lateral cuticular alae present. Oesophagus with bulb containing valvular apparatus.

Male: Body length 7.98—13.68 mm, maximum body width 0.20—0.43 mm. Vestibule 0.032—0.041 mm deep and 0.036—0.049 mm wide. Oesophagus 0.92—1.22 mm long, bulb 0.18—0.23 mm wide. Nerve ring 0.24—0.30 mm and excretory pore 0.30—0.51 mm from anterior end of body. Posterior end of body narrow, caudal alae indistinct. Of the 11 pairs of caudal papillae (three precloacal, two paracloacal and six postcloacal), 10 are proper caudal papillae;

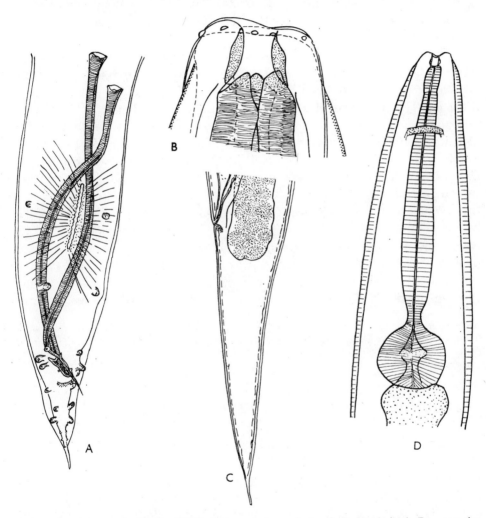

Fig. 51. *Subulura suctoria* (Molin, 1860). A — posterior end of male (ventral view); B — anterior end (detail); C — posterior end of female (lateral view); D — anterior end (overall view — dorsal view). A, B, C — After Kasimov (1946) in Kasimov (1956); D — After Baruš & Blažek (1970).

the others (fourth pair of postcloacal papillae) are outlet of caudal glands, shaped like small papillae. Precloacal pseudosucker, elliptical, formed of radial muscles without pseudochitinous support. Spicules equal, 0.98—1.38 mm long. Cloaca 0.19—0.28 mm from tail end, pseudosucker 0.49—0.74 mm. Gubernaculum 0.12—0.18 mm long.

Female: Body length 11.02—19.00 mm, maximum body width 0.32—0.51 mm. Vestibule 0.036—0.049 mm deep, 0.041—0.057 mm wide. Oesophagus 1.19—1.42 mm long, bulb 0.21—0.26 mm wide. Nerve ring 0.22—0.34 mm, excretory pore 0.45—0.55 mm and vulva 4.75—8.64 mm, respectively, from anterior end of body. Anus 0.72—1.13 mm from tail end. Eggs thin-shelled, 0.061—0.070 × 0.045—0.062 mm, containing coiled larva.

This species was described in Brazil from *Caprimulgus campestris* and was placed in the genus *Heterakis*. It was transferred to *Subulura* by Railliet & Henry (1912).

Biology: The biology of this species in the Hawaiian Islands, in Egypt and in Cuba was dealt with in detail in the papers by Alicata (1939), Cuckler & Alicata (1944), Abdou & Selim (1957, 1963), Baruš (1969, 1970 a, b) and Baruš & Blažek (1970). In intermediate hosts such as some *Coleoptera, Blattaria, Orthoptera* and *Dermaptera*, the larvae develop to the infective stage at temperatures higher than 20 °C. After infection of definitive hosts *(Gallus gallus* f. *dom.)* they become adult within 37 to 46 days but do not live more than nine months in the definitive host.

Note: Ortlepp (1937) suggested that the three species of *Subulura* Molin, 1860, namely, *S. suctoria* (Molin, 1860), *S. brumpti* (López-Neyra, 1922) and *S. differens* (Sonsino, 1890), are synonyms. Priority should be given to *S. suctoria*. In his experimental study of variability Baruš (1969) supported Ortlepp's assumption.

References: Abdou & Selim (1957, 1963); Alicata (1939); Baruš (1969*, 1970 a,b); Baruš & Blažek (1970); Cuckler & Alicata (1944); Kasimov & Feyzullaev (1965); Molin (1860 c)*; Ortlepp (1937); Railliet & Henry (1912); Skrjabin, Shikhobalova & Lagodovskaya (1964)*.

SUBORDER *SPIRURATA* RAILLIET, 1914

Mouth with two lips (pseudolabia), which in some forms are indistinct or absent. Cuticle often with spines, cordons or denticles, especially at the anterior end of body. Oesophagus long, divided into anterior muscular and posterior glandular parts. Vestibule present in most species. Two spicules usually distinctly unequal and dissimilar. Gubernaculum present or absent. Tail end of male usually spirally coiled. Lateral and caudal alae may be present. Vulva in anterior or posterior part of body. Eggs embryonated at deposition. All species utilize arthropods as intermediate hosts. Reservoir parasitism in which a paratenic host is involved is characteristic for many species (Ryšavý & Baruš 1965 analysed all forms).

Parasites of representatives of all classes of vertebrates. Located mostly in tissues of digestive tract. *Spirurata* are one of the most numerous suborders of nematodes and are widely distributed in birds. According to Yamaguti (1961), about 500 species of the world fauna of *Spirurata* parasitize birds. Fish-eating birds of the Palaearctic Region are parasitized by 74 species belonging to 36 genera and eight families.

KEY TO THE FAMILIES OF THE SUBORDER *SPIRURATA*

1   Head end without lips, mouth rounded or hexagonal, buccal cavity (vestibule) massive, strongly chitinized; parasites of orbital cavity, lachrymal ducts and conjunctival sacs . . . . *Thelaziidae*
—   Head end with distinct lateral lips (= pseudolabia); if lips are rudimentary, then buccal cavity is slightly chitinized; parasites of digestive tract and air sacs 2
2   Lateral pseudolabia large, trilobate; dorsal and ventral lips sometimes present . . . . *Spiruridae*
—   Pseudolabia, even if large, not trilobate 3
3   Lateral lips with denticles on their inner surface . . . . *Physalopteridae*
—   Lateral lips without denticles 4
4   Head end with complementary cuticular structures in form of cordons, posteriorly directed, horn-like appendages, radial processes or homologous formations 5
—   Head end without additional cuticular structures 7
5   Anterior part of body with typical cuticular cordons . . . . *Acuariidae*
—   Other type of cuticular ornamentation 6
6   Ornamentation of head end in form of collarette, its posterior margin smooth, rough or with small denticles . . . . *Streptocaridae*
—   Head end with complementary appendages in form of hood-shaped covers, paired posteriorly directed, horn-like appendages or cuticular swellings . . . *Schistorophidae*
7   Parasitic in air sacs of birds; bundle of peculiar „filiform papillae", or verrucose modifications of them, present on tail end of both sexes . . . . *Desmidocercidae*
—   Parasitic in proventriculus and gizzard; body of females deformed (globular, coiled in dense spiral), „filiform papillae" at posterior end of body absent . . . . *Tetrameridae*

Family *Spiruridae* Oerley, 1885

Mouth usually with trilobed lateral pseudolabia, dorsal and ventral lips often absent, sometimes present. Sclerotized vestibule present. Oesophagus long, divided into portions. Caudal alae of male present or absent. Spicules usually unequal and dissimilar. Gubernaculum usually present.

Parasites of birds and mammals.

According to Skrjabin & Sobolev (1963) the familly comprises 12 genera and more than 100 species. The representative of this family are divided into three subfamilies: *Spirurinae, Spiruracercinae* and *Cyrneinae*. Fish-eating birds of the Palaearctic Region are parasitized by four species of the subfamily *Cyrneinae*.

KEY TO THE GENERA OF THE FAMILY *SPIRURIDAE*

1 Teeth, if present, on anterior margin of pseudolabia do not fit together with those of the opposite side . . . . *Cyrnea*
— Teeth on anterior margin of pseudolabia very large, fit together deeply with those of the opposite side . . . . *Excisa*

Genus *Cyrnea* Seurat, 1914

Two lips — ventral and dorsal — deeply notched on median axis and each lip also forming two submedian masses, simple or divided into lobes. Two large lateral pseudolabia with prominent anterior margin and median lobe often armed with three or four large teeth. Four pairs of cephalic papillae, arranged in two circles on submedian lobes of lips. Vestibule well developed. Lateral alae present or absent. Vulva anterior or posterior to oesophagus. Spicules unequal, gubernaculum present. Parasites of various orders of birds.

Type species *C. eurycerca* Seurat, 1914.

The genus includes two subgenera, namely, *Cyrnea* and *Procyrnea*, which differ mainly in the position of the lateral teeth. In the former they are situated on the posterior part of the pseudolabia near the base of the buccal cavity or they may be absent; in the latter they are situated on the anterior margin of pseudolabia. The subgenus *Cyrnea* contains about 20 species and *Procyrnea* more than 25 species. At present only representatives of *Procyrnea* have been reported from fish-eating birds of the Palaearctic Region.

KEY TO THE GENUS *CYRNEA*

1 Lateral pseudolabia with three large, prominent teeth 2
— Lateral pseudolabia with four long teeth . . . . *C. monoptera*
2 Spicules long, left spicule more than 1.20 mm, right spicule more than 0.30 mm . . . .
   . . . . *C. ficheuri*
— Spicules considerably shorter; left spicule less than 0.70 mm, right spicule less than 0.25 mm . . . . *C. leptoptera*

Cyrnea ficheuri (Seurat, 1916)

Fig. 52
Host: *Bubulcus ibis;* found also in eagle owl.

Localizatin: gizzard.
Distribution: Africa (Algeria); outside the Palaearctic Region, North America.

Description: Cuticle thick, with transverse striations. Lateral alae absent. One pair of precervical papillae present. Two large lateral trilobed lips, their median lobe with three teeth, dorsal and ventral lobes each with visible process.
Male: Body length 8.4 mm, body width 0.34 mm. Length of tail 0.12 mm.

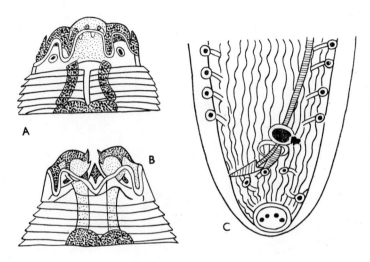

Fig. 52. *Cyrnea ficheuri* (Seurat, 1916). A — anterior end (lateral view); B — anterior end (dorso-ventral view); C — posterior end of male (ventral view). After Seurat (1916).

Two long caudal alae with transverse striations. Ventral part of body also transversely striated, 0.45 mm behind cloaca. Four pairs of pedunculate preanal papillae, one large unpaired papilla near anterior margin of cloaca and two pairs of large postanal papillae. Circular smooth area near caudal extremity with some small papillae. Length of left spicule 1.27 mm, right spicule 0.35 mm. Gubernaculum 0.07 mm long.
Female: Body length 12.8 mm, body width 0.39 mm. Length of tail 0.17 mm. Vulva 6.6 mm from anterior end, slightly salient. Didelphic. Mature eggs measure 0.05 × 0.02 mm and each contains a larva.

The species was described as *Habronema ficheuri* from specimens from *B. ibis* from Algeria. Chabaud (1958) transferred it to the genus *Cyrnea*.

References: Chabaud (1958); Seurat (1916 b)*; Skrjabin & Sobolev (1963)*.

Cyrnea leptoptera (Rudolphi. 1819)

Fig. 53
Host: *Haliaetus pelagicus;* found also in other *Falconiformes.*
Localization: proventriculus.
Distribution: Asia (U.S.S.R. — Far East). In birds other than those eating fish, in Europe, North Africa, South-East Asia, South America.

Description: Body extended, cuticle with transverse striations. Two lateral alae in anterior fourth of body. Cervical papillae in front of nerve ring. Head

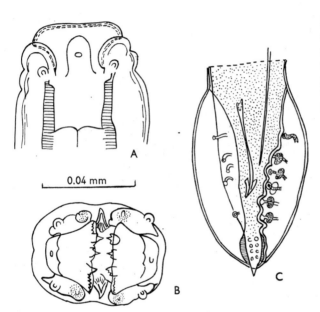

Fig. 53. *Cyrnea leptoptera* (Rudolphi, 1819). A — anterior end (lateral view); B — anterior end (apical view); C — posterior end of male (ventral view). A, B — After Chabaud (1958); C—After Seurat (1914).

end distinctly set off from body. Four very large, trilobed lateral lips with three large teeth; dorsal and ventral lips with protruding median comb and four papillae near margins of lips.

Male: Body length 7.1 mm, body width 0.72 mm. Left spicule 0.64 mm long, right spicule 0.20 mm long. Gubernaculum 0.03 mm long. Two caudal alae each divided into two parts, the dorsal part with fine transverse striations, the other part with longitudinal striations. Left ala thickened at edge; papillae ending in it are mushroom-shaped. Four pairs of preanal papillae, two pairs of asymmetrical postanal papillae and a group of eight sessile papillae near tail end.

Female: Body length 11.7—14.0 mm, body width 0.28—0.33 mm. Cervical papillae 0.12—0.13 mm, nerve ring 0.23—0.25 mm, and excretory pore 0.30—0.32 mm, respectively, from anterior end. Vestibule short, cylindrical,

0.02—0.03 mm in length and 0.02—0.03 mm in diameter. Muscular part of oesophagus 0.36—0.38 mm long and 0.03—0.04 mm wide, glandular part 2.55—2.71 mm long and 0.10—0.12 mm wide. Tail pointed, 0.28—0.30 mm long. One pair of papillae situated 0.07—0.08 mm from tail end. Vulva divides the body in ratio of 1 : 1.3—1.4. Eggs oval, 0.032—0.041 × 0.019—0.024 mm.

The species was described as *Spiroptera leptoptera* from specimens recovered from *Buteo buteo* in Europe. Chabaud (1958) transferred it to the genus *Cyrnea*.

References: Belogurov (1965); Chabaud (1958); Rudolphi (1819)*; Skrjabin & Sobolev (1963)*.

Cyrnea monoptera (Gendre, 1922)

Fig. 54
Host: *Bubulcus ibis;* originally found in an unidentified nocturnal bird of prey.
Localization: proventriculus.
Distribution: Africa (A.R.E.), outside the Palaearctic Region, in tropical Africa.

Description: Body attenuated anteriorly and widened posteriorly. One lateral ala originates 0.25—0.28 mm from anterior end, on left side of body, and extends to the level of end of glandular oesophagus. Cervical papillae asymmetrically arranged, the right being anterior to the left. Head end with four pseudolabia, the lateral ones trilobed, rounded or flattened, with four long teeth on inner side of middle lobe and the dorsal and ventral pseudolabia each with a central triangular tooth and two lateral semicircular lobes, each with large papilla.
Male: Body length 7.6—9.8 mm, body width 0.30—0.35 mm. Lateral alae 2.58 mm long. Length of tail 0.23—0.26 mm. Caudal alae long, thick. External or dorsal surface of alae transversely striated, inner surface covered with longitudinal parallel combs and plates, which also cover the ventral surface of body from a distance of 0.61 mm anterior to cloaca. Eleven pairs of caudal papillae present, of which four are preanal and seven postanal. There is, in addition, an unpaired large papilla in front of cloaca, near its left margin. Left spicule 1.56—1.68 mm long and right 0.38—0.46 mm long. Gubernaculum hardly discernible.
Female: Body length 12.3—13.6 mm, body width 0.43—0.44 mm. Lateral ala 2.98 mm long. Tail 0.18 mm long. Vulva near middle of body, sometimes somewhat posterior, not salient. Eggs 0.04 × 0.02 mm.

The species was described as *Habronema monoptera* from specimens recovered from a nocturnal bird of prey from equatorial Africa. Chabaud (1958) transferred it to the genus *Cyrnea*.

References: Chabaud (1958); Gendre (1922)*; Myers, Kuntz & Wells (1962); Skrjabin & Sobolev (1963)*.

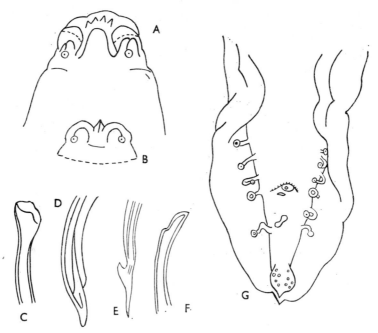

Fig. 54. *Cyrnea monoptera* (Gendre, 1922). A — anterior end (lateral view); B — anterior end (dorso-ventral view); C, F — proximal ends of spicules; D, E — distal ends of spicules; G — posterior end of male (ventral view). After Gendre (1922).

Genus *Excisa* Gendre, 1928

Two lips, ventral and dorsal, deeply notched, forming two independent lobes each. Eight cephalic papillae: latero-median near the base of each lobe; medio-median small and situated at the front. Median teeth absent. Two lateral pseudolabia, their anterior margin with three large teeth, deeply fitting together with those of the opposite side. Vestibule slightly chitinized. Cervical papillae behind nerve ring. Lateral alae absent. Vulva equatorial. Spicules unequal. Gubernaculum present.

Type species *E. excisa* (Molin, 1860).

The genus comprises three species, one of them parasitizing storks in the Palaearctic Region.

Excisa excisa (Molin, 1860)

Fig. 55
Host: *Ciconia ciconia*.
Localization: proventriculus.
Distribution: Europe (Spain, Czechoslovakia, U.S.S.R. — Byelorussia), Asia (U.S.S.R. —

Transcaucasus region and Middle Asia; Japan), Africa (Algeria). Outside the Palaearctic Region, found in *Ciconia* spp. in South America and tropical Africa.

Description:

Male: Body length 15.4—20.0 mm, maximum body width 0.40—0.60 mm. Ventral surface of posterior part of body covered with cuticular combs extending anteriorly from tail end. Nerve ring 0.33—0.36 mm from anterior end.

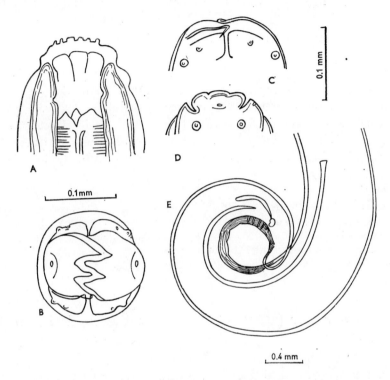

Fig. 55. *Excisa excisa* (Molin, 1860). A — anterior part of body (lateral view, optical section in medial level); B — anterior end (apical view); C — anterior end (dorso-ventral view); D — anterior end (lateral view); E — posterior end of male (lateral view). A, B, C, D — After Chabaud (1958); E — After Yamaguti (1941).

Vestibule 0.056—0.095 mm deep. Total length of oesophagus 3.78—5.2 mm, with muscular part 0.3—0.4 mm long. Tail blunt, ventrally coiled, 0.30—0.35 mm long. Caudal alae supported on each side with four preanal and two postanal papillae. There is, in addition a large unpaired papilla near anterior margin of cloaca and three postcloacal papillae arranged in triangle, the base of which is formed by the posterior margin of cloaca. Length of right spicule

121

0.5—0.63 mm, of left spicule 1.80—2.5 mm and of gubernaculum 0.078 mm. Female: Body length 18.8—26 mm, maximum body width 0.55—0.72 mm. Nerve ring 0.3—0.33 mm from anterior end. Total length of oesophagus 4.98—5.30 mm, with muscular part 0.32—0.40 mm long. Vulva at level of mid-body, somewhat anterior or posterior to it. Eggs 0.042—0.051 × 0.018 to 0.024 mm.

The species was described as *Spiroptera excisa* from specimens recovered from *C. maguari* in Brazil. It was transferred to the genus *Excisa* by Gendre (1928).

Notes: Seurat (1914 b) placed this species in the genus *Cyrnea*, Chitwood & Wehr (1934) did not regard the genus *Excisa* Gendre, 1928 as valid and considered it to be a synonym of *Hadjelia*. López-Neyra (1947) assigned this species to the genus *Seurocyrnea*, and Chabaud & Campana (1950 a) and Chabaud (1958) restored the independence of the genus *Excisa*.

According to Skrjabin & Sobolev (1963), the species *Physaloptera striata* Linstow, 1883 is a synonym of *E. excisa*. We are of the opinion that the species *Habronema sobolevi* Ryzhova & Dubov, 1955 from storks in Byelorussia, is also a synonym of *E. excisa*.

References: Chabaud (1958); Chabaud & Campana (1950 a); Chitwood & Wehr (1934); Feyzullaev (1963 a, b); Gendre (1928); Kasimov & Feyzullaev (1965); Linstow (1883*, 1886*); López-Neyra (1947 a)*; Macko (1963 b); Molin (1860 a)*; Ryzhova & Dubov (1955)*; Seurat (1914 b); Shakhtakhtinskaya & Sadykov (1967); Skrjabin (1915 a, 1923); Skrjabin & Sobolev (1963)*; Sultanov (1963); Yamaguti (1941)*.

Family *Acuariidae* (Railliet, Henry & Sisoff, 1912 subfam.) Seurat, 1913

Head end with two distinct lateral lips. Anterior part of body provided with cuticular cordons beginning on sides of lips or homologous structures. Walls of vestibule usually transversely striated. Oesophagus divided into muscular and glandular parts. Male with caudal alae supported by four pairs of preanal and various numbers of postanal papillae. Spicules usually unequal and dissimilar. Parasitic in digestive tract of birds. Some species are known to parasitize mammals.

According to Skrjabin, Sobolev & Ivashkin (1965) the family comprises 16 genera and 179 species. Recently Sergeeva (1968) transferred the genus *Rusguniella*, previously assigned to the family *Streptocaridae*, to the *Acuariidae*. The representatives of *Acuariidae* are divided into two subfamilies, namely, the *Acuariinae* and *Echinuriinae*. The fish-eating birds of the Palaearctic Region are parasitized by 37 species belonging to both subfamilies.

KEY TO THE SUBFAMILIES OF THE FAMILY *ACUARIIDAE*

1 Cuticle without longitudinal rows of spines . . . . *Acuariinae*
— Cuticle with longitudinal rows of spines . . . . *Echinuriinae*

Subfamily *Acuariinae* Railliet, Henry & Sisoff, 1912

The characteristic feature of this subfamily, which includes 15 genera, is the absence of longitudinal rows of spines on the surface of the cuticle.

Species of all genera were recorded in the fish-eating birds of the Palaearctic Region.

KEY TO THE GENERA OF THE SUBFAMILY *ACUARIINAE*

1 Cordons short, not extending posterior to head region 2
— Cordons long, extending posterior to head region 4
2 Cervical papillae tricuspid . . . . *Paracuaria*
— Cervical papillae simple 3
3 Vestibule short, approximately twice as long as cordons; vulva in middle part of body
. . . . *Rusguniella*
— Vestibule long, approximately three to four times as long as cordons; vulva close to anus
. . . . *Aviculariella*
4 Cervical papillae large, transformed into numerous denticles surrounding body from lateral sides in semicircle . . . . *Pectinospirura*
— Cervical papillae small, with one or three denticles 5
5 Posterior end of cordons not curving toward the head end, not anastomosing . . . 6
— Other formation of cordons 7
6 Male with precloacal ventral ridge, female tail with tuft of spines . . . .
. . . . *Skrjabinocerca*
— Male without precloacal ventral ridge, female tail without spines . . . . *Acuaria*
7 Posterior ends of cordons united in pairs, never curving towards anterior end . . . 8
— Posterior ends of cordons, united or not united in pairs, always curving towards anterior end 11
8 Characteristic cuticular collarette present in region of cordon junction . . . .
. . . . *Chevreuxia*
— Cuticular collarette absent 9
9 In addition to typical cordons there are two thick lateral cordons, originating posterior to cervical papillae and extending to posterior end of body . . . . . . *Skrjabinocara*
— Lateral cordons usually absent, if present, only slightly developed and indistinct . . 10
10 Cordons relatively narrow, without denticles and hooks on outer margin . . . . .
. . . . *Syncuaria*
— Cordons relatively large, with denticles or hooks on outer margin . . . *Decorataria*
11 Cordons recurrent but not anastomosing . . . . *Dispharynx*
— Cordons recurrent and anastomosing 12
12 Cordons forming a small loop immediately posterior to their origin 13
— Cordons not forming a loop 14

13	Cervical papillae simple or biscuspid	*Cosmocephalus*
—	Cervical papillae tricuspid	*Sexansocara*
14	Cordons without spines, uniform width throughout length; females didelphic	*Synhimantus*
—	Cordons with spines, widening gradually posteriorly; females monodelphic	*Desportesius*

Genus *Acuaria* Bremser, 1841

Cuticular cordons, originating on both sides of lips, extend to posterior end, not recurrent, anastomosing or curving. Tail end of male with lateral alae. Four preanal papillae, variable number of postanal papillae. Vulva usually in posterior third of body.

Type species *A. anthuris* (Rudolphi, 1819).

This genus is the most numerous of all genera belonging to the family *Acuariidae*. It includes about 70 species, mainly parasites of *Passeriformes* and *Galliformes*. Only one species was recorded in the fish-eating birds of the Palaearctic Region.

Acuaria phalacrocoracis (Smogorzhevskaya, 1961)

Fig. 56
Host: *Phalacrocorax aristotelis*.
Localization: stomach.
Distribution: Europe (U.S.S.R. — Crimea).

Description:
Male: Body length 6.3—7.7 mm. Length of cordons 0.35—0.47 mm. Vestibule 0.12 mm long, muscular part of oesophagus 0.33 mm long, glandular part 1.92 mm long. Cervical papillae bicuspid, 0.34—0.36 mm from anterior end of body. Caudal alae 0.021—0.024 mm wide. Four pairs of pedunculate preanal papillae and five pairs of postanal papillae (four pedunculate, one sessile). Large spicule 0.33—0.36 mm long, small spicule 0.114 mm long.
Female: Body length 13—19 mm. Length of cordons 0.672—1.36 mm. Vestibule 0.12—0.22 mm long, muscular part of oesophagus 0.44—0.45 mm long, glandular part 2.93—3.08 mm long. Cervical papilla 0.51—0.54 mm from anterior end. Vulva 0.68—1.21 mm from posterior end. Eggs 0.034—0.036 × × 0.020—0.022 mm.

The species was originally assigned to the genus *Cheilospirura*. Skrjabin, Sobolev & Ivashkin (1965) transferred it to the genus *Acuaria*. The species has not been found since the original description.

References: Skrjabin, Sobolev & Ivashkin (1965)*, Smogorzhevskaya (1961*, 1962 b).

Genus *Aviculariella* Wehr, 1931

Cordons short, uniting in pairs on dorsal and ventral sides. Inner borders of cordons denticulate. Cuticle inflated immediately behind transverse arch of cordons. Cervical papillae simple, situated at level of posterior part of vestibule. Caudal alae present in male. Spicules unequal, dissimilar. Vulva near anus.

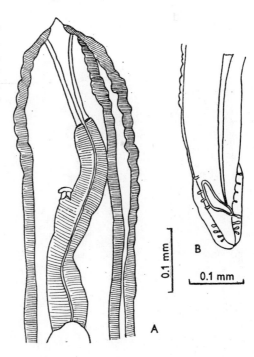

Fig. 56. *Acuaria phalacrocoracis* (Smogorzhevskaya, 1961). A — anterior end (lateral view); B — posterior end of male (ventral view). After Smogorzhevskaya (1961).

Type species *A. alcyona* Wehr, 1931.

The genus includes three species, two of them (including the type species) parasitic in kingfishers. Only one species has been recorded in the Palaearctic Region.

Aviculariella alcedonis (Yamaguti & Mitunaga, 1943)

Fig. 57
Host: *Alcedo atthis*.
Localization: under the cuticle of gizzard.
Distribution: Asia (U.S.S.R. — Far East; China). Outside the Palaearctic Region it has been reported from kingfishers in South-East Asia.

Description: Mouth surrounded by two lips, from base of which arise rather large (up to 0.0035 mm) transversely striated cordons, forming, on dorsal and ventral sides, approximately isosceles triangles. Cervical papillae approximately at level of nerve ring. Narrow lateral alae arise immediately posterior to cervical papillae. Vestibule long, oesophagus divided.
Male: Body length 7.2—10.0 mm, maximum body width 0.08—0.144 mm. Cordons 0.04—0.053 mm long. Vestibule 0.106—0.20 mm long. Length of muscular part of oesophagus 0.41—0.605 mm, of glandular part 1.13—2.8 mm.

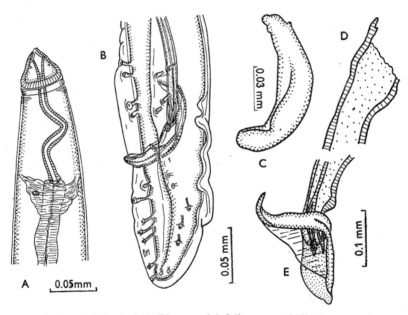

Fig. 57. *Aviculariella alcedonis* (Yamaguti & Mitunaga, 1943). A — anterior end (lateral view); B — posterior end of male (lateral view); C — small spicule; D — proximal end of long spicule; E — distal end of long spicule. A, B — After Ryzhikov & Khokhlova (1965), C, D, E — After Smetanina & Alekseev (1968).

Nerve ring 0.17—0.23 mm, cervical papillae 0.132—0.174 mm, respectively, from anterior end. Tail end conical, 0.096—0.155 mm long, with caudal alae. There are five pairs of pedunculate preanal papillae and seven pairs of postanal papillae, the last two of which are very small and sessile. Left spicule cylindrical, 0.28—0.40 mm long, its last third triangular; distal end with hook-like process 0.03 mm long, and three finger-like processes joined by a membrane. Right spicule 0.09—0.11 mm long, cymbiform.
Female: Body length 18.6—24.5 mm, maximum body width 0.14—0.17 mm. Cordons 0.06—0.07 mm long. Vestibule 0.13—0.20 mm long. Length of

muscular part of oesophagus 0.45—0.67 mm, of glandular part 2.24—2.90 mm. Nerve ring 0.165—0.18 mm from head end. Vulva in posterior part of body, 0.30—0.40 mm from posterior end. Vagina thick-walled, 0.35 mm long. Eggs 0.073—0.078 × 0.039—0.042 mm.

This species was described on the basis of specimens recovered from *A. atthis* in China (Taiwan Isle) and was originally assigned to the genus *Rusguniella*. It was transferred to the genus *Aviculariella* by Wang (1966).

Notes: We also regard *A. collaricephala* (= *Alcedospirura collaricephala*), described by Oshmarin (1959) on the basis of specimens from *Alcedo atthis* from Primorye Territory (U.S.S.R.), as a synonym of *Aviculariella alcedonis*. We obtained the nematodes from the same host from the Amur region, compared them with the specimens previously described and concluded that all of them were *A. alcedonis*.

References: Alekseev & Smetanina (1968)*; Oshmarin (1959*, 1963); Sergeeva (1968 b); Skrjabin, Sobolev & Ivashkin (1965)*; Smetanina & Alekseev (1967)*; Wang (1966)*; Yamaguti & Mitunaga (1943)*.

Genus *Chevreuxia* Seurat, 1918

Wide straight cordons unite in pairs on lateral sides of body. Cuticular collarette with distinct transverse striations encircles body at junction of cordons. Cervical papillae small, simple. Spicules unequal, dissimilar. Vulva in posterior part of body.

Type species *Ch. revoluta* (Rudolphi, 1819).

The genus includes four species, three of them recorded in fish-eating birds. Only one species is known from the Palaearctic Region.

Chevreuxia revoluta (Rudolphi, 1819)

Fig. 58

Host: *Larus ridibundus*. Characteristic parasite of *Himantopus himantopus*, only once reported from gull.
Localization: under the cuticle of gizzard.
Distribution: Asia (U.S.S.R. — West Siberia). In *H. himantopus* in many areas of the Palaearctic Region.

Description:
Male: Body length 6.4—6.53 mm, maximum body width 0.14—0.16 mm. Cordons 0.4 mm long. Vestibule 0.16 × 0.012 mm, muscular part of oesophagus 0.74 × 0.036 mm, glandular part 2.2 × 0.080 mm. Caudal alae supported by four pairs of preanal and two pairs of postanal papillae. Large spicule 0.7—0.75 mm long, small spicule 0.092 mm long.

Female: Body length 8.3—18.3 mm, body width 0.18—0.265 mm. Cordons 0.59—0.765 mm long. Vestibule 0.156—0.24 mm long, muscular part of oesophagus 0.50—0.96 mm long and glandular part 2.34 mm. Collarette 0.21 mm long. Vulva 4.86—6.22 mm from posterior end. Tail 0.17—0.18 mm long. Eggs 0.32—0.036 × 0.018—0.022 mm.

The species was described from specimens from *H. himantopus* in Europe and was originally assigned to the genus *Spiroptera*. It was transferred to *Chevreuxia* by Seurat (1918).

References: Cram (1927)*; Rudolphi (1819)*; Serkova (1948)*; Seurat (1918); Skrjabin, Sobolev & Ivashkin (1965)*.

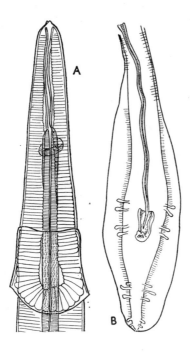

Fig. 58. *Chevreuxia revoluta* (Rudolphi, 1819). A — anterior end (lateral view); B — posterior end of male (ventral view). After Sobolev (1943).

Genus *Cosmocephalus* Molin, 1858

Anterior part of body widened, provided with cordons which originate at base of lips, run posteriorly, turn ventrally or dorsally and then run anteriorly, forming a characteristic loop. From this the cordons then turn posteriorly and some distance from anterior end turn laterally towards one another and run anteriorly, uniting in middle of lateral fields. Lateral alae present.

Type species *C. obvelatus* (Creplin, 1825).

The genus includes ten species; five of them have been recorded from fish-eating birds of the Palaearctic Region.

When erecting the genus, *C. diesingi* was originally designated type species. Since a detailed description of this species was unavailable, Skrjabin, Shikhobalova & Sobolev (1949) treated it as *species inquirenda* and designated *C. obvelatus* as type. This species had been described earlier than other species of this genus. We are of the same opinion as these authors.

KEY TO THE SPECIES OF THE GENUS *COSMOCEPHALUS*

1. Length of cordons half the width of body behind cordons; cervical papillae not projecting . . . . *C. aduncus*
— Length of cordons twice the width of body behind cordons; cervical papillae projecting 2
2. Vestibule long, reaching the level of cervical papillae 3
— Vestibule short, ending somewhat anterior to cervical papillae 4
3. Cordons scalloped on their inner border . . . . *C. obvelatus*
— Cordons with smooth inner border . . . . *C. faridi*
4. Left spicule more than four times as long as right; only nine pairs of pedunculate caudal papillae present, sessile papillae absent . . . . *C. jaenschi*
— Left spicule less than four times as long as right one; in addition to pedunculate papillae, two pairs of sessile caudal papillae present . . . . *C. imperialis*

Cosmocephalus obvelatus (Creplin, 1825)

Fig. 59
Hosts: *Gavia arctica, Podiceps cristatus, P. auritus, P. nigricollis, P. griseigena, Phalacrocorax pygmaeus, Egretta garzetta, Platalea leucorodia, Mergus serrator, Stercorarius longicaudatus, S. parasiticus, Rissa tridactyla, Xema sabini, Larus argentatus, L. canus, L. crassirostris, L. genei, L. glaucescens, L. fuscus, L. hyperboreus, L. ichthyaetus, L. marinus, L. medius, L. melanocephalus, L. minutus, L. ridibundus, L. schistisagus, Chlidonias nigra, Ch. hybrida, Ch. leucoptera, Hydroprogne tschegrava, Gelochelidon nilotica, Sterna hirundo, S. albifrons, S. paradisea, S. sandvicensis, Alca torda, Uria aalge, Cepphus carbo* and *Cerorhinca monocerata*.
Localization: oesophagus.
Distribution: Europe (England, France, Italy, Denmark, G.D.R., F.R.G., Czechoslovakia, Bulgaria, U.S.S.R. — Pribaltic Territory, Ukraine, northern and central regions of Russia, the Volga Region); Asia (U.S.S.R. —West and South Siberia, Far East, Transcaucasus Region, republics of Middle Asia and Kazakhstan); Africa (Algeria). Outside the Palaearctic Region, this species has been reported from South-East Asia, North America, Equatorial Africa and Australia.

Description: Cordons with scalloped inner borders. Cervical papillae bicuspid, situated at level of beginning of muscular part of oesophagus. Pharynx long, reaching level of cervical papillae.
Male: Body length 4.6—12.2 mm, maximum body width 0.153—0.29 mm. Cordons 0.24—0.40 mm long. Vestibule 0.29—0.36 mm long; muscular part of oesophagus 0.20—0.98 mm, glandular part 2.2—3.2 mm long. Cervical

papillae 0.43 mm from anterior end (in specimens 12.2 mm long). Tail 0.24 to 0.42 mm long. Ten pairs of pedunculate papillae present, of which four pairs are preanal and six postanal; one additional pair of sessile papillae present at end of tail. Spicules unequal, dissimilar; right massive, 0.12—0.155 mm long, left more elongated and slender, 0.42—0.63 mm long.

Female: Body length 7.75—15.6 mm, maximum body width 0.20—0.40 mm. Cordons 0.307—0.44 mm long. Vestibule 0.5 mm long; muscular part of oesophagus 1.14 mm long, glandular part 3.26 mm long. Vulva near middle of body, 3.6—8.92 mm from posterior end. Tail end with characteristic knob-like thickening. Eggs 0.026—0.040 × 0.018—0.024 mm.

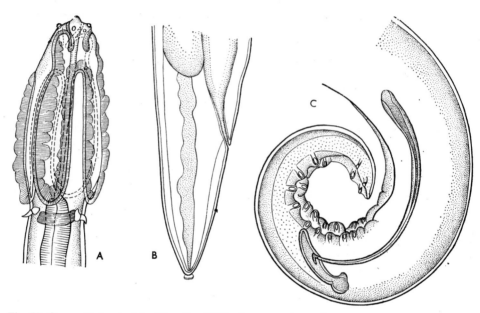

Fig. 59. *Cosmocephalus obvelatus* (Creplin, 1825). A — anterior end (lateral view); B — posterior end of female (lateral view); C — posterior end of male (lateral view). Original.

The description was based on specimens recovered from gulls in Central Europe. The species, originally assigned to the genus *Spiroptera* and later to some other genera, was transferred to the genus *Cosmocephalus* by Seurat (1916).

References: Ablasov (1957); Ablasov & Chibichenko (1961, 1962); Agapova & Zhatkanbaeva (1971); Akhyuman (1966); Alekseev & Smetanina (1968); Babaev (1970); Bakke (1972); Bakke & Baruš (1976); Baylis (1939); Belogurov, Leonov & Zueva (1968); Belopolskaya (1952); Braun (1892); Chabaud (1954); Cram (1927); Creplin (1825*, 1829, 1845—1846); Creutz & Gottschalk (1969); Daiya (1971); Diesing (1851); Dollfus et al. (1961); Ellis & Williams (1973); Golikova (1959); Golovin (1964); Gubanov & Sergeeva (1971); Guildal (1968); Gurlt (1845); Gushanskaya (1950 c); Iksanov & Dikambaeva (1962); Jennings & Soulsby (1957, 1958); Kibakin (1965); Kosupko (1962, 1963); Krotov & Delyamure (1952); Kulachkova & Kochetova (1964); Kurashvili (1950, 1953, 1957, 1961); Leonov (1958); Leonov & Belogurov (1963); Leonov & Shvetsova (1970); Linstow (1878, 1889); McIntosh (1927); Macko (1964 a, d); Michelson (1968); Molin (1859); Mozgina (1967, 1969); Muehling (1898); Mukhamadiev (1966); Nicoll (1927); Oshmarin (1963); Parona (1894); Pemberton (1963); Sailov (1963, 1965 b, 1966, 1970); Sergeeva (1968 a, 1969); Seurat (1916 a, 1919*); Shakhtakhtinskaya (1959 a, b); Shigin (1961); Skrjabin (1923); Skrjabin, Sobolev & Ivashkin (1965)*; Smetanina (1972); Smetanina & Alekseev (1967); Smogorzhevskaya (1964, 1967); Solonitsin (1928 a, b); Sonin & Larchenko (1974); Sprehn (1962); Threlfall (1965 b, 1967); Tsimbalyuk & Belogurov (1964); Turemuratov (1962 a, 1965 a); Vaidova (1965, 1970); Vasilkova (1926); Yigis (1962); Zhatkanbaeva (1971); Zhelyazkova-Paspaleva (1962 a).

Cosmocephalus aduncus (Creplin, 1846)

Fig. 60
Hosts: *Gavia stellata, Podiceps cristatus, P. auritus, P. nigricollis, P. ruficollis, Phalacrocorax carbo, Larus canus, L. argentatus, L. fuscus, L. genei, L. ichthyaetus, L. marinus, L. medius, L. ridibundus, Chlidonias hybrida, Hydroprogne tschegrava, Gelochelidon nilotica, Sterna hirundo* and *S. sandvicensis*.

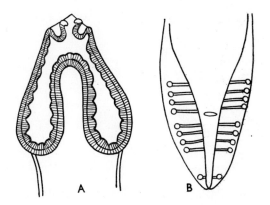

Fig. 60. *Cosmocephalus aduncus* (Creplin, 1846). A — anterior end (lateral view); B — posterior end of male (ventral view). After Stossich (1892).

Localization: oesophagus.
Distribution: Europe (England, Sweden, Poland, G.D.R., Rumania, Italy, U.S.S.R. — northern and central areas of Russia, the Volga Region, Ukraine), Asia (U.S.S.R.—republics of Transcaucasus Region and Middle Asia, Kazakhstan, Far East.) Outside the Palaearctic Region, North America.

Description:

Male: Body length 8—11.05 mm, maximum body width 0.2—0.31 mm. Vestibule 0.4—0.42 mm long; muscular part of oesophagus 0.77 mm long, glandular part 3.58—3.67 mm long. Tail 0.34—0.4 mm long. Large spicule 0.52—0.7 mm long, small spicule 0.15—0.18 mm long. There are four pairs of preanal and five pairs of postnatal papillae.
Female: Body length 10—17.3 mm, maximum body width 0.38—0.43 mm.

Vestibule 0.44—0.54 mm long; muscular part of oesophagus 0.75—1.21 mm long, glandular part 3.6—4.24 mm long. Cordons 0.44—0.49 mm long. Cervical papillae 0.47—0.52 mm from anterior end. Tail 0.21—0.24 mm long. Vulva 6.29—9 mm from posterior end. Eggs 0.032—0.039 × 0.018 to 0.023 mm.

The first description of this species was based on specimens recovered from *Gavia stellata* in Central Europe. The species was originally assigned to the genus *Spiroptera* and was transferred to *Cosmocephalus* by Yorke & Maplestone (1926).

References: Akhumyan (1966); Babaev (1970); Baird (1853); Bakke (1972); Belopolskaya (1952); Bezubik (1956); Chiriac (1965); Cram (1927)*; Creplin (1845—1846)*; Creutz & Gottschalk (1969); Diesing (1851); Golovin (1964); Guildal (1966); Gushanskaya (1950 c);

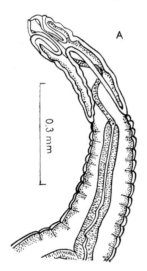
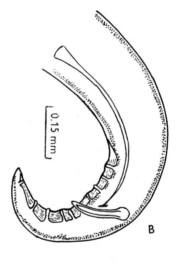

Fig. 61. *Cosmocephalus faridi* Khalil, 1931. A — anterior end (lateral view); B — posterior end (lateral view). After Khalil (1931).

Gvozdev & Kasymzhanova (1965); Iksanov & Dikambaeva (1962); Kibakin (1965); Kibakin, Dobrynin, Skladchikov & Zhuchenko (1963); Kurochkin & Zablotsky (1961); Leonov (1958); Linstow (1878, 1889); Molin (1860 b); Muehling (1898); Pemberton (1963); Sailov (1962, 1963, 1965 b, 1966, 1970); Sergeeva (1968 a); Shakhtakhtinskaya (1959 a, b); Skrjabin (1923); Skrjabin, Sobolev & Ivashkin (1965)*; Smogorzhevskaya (1964); Sprehn (1962); Stossich, (1829 a* 1895, 1902); Sultanov (1959, 1963); Threlfall (1966 c); Turemuratov (1962 a, 1965 a); Vasilkova (1926*, 1927); Yorke & Maplestone (1926); Zhatkanbaeva (1971).

Cosmocephalus faridi Khalil, 1931

Fig. 61
Host: *Pelecanus onocrotalus*.
Localization: gizzard.
Distribution: Africa (Egypt), Outside the Palaearctic Region recorded in tropical Africa.

Description: Cordons not scalloped, running posteriorly as far as 0.3—0.4 mm from anterior end. Cervical papillae simple, short and thick, immediately behind cordons. Excretory pore posterior to level of cervical papillae, nerve ring surrounding muscular part of oesophagus a short distance from its origin.
Male: Body length 7.6 mm. Oesophagus 0.93 mm long. Cloaca 0.27 mm from tail end. Caudal alae supported by nine pairs of caudal papillae, four of which are preanal and five postanal, situated close to one another, except the distal pair which is subterminal. Long spicule 0.54 mm long, thin, its end bent in the form of a fish-hook. Short spicule thick, 0.16 mm long.
Female: Body length 5.7 mm, oesophagus 0.68 mm long. Anus 0.18 mm from tail end. Tail straight. Vulva in mid-body or slightly nearer tail end.

The description was based on specimens recovered from *P. onocrotalus* in Egypt.

References: Khalil (1931)*; Skrjabin, Sobolev & Ivashkin (1965)*.

Cosmocephalus imperialis Morishita, 1930

Fig. 62
Host: *Uria aalge*.
Localization: proventriculus.
Distribution: Asia (U.S.S.R. — Sakhalin Island). Unknown outside the Palaearctic Region.

Description: Four long cordons with scalloped inner border. Tip of lateral loops of cordons 0.11—0.13 mm from anterior end of body in male and about 0.13 mm or more in female. Cervical papillae salient, bicuspid. Lateral alae originate immediately posterior to papillae, and terminate in female at, or slightly posterior to, level of oesophageal-intestinal junction and in male extend nearly whole length of body.
Male: Body length 6.3—7.3 mm, maximum body width 0.29 mm. Posterior margin of cordons 0.45—0.47 mm, cervical papillae 0.50—0.56 mm, nerve ring 0.48—0.52 mm and excretory pore 0.58—0.60 mm from anterior end of body. Cordons 0.23—0.27 mm wide (dorsally) and 0.29 mm (laterally). Vestibule 0.43 mm long, muscular part of oesophagus 0.8—0.9 mm long, glandular part 3.1—3.2 mm long. Tail 0.40—0.44 mm long. Spicules 0.58 to 0.66 m and 0.18—0.28 mm long. Nine pairs of pedunculate papillae present, four of which are preanal and five postanal, plus two pairs of small papillae near tail end.
Female: Body length 16—18 mm, maximum body width 0.47—0.50 mm. Posterior margin of cordons 0.54—0.65 mm, cervical papillae 0.56—0.65 mm, nerve ring 0.56—0.61 mm and excretory pore 0.65—0.74 mm from anterior end of body. Cordons 0.40—0.45 mm wide (dorsally) and 0.45—0.54 mm

(laterally). Vestibule 0.45—0.54 mm long, muscular part of oesophagus 1.10—1.1 mm long, glandular part 3.3—3.7 mm long. Vulva 8.5—10.2 mm from posterior end. Length of tail 0.25—0.3 mm. Eggs 0.035—0.036 × 0.017 mm.

References: Morishita (1930)*; Skrjabin, Sobolev & Ivashkin (1965)*.

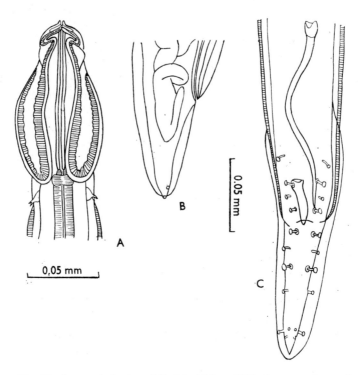

Fig. 62. *Cosmocephalus imperialis* Morishita, 1930. A — anterior end (dorsal view); B — posterior end of female (lateral view); C — posterior end of male (ventral view). After Morishita (1930).

Cosmocephalus jaenschi Johnston & Mawson, 1941

Fig. 63
Host: *Phalacrocorax carbo* and *P. pygmaeus*.
Localization: under cuticle of gizzard.
Discribution: Europe (Rumania), Asia (U.S.S.R. — Uzbekistan). Outside the Palaearctic Region recorded from gulls and cormorants in Australia.

Description:
Male: Body length about 10.5 mm. Lips with two large papillae. Rounded cuticular projections situated dorsally and ventrally between cordons. Cordons

salient, scalloped on inner border, forming a narrow loop. Cervical papillae tricuspid, 0.46 mm from head end. Vestibule 0.39 × 0.020 mm. Length of muscular part of oesophagus 0.9 mm and glandular part 3.7 mm. Nerve ring and excretory pore 0.45—0.53 mm from head end. Caudal alae supported by four pairs of preanal and five pairs of postanal papillae, the last pair of which is more massive than the others. Length of spicules 0.61 mm and 0.15 mm. Length of tail 0.29 mm.

The species was described on the basis of male specimens recovered from

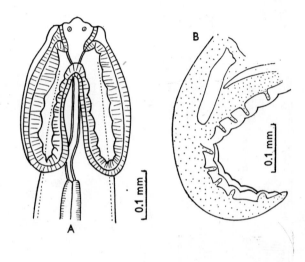

Fig. 63. *Cosmocephalus jaenschi* Johnston & Mawson, 1941. A — anterior end (lateral view); B — posterior end of male (lateral view). After Johnston & Mawson (1941).

Ph. carbo in Australia. It has since been reported several times, but the description of the female is still lacking.

References: Chiriac (1965); Johnston & Mawson (1941)*; Skrjabin, Sobolev & Ivashkin (1965)*; Sultanov (1959 a, b, 1963).

Genus *Decorataria* Skrjabin, Sobolev & Ivashkin, 1965

Nematodes with very large cuticular cordons, covering nearly the whole anterior part of body and uniting at level of origin of glandular part of oesophagus. External border of cordons scalloped. Cervical papillae simple. Females monodelphic.

Type species *D. decorata* (Cram, 1927).

The genus includes two species. Both of them parasitize fish-eating birds. One species has been reported from the territory of the Palaearctic Region.

Decorataria decorata (Cram, 1927)

Fig. 64
Hosts: *Podiceps cristatus, P. nigricollis, P. griseigena, P. ruficollis* and *Podiceps* sp.
Localization: under cuticle of gizzard.
Distribution: Europe (Czechoslovakia, U.S.S.R. — Estonia, northern and central regions, delta of the Volga River, Ukraine), Asia (U.S.S.R. — republics of Transcaucasus Region and Central Asia, Kazakhstan, Ural, West and East Siberia, Far East).

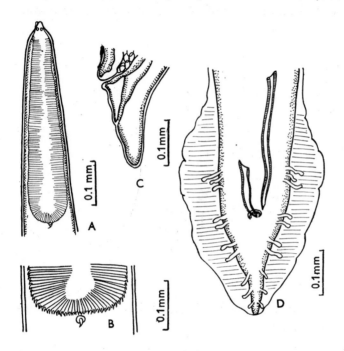

Fig. 64. *Decorataria decorat;* (Cram, 1927). A — anterior end (lateral view); B — detail of cordons a C — posterior end of female (lateral view); D — posterior end of male (ventral view). After Sobolev (1943).

Description:
Male: Body length 9.2—15 mm, maximum body width 0.3—0.4 mm. Vestibule 0.206—0.4 mm long, muscular part of oesophagus 0.41—0.81 mm, glandular part 2.93—4.62 mm long. Cordons 1.16—2.023 mm long. Tail 0.24—0.34 mm long. Caudal alae wide, transversely striated. Five to nine pairs of caudal papillae plus one unpaired papilla. Left spicule 0.42—0.55 mm, right spicule 0.13—0.23 mm long.
Female: Body length 12.8—17.5 mm, maximum body width 0.29—0.408 mm. Cordons 1.42—2.64 mm long. Vestibule 0.324—0.4 mm long, muscular part of oesophagus 0.694—0.9 mm long, glandular part 2.6 to 3.8 mm long. Tail 0.199—0.238 mm long. Vulva 0.35—0.462 mm from posterior end. Eggs 0.027—0.036 × 0.015—0.022 mm.

The species was described from specimens recovered from *P. auritus* in North America and was originally assigned to the genus *Echinuria*. It was

transferred to the genus *Decorataria* by Skrjabin, Sobolev & Ivashkin (1965).
Notes : Following Sobolev (1943 a), we regard *Syncuria ciconiae*, described from specimens found by Solonitsin (1928) in *P. ruficollis*, as a synonym of *D. decorata* and, according to Baruš & Zajíček (1967), the nematodes described by Škarda (1964) as *Syncuaria* sp. are also this species.

References: Baruš & Zajíček (1967); Borgarenko (1970)*; Cram (1927)*; Daiya (1967 a); Golikova (1959); Golovin (1964); Gushanskaya (1950 c); Gvozdev & Kasymzhanova (1965); Iksanov & Dikambaeva (1962); Kosupko (1963); Kurashvilli (1957, 1961); Okorokov (1957, 1964); Okorokov & Tkachev (1969); Panin (1960); Sailov (1962, 1965 b, 1966); Serkova (1948)*; Shakhtakhtinskaya (1959 a); Shigin (1957); Škarda (1964); Skrjabin, Sobolev & Ivashkin (1965)*; Smetanina (1972); Smogorzhevskaya (1962 a, 1964); Sobolev (1943 a)*; Solonitsin (1928 a, b); Turemuratov (1965 a); Vaidova (1965); Yigis (1962); Zhatkanbaeva (1964, 1965, 1966 a, b).

Genus *Desportesius* Skrjabin, Sobolev & Ivashkin, 1965

Cervical cordons with spines, running posteriorly from point of origin, widening gradually, then turning anteriorly and uniting on lateral sides of body. Vulva in posterior part of body. Monodelphic.

Type species *D. invaginatus* (Linstow, 1901).

The genus includes nine species, usually parasites of birds of the order *Ciconiiformes*. Eight species have been recorded in fish-eating birds of the Palaearctic Region.

Two of these species, namely, *D. groffi* and *D. raillieti*, were described on the basis of female specimens only. We have omitted these from the Key, because it is hardly possible to differentiate them from other species on the basis of their morphological features.

It should be noted also that some other species *(D. invaginatus, D. equispiculatus* and *D. spinulatus)* are very close to one another morphologically and in their hosts and distribution. It is possible that a more detailed study will show them to be identical.

KEY TO THE SPECIES OF THE GENUS *DESPORTESIUS*

```
1   Spicules nearly equal . . . . . . . . . . . . . . . . . . . . . . . . . 2
—   Spicules markedly unequal . . . . . . . . . . . . . . . . . . . . . . 4
2   Preanal papillae symmetrical . . . . . . . . . . . . . . . . D. invaginatus
—   Preanal papillae asymmetrical . . . . . . . . . . . . . . . . . . . . 3
3   Spicules 0.76—0.82 and 0.70—0.79 mm long . . . . . . . . . D. equispiculatus
—   Spicules 0.65—0.73 and 0.42—0.6 mm long . . . . . . . . . D. spinulatus
4   Eggs less than 0.02 mm long . . . . . . . . . . . . . . . . D. orientalis
—   Eggs more than 0.02 mm long . . . . . . . . . . . . . . . . . . . . 5
5   Spicules 0.207—0.232 and 0.679—0.795 mm long . . . . . . . . D. sagittatus
—   Spicules 0.060 and 0.110 mm long . . . . . . . . . . . . . D. brevicaudatus
```

Desportesius invaginatus (Linstow, 1901)

Fig. 65 A—C
Hosts: *Ardea cinerea, A. purpurea, Egretta alba, E. garzetta, Bubulcus ibis* and *B. lucidus*.
Localization: oesophagus and under cuticle of gizzard.
Distribution: Europe (France, F.R.G., U.S.S.R. — Ukraine, Volga Delta), Asia (U.S.S.R.—Transcaucasus Region, republics of Middle Asia, East Siberia, Far East; Japan), Africa (Algeria, Egypt). Outside the Palaearctic Region, India, North and South America.

Description:
Male: Body length 8.4—12 mm, body width 0.24—0.29 mm. Cordons 0.49—0.55 mm long. Excretory pore 0.4—0.41 mm, nerve ring 0.26—0.3 mm and cervical papillae 0.57—0.66 mm from anterior end. Vestibule 0.24—0.27 mm long. Muscular part of oesophagus 0.81—0.97 mm, glandular part 2.53 to 3.2 mm long. Caudal alae wide, 1.32—1.56 mm long. Left spicule 0.47 to 0.53 mm long, right spicule 0.62—0.9 mm long. Four pairs of preanal and five pairs of postanal papillae present.
Female: Body length 9.5—12.8 mm, body width 0.31—0.41 mm. Cordons 0.61—0.68 mm long. Excretory pore 0.45—0.49 mm, nerve ring 0.3—0.37 mm and cervical papillae 0.77—0.9 mm from anterior end. Vestibule 0.27—0.3 mm

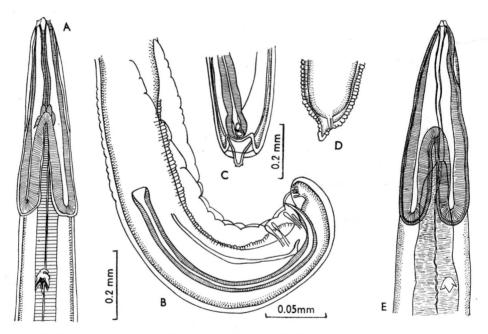

Fig. 65. *Desportesius invaginatus* (Linstow, 1901) — A, B, C; *D. brevicaudatus* (Dujardin, 1845) — D, E. A — anterior end (lateral view); B — posterior end of male (lateral view); C — posterior end of female; D — posterior end of female; E — anterior end (lateral view). A, B, C — After Ali (1968); D, E — After Skrjabin (1915).

long, muscular part of oesophagus 0.93—1.01 mm long, glandular part 2.28—2.29 mm long. Vulva immediately in front of anus. Eggs 0.025—0.029 × × 0.018—0.02 mm.

The species was described from the birds of South Africa and was originally placed in the genus *Dispharagus*. It was transferred by Skrjabin, Sobolev & Ivashkin (1965) to the genus *Desportesius*.

References: Babaev (1970); Bashkirova (1960); Cram (1927)*; Dubinin & Dubinina (1940); Feyzullaev (1963 a); Kasimov & Feyzullaev (1965); Kosupko (1963); Kurashvili (1957)*; Leonov (1960 a); Linstow (1901 a)*; Myers, Kuntz & Wells (1972); Oshmarin (1963, 1965); Sailov (1962, 1965 b); Seurat ((1915); Shakhtakhtinskaya (1959 a, b); Skrjabin, Sobolev & Ivashkin (1965)*; Smetanina (1972); Smogorzhevskaya (1962 a, 1964); Stossich (1902)*.

Desportesius brevicaudatus (Dujardin, 1845)

Fig. 65 D, E
Hosts: *Ardea cinerea, Egretta garzetta, E. alba, Botaurus stellaris, Ixobrynchus minutus* and *Ciconia ciconia* (experimentally).
Localization: gizzard.
Distribution: Europe (France, F.R.G., Poland, Rumania, U.S.S.R. — Ukraine), Asia (Tajikistan).

Description:
Male: Body length 10 mm, maximum body width 0.16 mm. Cordons 0.228 mm long. Tail 0.063 mm long. Six to seven pairs of caudal papillae present. Small spicule 0.06 mm long, large spicule 0.11 mm long.
Female: Body length 7.5—11.6 mm, body width 0.29 mm. Cordons 0.31 to 0.425 mm long. Cervical papillae 0.56 mm from anterior end. Vestibule 0.21 mm long, muscular part of oesophagus 0.85 mm long. Tail 0.068 mm long. Vulva 0.64—0.7 mm from posterior end. Eggs 0.02—0.03 × 0.025 mm.

The species was described from specimens recovered from *B. stellaris* in France and was originally assigned to the genus *Dispharagus*. It was transferred to the genus *Desportesius* by Skrjabin, Sobolev & Ivashkin (1965).
Biology: The life-cycle of this species was studied by Mozgovoy, Popova & Semenova (1965). *D. brevicaudatus* utilizes intermediate and paratenic or transport hosts during its development. Larvae of *Odonata* serve as intermediate hosts and fish (catfish, pike, tench, stickleback) as paratenic hosts. Young forms of the nematodes were found in stork and common heron seven days after ingestion of infective larvae.

References: Bezubik (1956); Cram (1927)*; Creplin (1846); Diesing (1851); Dujardin (1845)*; Dubinina & Serkova (1951); Gurlt (1845); Iordăchescu-Lăzărescu (1963); Mozgovoy, Popova & Semenova (1965); Skrjabin (1917*, 1923*); Skrjabin, Sobolev & Ivashkin (1965)*; Smogorzhevskaya (1962 a, 1964); Stossich (1892 a)*; Walter (1866).

Desportesius equispiculatus (Wu & Liu, 1943)

Host: *Egretta garzetta*.
Localization: stomach.
Distribution: Europe (Rumania), Asia (China); outside the Palaearctic Region not reported.

Description:
Male: Body length 9.6—12 mm, body width 0.21—0.23 mm. Cordons 0.53 to 0.55 mm long. Vestibule 0.27—0.28 mm, muscular part of oesophagus 0.94—0.98 mm and glandular part 2.75—3.1 mm long. Nerve ring 0.285 to 0.30 mm from anterior end. Caudal alae 1.6—1.8 mm long. Spicules nearly equal, 0.76—0.82 mm and 0.70—0.79 mm long. Nine to ten pairs of caudal papillae present. Preanal papillae asymmetrical.
Female: Body length 11.1—15.1 mm, body width 0.245—0.374 mm. Cordons 0.56—0.62 mm long, Vestibule 0.27—0.30 mm, muscular part of oesophagus 0.84—1.1 mm and glandular part 2.8—3.71 mm long. Nerve ring 0.29 to 0.33 mm from anterior end. Tail 0.038—0.05 mm long. Eggs 0.027—0.032 × × 0.019—0.02 mm.

The species was described from specimens recovered from *E. garzetta* in China and was originally assigned to the genus *Synhimantus*. It was transferred to the genus *Desportesius* by Skrjabin, Sobolev & Ivashkin (1965). (Dravings of this species have not appeared in the literature.)

References: Chiriac (1965); Skrjabin, Sobolev & Ivashkin (1965)*; Wu & Liu (1943)*.

Desportesius groffi (Li, 1934)

Fig. 66
Host: *Nycticorax nycticorax*. Recorded also from *Galliformes*.
Localization: gizzard.
Distribution: Asia (Japan).

Description: Only female known. Body length 5.85—7.9 mm, body width 0.14—0.25 mm. Cordons turn anteriorly 0.26—0.30 mm from anterior end and anastomose 0.1—0.13 mm from anterior end. Cervical papillae and nerve ring 0.29—0.45 mm and 0.20—0.22 mm, respectively, from anterior end. Vestibule 0.15—0.23 mm, muscular part of oesophagus 0.55—0.85 mm and glandular part 1.9—2.4 mm long. Tail 0.054—0.09 mm long. Vulva 0.10 to 0.12 mm from posterior end. Eggs 0.018—0.020 × 0.010—0.013 mm.

The species was described from specimens recovered from *Bambusicola thoracica* in the southern regions of China and was originally placed in the genus *Synhimantus*. It was transferred to the genus *Desportesius* by Skrjabin, Sobolev & Ivashkin (1965).

References: Li (1934)*; Skrjabin, Sobolev & Ivashkin (1965)*; Yamaguti (1935)*.

Fig. 66. *Desportesius groffi* (Li, 1934). A — anterior end (lateral view); B — anterior end (apical view); C — posterior end of female (lateral view). After Li (1934).

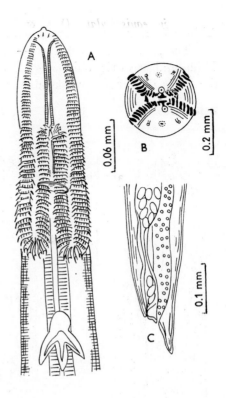

Desportesius orientalis (Wu, 1933)

Fig. 67
Host: *Botaurus stellaris*.
Localization: stomach.
Distribution: Asia (China). Not recorded outside the Palaearctic Region.

Description:

Male: Body length 5.59—6.35 mm, body width 0.13—0.15 mm. Vestibule 0.14—0.20 mm, muscular part of oesophagus 0.46—0.60 mm and glandular part 2.28—2.51 mm long. Anterior and posterior bends of cordons 0.112 to 0.116 mm and 0.22—0.25 mm, respectively, from anterior end. Cervical papillae 0.26—0.36 mm and excretory pore 0.95—1.24 mm from anterior end. Caudal alae well developed, 0.82—1.00 mm long. Tail 0.03—0.06 mm long. Spicules 0.36 mm and 0.08 mm long. Three or four pairs of preanal and five pairs of postanal papillae present.

Female: Body length 8.18—10.43 mm, body width 0.25—0.28 mm. Vestibule 0.176—0.236 mm, muscular part of oesophagus 0.592—1.022 mm and glandular part 1.75—2.95 mm long. Nerve ring 0.22—0.28 mm and cervical papillae 0.352—0.432 mm from anterior end. Anterior and posterior bends of cordons 0.11—0.14 mm and 0.27—0.31 mm, respectively, from anterior end. Vulva

8.13—8.76 mm from anterior end. Eggs 0.016—0.20 × 0.020—0.024 mm.

The species was described from specimens recovered from *B. stellaris* in China and was originally assigned to the genus *Acuaria*. It was transferred to the genus *Desportesius* by Skrjabin, Sobolev & Ivashkin (1965).

References: Hsü, W. N. (1957)*; Skrjabin, Sobolev & Ivashkin (1965)*; Wu (1933)*.

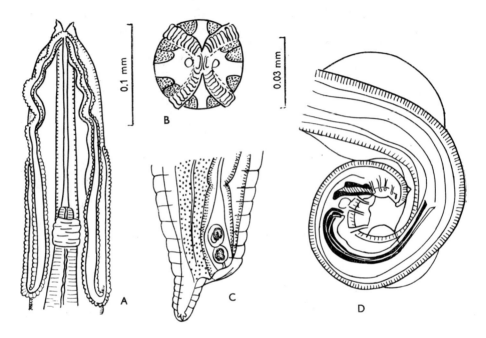

Fig. 67. *Desportesius orientalis* (Wu, 1933). A — anterior end (dorso-ventral view); B — anterior end (apical view); C — posterior end of female (lateral view); D — posterior end of male (lateral view). After Wu (1933).

Desportesius raillieti (Skrjabin, 1924)

Host: *Pelecanus onocrotalus*.
Localization: mucosa of the pouch under beak.
Distribution: Europe (Rumania). Outside the Palaearctic Region in Africa.

Description: Only female described. Body length 13—15 mm, maximum body width 0.7 mm. Vestibule 0.425 mm, muscular part of oesophagus 0.56 mm and glandular part 3.57 mm long. Tail 0.17 mm long. Cordons turning anteriorly 0.84 mm and anastomosing 0.28 mm from anterior end. Vulva 0.51 mm from posterior end. Eggs 0.038—0.02 mm.

The species was described from specimens recovered from *Pelecanus* sp.

in Somaliland and was originally placed in the genus *Acuaria*. It was transferred to the genus *Desportesius* by Skrjabin, Sobolev & Ivashkin (1965). *D. raillieti* was also reported from *P. onocrotalus* in Rumania (Chiriac 1965) but unfortunately this author neither described nor illustrated the nematodes recovered.

References: Chiriac (1965); Skrjabin (1924)*; Skrjabin, Sobolev & Ivashkin (1965)*.

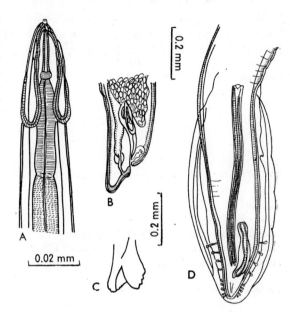

Fig. 68. *Desportesius sagittatus* (Rudolphi, 1809). A — anterior end (dorso-ventral view); B — posterior end of female (lateral view); C — distal end of long spicule; D — posterior end of male (ventral view). After Macko (1964).

Desportesius sagittatus (Rudolphi, 1809)

Fig. 68
Hosts: *Ardea purpurea*, *Ciconia nigra* and *Nycticorax nycticorax*.
Localization: gizzard.
Distribution: Europe (Rumania, Bulgaria, Czechoslovakia), Asia (U.S.S.R. — Azerbaijan, Far East). Not reported outside the Palaearctic Region.

Description: In addition to cervical cordons, lateral cordons are present which originate behind cervical papillae and run posteriorly, disappearing gradually in caudal region. The presence of these cordons was pointed out by Macko (1964).
Male: Body length 4.9—7.7 mm, maximum body width 0.185—0.219 mm. Cervical cordons 0.274—0.531 mm long. Vestibule 0.15—0.205 mm, muscular part of oesophagus 0.384—0.490 mm and glandular part 1.676—2.104 mm long. Caudal alae 0.626—0.819 mm long. There are nine pairs of caudal papillae, four pairs preanal and five pairs postanal. Small spicule 0.207—0.232 mm, large spicule 0.679—0.795 mm long.

Female: Body length 6.7—9 mm, maximum body width 0.274—0.294 mm. Cervical cordons 0.368—0.521 mm long. Vestibule 0.13—0.192 mm, muscular part of oesophagus 0.343—0.445 mm and glandular part 1.39—2.308 mm long. Tail 0.058—0.075 mm long. Vulva 0.102—0.137 mm from posterior end. Eggs 0.027—0.033 × 0.016—0.018 mm.

The species was described on the basis of specimens recovered from *Ardea purpurea* in Central Europe. It was originally assigned to the genus *Ascaris* and later transferred several times to other genera; it was transferred to the genus *Desportesius* by Skrjabin, Sobolev & Ivashkin (1965).

References: Cram (1927)*; Diesing (1851); Dujardin (1845); Feyzullaev (1963 a); Gurlt (1845); Iordăchescu (1962)*; Macko (1964 c)*; Molin (1860 b); Oshmarin & Parukhin (1963); Rudolphi (1809*, 1819); Skrjabin (1923); Skrjabin, Sobolev & Ivashkin (1965)*; Stossich (1892 a)*; Zhelyazkova-Paspaleva (1962 b).

Desportesius spinulatus (Chabaud & Campana, 1949)

Fig. 69
Host: *Egretta garzetta*.
Localization: under mucous coat of crop.
Distribution: Europe (France). Outside the Palaearctic Region in *Ciconiiformes* from Africa and South-East Asia.

Description:
Male: Body length 6.3—8.9 mm. Cordons turn anteriorly 0.51 mm and anastomose 0.25 mm from anterior end. Vestibule 0.25 mm, muscular part of oesophagus 0.69 mm and glandular part 2.1 mm long. Nerve ring 0.27 mm, excretory pore 0.41 mm and cervical papillae 0.65 mm from anterior end. Large spicule 0.65—0.73 mm, small one 0.42—0.60 mm long. Preanal papillae in two groups, one, situated near cloaca, with two symmetrical pairs; the other, situated anteriorly, with only three papillae, two on the right and one (larger and longer) on the left. There are five pairs of postanal papillae. Phasmids subterminal.

Female: Body length 8.0—9.6 mm. Cordons turn anteriorly 0.57 mm and anastomose 0.28 mm from anterior end. Vestibule 0.26 mm, muscular part of oesophagus 0.73 mm and glandular part 2.8 mm long. Nerve ring 0.28 mm, excretory pore 0.46 mm and cervical papillae 0.655 mm from anterior end. Cuticle of posterior part inflated and distinctly elevated above body; swelling commences 0.5—0.9 mm from posterior end and disappears at level of vulva and anus; it forms a large dome-like structure about 0.1 mm deep in which the tail originates. Tail 0.08—0.09 mm long with a spur composed of one pair of small chitinous formations about 0.02 mm long at base. Vulva opens immediately in front of anus. Eggs 0.028—0.030 × 0.018—0.020 mm.

The nematodes of this species were described as subspecies of *Synhimantus equispiculatus* Wu & Liu 1943. The description was based on material from *E. garzetta* from France. Later, the authors raised this subspecies to the rank of species and assigned it to the newly erected subgenus *Desportesius* in the genus *Synhimantus*. Skrjabin, Sobolev & Ivashkin (1965) raised the subgenus *Desportesius* to generic rank.

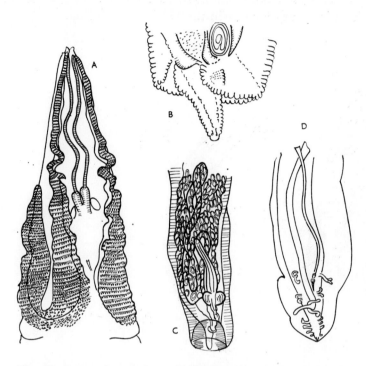

Fig. 69. *Desportesius spinulatus* (Chabaud & Campana, 1949). A — anterior end (dorso-ventral view); B — posterior end of female (detail); C — tail of female; D — posterior end of male (ventral view). After Chabaud & Campana (1949).

Biology: Ostracods *(Cyprinotus salinus* and *Pionocypris vidus)* serve as intermediate hosts. *D. spinulatus* may also employ paratenic hosts (tadpoles) in which the larvae encapsulate (Chabaud 1950, 1954).

Notes: In the opinion of Schmidt & Kuntz (1971), with which we agree, the nematodes found in *Egretta garzetta* in Vietnam and designated as *Cosmocephalus* sp. Ryzhikov & Khokhlova, 1965, are *D. spinulatus*.

References: Chabaud (1950, 1954); Chabaud & Campana (1949 b)*; Schmidt & Kuntz (1971); Skrjabin, Sobolev & Ivashkin (1965)*.

Genus *Dispharynx* Railliet, Henry & Sisoff, 1912

Posterior ends of cuticular cordons recurrent but not anastomosing. Spicules unequal and dissimilar. Parasitic in oesophagus and proventriculus of birds.

Type species *D. nasuta* (Rudolphi, 1819).

The genus includes more than 20 species. Five species have been recorded in fish-eating birds but only one in the Palaearctic Region.

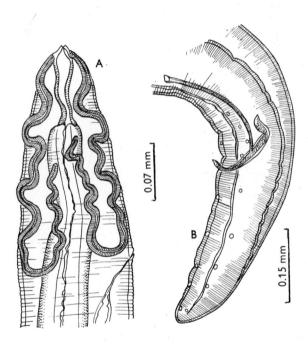

Fig. 70. *Dispharynx nasuta* (Rudolphi, 1819). A — anterior end (lateral view); B — posterior end of male (lateral view). After Chertkova (1949).

Dispharynx nasuta (Rudolphi, 1819)

Fig. 70

Hosts: *Ciconia* sp., *Larus genei* and *L. ridibundus*. Not characteristic of fish-eating birds, usually parasitizing *Galliformes* and *Passeriformes*.
Localization: oesophagus, proventriculus, gut.
Distribution: Europe (U.S.S.R. — the Sea of Azov, Danube Delta), Asia (U.S.S.R. — Tajikistan). Outside the Palaearctic Region, North and South America and Africa.

Description:

Male: Body length 4.1—8 mm. Cordons 0.145—0.42 mm long. Nerve ring 0.225—0.3 mm, and cervical papillae 0.21—0.36 mm from anterior end. Length of vestibule 0.093—0.185 mm, muscular part of oesophagus 0.41—0.88

mm and glandular part 1.46—2.068 mm. Spicules 0.395—0.482 and 0.12—0.0.189 mm long; nine pairs of caudal papillae present, four preanal, five postanal.

Female: Body length 3.47—9.21 mm. Cordons 0.115—0.56 mm long. Nerve ring 0.17—0.35 mm and cervical papillae 0.155—0.54 mm from anterior end. Length of vestibule 0.09—0.14 mm, muscular part of oesophagus 0.33—1.05 mm and glandular part 1.61—1.91 mm. Vulva 1.0—1.66 mm from posterior end. Eggs 0.028—0.038 × 0.016—0.018 mm.

The description was based on material from *Passeriformes* in Austria. The species was originally placed in the genus *Spiroptera*, and then in many other genera. It was transferred to the genus *Dispharynx* by Railliet, Henry & Sisoff (1912).

Biology: According to Cram (1931) the intermediate hosts of this species are isopods *(Porcellio scaber, P. laevis, Armadillidium vulgare)*. The larva develops into the infective stage within 26 days in the intermediate host and in the definitive host *(Galliformes)* the parasites become adult within 27 days.

Notes: In agreement with the opinion of many authors (Goble & Kutz 1945; Chertkova & Petrov 1961; Skrjabin, Sobolev & Ivashkin 1965 and others) we regard *D. spiralis* (Molin, 1858) as a synonym of *D. nasuta*. The bases of the synonymy are given in the papers of the authors referred to above. This species was also recorded under the name *D. spiralis* (Gushanskaya 1950, Kulachkova 1950) from *Ciconia* sp. and *Larus ridibundus*.

References: Chertkova (1961)*; Cram (1927*, 1931); Goble & Kutz (1945;) Gushanskaya (1950 c); Kulachkova (1950); Railliet, Henry & Sisoff (1912); Rudolphi (1819)*; Sergeeva (1968 a, 1969); Skrjabin, Sobolev & Ivashkin (1965)*; Smogorzhevskaya (1964).

Genus *Paracuaria* Rao, 1951

Cordons very short, straight, in shallow groove-like depressions, not extending beyond apical region of nematode. Cervical papillae multicuspid. Lateral alae absent. Spicules unequal and dissimilar.

Type species — *P. tridentata* (Linstow, 1877).

The genus includes two species; both of them have been recorded in fish-eating birds of the Palaearctic Region.

KEY TO THE SPECIES OF THE GENUS *PARACUARIA*

1 Pharynx short (0.03—0.07 mm); usually parasites of *Nyrocinae (Melanitta, Somateria, Aythia)* . . . . *P. formosensis*
— Pharynx long (0.09—0.18 mm); usually parasites of gulls and divers . . . . *P. tridentata*

Paracuaria tridentata (Linstow, 1877)

Fig. 71
Hosts: *Gavia arctica, G. stellata, Podiceps griseigena, Fulmarus glacialis, Phalacrocorax pelagicus, Botaurus stellaris, Stercorarius parasiticus, Pagophila eburnea, Rissa tridactyla, R. brevirostris, Rhodosthethia rosea, Larus argentatus, L. crassirostris, L. fuscus, L. genei, L. glaucescens, L. hyperboreus, L. ichthyaetus, L. marinus, L. melanocephalus, L. minutus, L. ridibundus, L. schistisagus, Chlidonias leucoptera, Hydroprogne tschegrava, Sterna hirundo, S. albifrons, S. sandvicensis, Aethia pygmaea* and *Cyclorrhynchus psittacula*.
Localization: oesophagus, intestine and under cuticle of gizzard.
Distribution: Europe (England, France, Denmark, G.D.R., Iceland, U.S.S.R. — Ukraine, Byelorussia, Pribaltic Region, North and Central Regions), Asia (U.S.S.R. — West and East Siberia, Far East, republics of Middle Asia and Kazakhstan). Outside the Palaearctic Region, in gulls from Canada and Cuba.

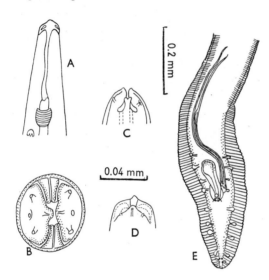

Fig. 71. *Paracuaria tridentata* (Linstow, 1877). A — anterior end (lateral view); B — anterior end (apical view); C, D — head end (dorso-ventral and lateral view); E — posterior end of male (ventral view); A — After Rao (1951), B to E — After Chabaud and Czaplinski (1961).

Description: Head with two large pointed lateral pseudolabia and two small, narrow median lips with deeply incised anterior margins and rather wide bases which penetrate under the bases of pseudolabia. Median lips barely two thirds of length of pseudolabia. Lips and pseudolabia united to tips by transparent membranes which resemble four short cordons which turn outwards. Cervical papillae tricuspid, slightly posterior to nerve ring.
Male: Body length 5.9—12.0 mm, maximum body width 0.10—0.125 mm. Length of vestibule 0.09—0.18 mm, of anterior part of oesophagus 0.50—1.0 mm and posterior part of oesophagus 1.25—1.80 mm. Nerve ring 0.14 to 0.185 mm and cervical papillae 0.18—0.24 mm from anterior end. Tail 0.19 mm long, caudal alae well developed. There are ten pairs of pedunculate papillae (four pairs precloacal, six pairs postcloacal) and a single median papilla situated in front of anterior lip of cloaca. Right spicule short and thick,

0.08—0.13 mm long; left spicule 0.30—0.47 mm long, with complex distal end.

Female: Body length 7.0—25.0 mm, maximum body width 0.08—0.23 mm. Length of vestibule 0.11—0.18 mm, anterior part of oesophagus 0.62—1.14 mm and posterior part of oesophagus 1.30—2.84 mm. Nerve ring 0.05—0.21 mm and deirids 0.08—0.27 mm from anterior end. Vulva salient, 4.93—11.30 mm from anterior end. Tail 0.11—0.195 mm long. Eggs 0.03—0.04 × 0.01—0.02 mm.

The species was described from specimens recovered from *G. arctica* in Central Europe and was originally placed in the genus *Filaria*. Later it was transferred to the genus *Spiroptera*, then to *Streptocara*. Leonov, Tsimbalyuk & Belogurov (1963) transferred it to the genus *Paracuaria*.

Biology: Shmytova (1967) found larvae of *Acuariidae* in *Pimelia subglobosa* and *Tentiria taurica* collected where gulls often occurred. Having fed the larvae to gulls, the author recovered adult *P. tridentata* from them.

Notes: In agreement with the opinion of Leonov, Tsimbalyuk & Belogurov (1963), we regard the species *P. macdonaldi* (described from gulls in Canada by Rao in 1951) as a synonym of *P. tridentata* but *Paracuaria macdonaldi* was the type and only species of the genus. After its synonymization, Kurochkin & Ryzhikov (1964) proposed *P. tridentata* as type species of the genus. Baruš (1967) agreed with these authors and redescribed *P. tridentata* from Linstow's material. Skrjabin, Sobolev & Ivashkin (1965) were of a different opinion and, while recognizing the identity of these two species, regarded *P. macdonaldi* as type species, placing *P. tridentata* among other synonyms of this species. We do not agree with this.

We consider the species *Streptocara transcaucasica* Solonitsin, 1928 (according to Ryzhikov 1966) and *Streptocara rissae* Kreis, 1958 (according to Daiya 1967 a) to be synonyms of *P. tridentata*.

References: Ablasov & Chibichenko (1961, 1962); Agapova & Zhatkanbaeva (1971); Alekseev & Smetanina (1968); Babaev (1970); Bakke (1972); Bakke & Baruš (1976); Baruš (1967)*; Baylis (1939); Belogurov, Leonov & Zueva (1968); Belopolskaya (1952); Chabaud & Czaplinski (1961 a*, b*); Creutz & Gottschalk (1969); Daiya (1967 a); Ellis & Williams (1973); Gubanov & Sergeeva (1968); Guildal (1964, 1966, 1968); Gvozdev & Kasymzhanova (1965); Kontrimavichus & Bakhmeteva (1960); Kreis (1958)*; Krivonogova (1963); Kulachkova & Kochetova (1964); Kurochkin & Ryzhikov (1964)*; Kurochkin & Zablotsky (1961); Leonov (1958); Leonov & Shvetsova (1970); Leonov, Tsimbalyuk & Belogurov (1963)*; Linstow (1877 a, b*); Mashtakov (1964); Michelson (1968); Mozgina (1967, 1969); Mukhamadiev (1966); Oshmarin & Parukhin (1963); Panova (1926); Pemberton (1963); Rao (1951)*; Sergeeva (1968 a, 1969); Serkova (1948); Shigin (1961); Shmytova (1967); Skrjabin, Sobolev & Ivashkin (1965)*; Smetanina (1972); Smetanina & Alekseev (1967); Smogorzhevskaya (1964); Smogorzhevskaya, Kornyushin, Iskova & Eminov (1965); Solonitsin (1928 a, b); Sonin & Larchenko (1974); Sultanov, Ryzhikov & Kozlov (1960); Tsimbalyuk & Belogurov (1964); Tsimbalyuk, Leonov & Belogurov (1963); Turemuratov (1962 a) Vasilkova (1926, 1927); Yigis (1962).

Paracuaria formosensis (Sugimoto, 1930)

Fig. 72
Hosts: *Mergus merganser*, *M. serrator* and *M. squamatus*. Characteristic of the genera *Melanitta*, *Somateria* and *Aythya*. Found only rarely in *Mergus* spp.
Localization: under cuticle of gizzard.
Distribution: Asia (U.S.S.R. — East Siberia, Far East). Outside the Palaearctic Region in *Anseriformes* of North America and South-East Asia.

Description: Two very indistinct cordons on each side running posteriorly from mouth opening. Cervical papillae posterior to end of pharynx, usually with three to four, rarely with six denticles.

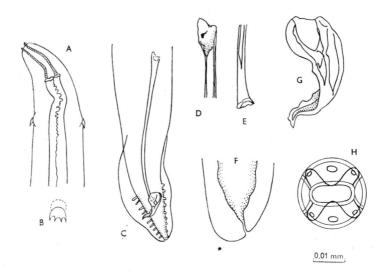

Fig. 72. *Paracuaria formosensis* (Sugimoto, 1930). A — anterior end (dorso-ventral view); B — cervical papilla; C — posterior end of male (ventral view); D — distal end of long spicule; E — proximal end of long spicule; F — posterior end of female (lateral view); G — small spicule; H — anterior end (apical view); A to G—After Ryzhikov (1960), H—After Ryzhikov & Kurochkin (1964).

Male: Body length 5.14—7.83 mm, maximum body width 0.14—0.20 mm. Cervical papillae 0.11—0.24 mm, nerve ring 0.135—0.172 mm and excretory pore 0.188—0.224 mm from anterior end. Cervical papillae 0.006—0.008 × × 0.008—0.011 mm. Length of vestibule 0.03—0.06 mm, muscular part of oesophagus 0.31—1.0 mm, and glandular part 1.56—3.70 mm. Caudal alae 0.331—0.424 mm long. Tail 0.073—0.08 mm. Ten pairs of caudal papillae present, four preanal and six postanal; last pair of papillae very small and not always visible. Left spicule 0.28—0.46 mm long, its proximal end swollen, distal

end with a small asymmetrical process; right spicule 0.062—0.12 mm long, cymbiform, with narrow distal end on ventral side of which is a small denticle.
Female: Body length 8.16—18 mm, maximum body width 0.17—0.52 mm. Cervical papillae 0.108—0.22 mm, nerve ring 0.098—0.185 mm and excretory pore 0.22—0.27 mm from anterior end. Cervical papillae 0.010—0.011 × 0.011 to 0.016 mm. Length of vestibule 0.028—0.065 mm, muscular part of oesophagus 0.66—1.2 mm, and glandular part 1.84—3.1 mm. Vulva 0.52—0.64 mm from anterior end. Tail 0.049—0.06 mm long. Eggs 0.03—0.04 × × 0.018—0.032 mm.

The species was described from specimens recovered from *Anas platyrhyncha* and *Cairina moschata* in China (Taiwan Island) and was originally placed in the genus *Streptocara*.

Notes: After they had revised the genus *Paracuaria*, Leonov, Tsimbalyuk & Belogurov (1963) concluded that *Streptocara somateriae* Ryzhikov, 1960 corresponded morphologically with the diagnosis of the genus *Paracuaria* and transferred it to this genus. Gibson (1968) analysed the species composition of the genus *Streptocara* and synonymized the species *P. somateriae* with *S. formosensis*. We agree with Gibson's opinion that these two species are synonyms but believe, however, that *S. formosensis* should be included in the genus *Paracuaria*, as the arrangement of its anterior end corresponds with the diagnosis of this genus.

References: Daiya (1967 b); Gibson (1968)*; Kurochkin & Ryzhikov (1964)*; Leonov Tsimbalyuk & Belogurov (1963); Ryzhikov (1960)*; Skrjabin, Sobolev & Ivashkin (1965)* Sugimoto (1930)*; Tolkacheva (1967); Tsimbalyuk (1965).

Genus *Pectinospirura* Wehr, 1933

Cervical papillae very large, transformed into numerous denticles, surrounding body from lateral sides in form of semi-circles. Cordons in front of cervical papillae, recurrent and anastomosing. Spicules unequal and dissimilar.

Type species *P. argentata* Wehr, 1933.

The genus includes two species, namely, *P. argentata* parasitizing gulls in north America and Australia and *P. multidentata* reported from the Palaearctic Region.

Pectinospirura multidentata Sobolev, 1943

Fig. 73
Host: *Larus argentatus*. Reported also in *Charadriiformes*.
Localization: intestine.
Distribution: Asia (U.S.S.R. — West Siberia, Far East, Uzbekistan).

Description: 30—38 denticles on each cervical papilla.
Male: Body length 3.79—5.59 mm, maximum body width 0.07—0.072 mm. Length of vestibule 0.09—0.20 mm, muscular part of oesophagus 0.389—0.490 mm, and glandular part 1.90—2.435 mm. Cordons extend 0.34—0.714 mm posteriorly. Cervical papillae 0.935 mm from anterior end, 0.216 mm wide. There are nine pairs of caudal papillae, four of which are preanal and five

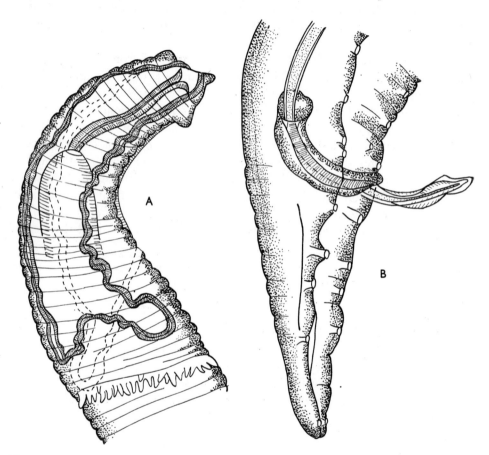

Fig. 73. *Pectinospirura multidentata* (Sobolev, 1943). A — anterior end (lateral view); B — posterior end of male (lateral view). After Sergeeva (1969).

postanal. Spicules unequal and dissimilar, large 0.75 to 0.80 mm long, small 0.15—0.174 mm long.

The species was originally described from a male recovered from *Terekia cinerea* in the Volga Region. The female was found later, but not described, by Krivonogova (1963) in *L. argentatus* from the Amur River.

Notes: Sergeeva (1968) regarded the species *P. sobolevi* Turemuratov, 1965, described from *L. argentatus* from the Aral Sea, as a synonym of *P. multidentata*. We agree with this opinion.

References: Krivonogova (1963)*; Sergeeva (1968 a, 1969); Skrjabin, Sobolev & Ivashkin (1965)*; Sobolev (1943 b, 1947*); Turemuratov (1965 b)*.

Genus *Rusguniella* Seurat, 1919

Two falciform cordons, present on anterior end, arising in depressions between lips and surrounding bases of lips on lateral sides in form of collarette. Lateral alae present or absent. Cervical papillae setiform. Spicules unequal and dissimilar. Vulva with salient lips, postequatorial.

Type species *R. elongata* (Rudolphi, 1819).

The genus includes seven species; two of them have been reported from fish-eating birds of the Palaearctic Region.

KEY TO THE SPECIES OF THE GENUS *RUSGUNIELLA*

1 Lateral alae present . . . . *R. elongata*
— Lateral alae absent . . . . *R. wedli*

Rusguniella elongata (Rudolphi, 1819)

Fig. 74 A, B

Hosts: *Podiceps cristatus, P. griseigena, Ardea cinerea, A. purpurea, Ixobrychus sinensis, Larus canus, L. argentatus, L. fuscus, L. genei, L. minutus, L. ridibundus, Xema sabini, Chlidonias nigra, Sterna hirundo,* and *S. albifrons.* Recorded also in members of *Anatidae* and *Charadriiformes.*

Localization: oesophagus and gizzard.

Distribution: Europe (countries of West and Central Europe, U.S.S.R. — Pribaltic Region, Beylorussia, Central regions of Russia, Ukraine), Asia (U.S.S.R. — West Siberia, Far East, republics of Middle Asia), Africa (Algeria). Outside the Palaearctic Region, South-East Asia.

Description:

Male: Body length 7.5—13 mm, maximum body width 0.15—0.23 mm. Lateral parts of cordons 0.05—0.08 mm, nerve ring 0.16—0.20 mm and excretory pore 0.28—0.31 mm from anterior end. Cervical papillae 0.10 to 0.15 mm from anterior end, sometimes symmetrical. Lateral alae ending 0.07—0.09 mm from posterior end. Length of vestibule 0.07—0.12 mm, muscular part of oesophagus 0.43—0.58 mm, and glandular part 2.05 to 3.4 mm. Nine pairs of caudal papillae present, of which four pairs are preanal and five postanal. Phasmids terminal. Large spicule 0.47—0.67 mm long, thin, tapering from proximal to distal end with uncinate process at tip. Smaller spicule cymbiform, 0.08—0.12 mm with widened proximal and bluntly rounded distal end.

Female: Body length 24—48 mm, maximum body width 0.25—0.54 mm. Lateral parts of cordons 0.05—0.13 mm from anterior end. Cervical papillae 0.11—0.19 mm from anterior end, asymmetrical. Length of vestibule 0.08 to 0.16 mm, length of muscular part of oesophagus 0.65—0.90 mm, glandular part 1.8—3.0 mm. Vulva 13—20 mm from anterior end. Nerve ring 0.09 to

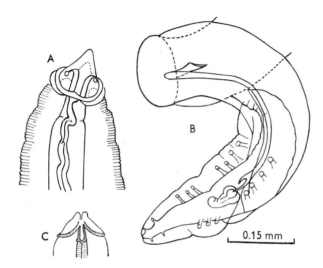

Fig. 74. *Rusguniella elongata* (Rudolphi, 1819) — A, B; *R. wedli* Williams, 1929 — C. A — anterior end (lateral view); B — posterior end of male (ventral view); C — anterior end of female (dorso-ventral view). A, B — After Ryzhikov (1966); C — After Williams (1929).

0.14 mm and excretory pore 0.26—0.42 mm from anterior end. Tail 0.16 to 0.45 mm long. Eggs 0.03—0.04 × 0.02 mm.

The description was based on a female specimen from *Ch. nigra* from Central Europe and the species was assigned to the genus *Spiroptera*. In the years following the species was placed in several genera and Seurat (1919) transferred it to the genus *Rusguniella*. The male of *R. elongata* was described by Gilbert (1930) from specimens recovered from the same definitive host caught in Byelorussia.

Notes: Sergeeva (1968 a) regarded *R. arctica* Ryzhikov, 1960 (recovered from *Anseriformes* in Siberia), *R. skrjabini* Chuan, 1961 and *R. tringae* Wang, 1966 (from *Charadriiformes* from the Far East and China) as synonyms of *R. elongata*. We agree with the opinion of this author.

References: Babaev (1970); Bakke (1972); Borgarenko (1970); Daiya (1971); Diesing (1851); Dujardin (1845); Gilbert (1930)*; Golikova (1959); Golovin (1964); Gurlt (1845); Kulachkova (1950); Leonov (1958); Leonov & Shvetsova (1970); Linstow (1878, 1909); Michelson (1968); Molin (1860 b); Parona (1902); Rudolphi (1819)*; Semenov (1927); Sergeeva (1968 a, b*, 1969); Serkova (1948); Seurat (1919)*; Shigin (1961); Skrjabin (1923); Skrjabin, Sobolev & Ivashkin (1965)*; Smetanina (1972); Smogorzhevskaya (1962 a, 1964); Smogorzhevskaya, Kornyushin, Iskova & Eminov (1965); Stossich (1891 b); Yigis (1962).

Rusguniella wedli Williams, 1929

Fig. 74 C
Hosts: *Podiceps cristatus, P. nigricollis* and *P. ruficollis*.
Localization: proventriculus, under parietal layer of peritoneum, muscles of femur and kidneys.
Distribution: Europe and Asia (U.S.S.R. — Ukraine and Transcaucasus).

Description: Lateral lips well developed. Head with a pair of narrow, granular, crescent-shaped cordons arising at angles of insertion of lips. Lateral alae absent.
Male: Very poorly known. According to Wedl's (1856) description, the posterior end is twisted, but the structure of the male genital apparatus is not accurately determined and no measurements are given.
Female: Body slightly attenuated anteriorly and posteriorly, length 30—40 mm, width 0.5 mm. Postanal region digitiform, curved. Eggs 0.038×0.024 mm, oval, thick-shelled, embryonated.

The species was described by Williams (1929) on the basis of Wedl's material.

Notes: Wedl (1856) briefly described the nematodes he found in *P. nigricollis* and assigned them to the genus *Dispharagus* without specific determination, but gave no locality in his paper. Seurat (1919) considered the nematodes described by Wedl to be identical with the species *R. elongata*. Williams (1929) did not share Seurat's opinion and separated these nematodes as an independent species, *R. wedli* Williams, 1929, considering the main difference between these species to be the absence of lateral alae in *R. wedli*. After the paper by Williams no further reports of this species appeared in the literature for a long time. *R. wedli* was reported again only recently in grebes in the territory of the U.S.S.R. (Vaidova 1965; Sailov 1966; Sergienko 1963, 1972) but these authors neither described nor figured the nematodes recovered. In our opinion, Seurat (1919) correctly identified the parasites described by Wedl as *R. elongata*. For a definitive elucidation of this problem, however, a larger number of specimens of *Rusguniella* from grebes should be studied in order to determine the importance of the presence or absence of lateral alae in given species.

References: Sailov (1966); Sergienko (1963, 1972); Vaidova (1965); Wedl (1856)*; Williams (1929)*.

Genus *Sexansocara* Sobolev & Sudarikov, 1939

Cordons situated in the anterior region, not extending beyond the length of pharynx, convoluted, each cordon forming a small bend at its origin, two larger bends in the middle part, first anteriorly and then posteriorly, uniting on lateral sides of body, with ends of each pair forming another loop. Cervical papillae tricuspid, situated behind cordons. Lateral alae originate behind cervical papillae.

Type and single species *S. skrjabini* Sobolev & Sudarikov, 1939.

Sexansocara skrjabini Sobolev & Sudarikov, 1939

Fig. 75
Host: *Pandion haliaetus.*
Localization: oesophagus.
Distribution: Europe and Asia (U.S.S.R. — Ukraine, the Volga Region, Azerbaijan, Georgia, Far East).

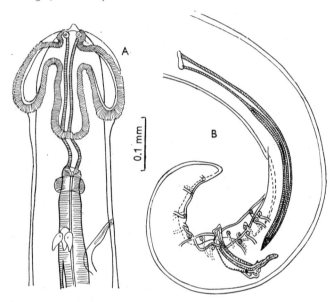

Fig. 75. *Sexansocara skrjabini* Sobolev & Sudarikov, 1939. A — anterior end (lateral view); B — posterior end of male (lateral view). After Sobolev & Sudarikov (1939).

Description: Nematodes of medium size. Pharynx long and narrow. Nerve ring surrounding oesophagus immediately behind its origin.
Male: Body length 9.68 mm. Width of anterior swollen part 0.19 mm, and middle part 0.17 mm. Cordons 0.16 mm long. Nerve ring 0.30 mm and cervical papillae 0.34 mm from anterior end. Length of vestibule 0.29 mm, of muscular part of oesophagus 0.77 mm, and glandular part 3.28 mm. Caudal alae 0.50 mm long and 0.04 mm wide. Four pairs of long, pedunculate preanal papillae and five pairs of postanal papillae present. Tail 0.30 mm long. Right spicule 0.11 mm long, short, thick, slightly bent, and rather complicated conformation; left spicule 0.42 mm long, with distal end attenuated and pointed.
Female: Body length 11.00—16.79 mm. Width of anterior swollen part 0.20—0.27 mm, and middle part 0.17—0.25 mm. Cordons 0.17—0.20 mm

long. Nerve ring 0.32—0.34 mm and cervical papillae 0.37—0.41 mm from anterior end. Length of vestibule 0.31 mm, of muscular oesophagus 0.82 to 0.84 mm, and glandular oesophagus 3.96—4.00 mm. Vulva 7.64—8.06 mm from anterior end. Tail 0.21—0.22 mm long, with rounded end. Eggs 0.04×0.02 mm.

The description was based on specimens found in *P. haliaetus* from the Middle Volga basin.

References: Kurashvili (1953, 1956, 1957); Oshmarin & Parukhin (1960); Samedov 1967 a, b, 1969); Skrjabin, Sobolev & Ivashkin (1965)*; Smogorzhevskaya (1962 a, 1964); Sobolev & Sudarikov (1939)*.

Genus *Skrjabinocara* Kurashvili, 1941

In addition to cordons situated on the head region, lateral cordon bands occur which originate behind the cervical papillae and end near the tip of the tail. Cordons of head end extend from behind lips and unite in pairs, not recurrent, composed of individual plates with sharp external margins. Caudal alae of males well developed. Vulva near anal opening.

Type species *S. squamata* (Linstow, 1883).

The opinions of different authors are not consistent as to the species composition of this genus. According to Skrjabin, Sobolev & Ivashkin (1965) the genus includes eight species but Kurochkin (1958) had synonymized four of them *(S. schikhobalovi, S. skrjabini, S. timofejevi* and *S. victori)* with the type species. He arrived at this conclusion from a study of a large number of specimens and we agree with his opinion.

The genus *Skrjabinocara* thus comprises four species only, namely, *S. squamata, S. buckleyi, S. parvepapillata* and *S. rostombekovi*. Two of these species parasitize fish-eating birds in the Palaearctic Region.

KEY TO THE SPECIES OF THE GENUS *SKRJABINOCARA*

1 Parasites of cormorants. Cervical papillae large (0.04—0.06 mm in diameter) . . . . *S. squamata*
— Parasites of storks. Cervical papillae small (0.005 mm in diameter) . . . . *S. parvepapillata*

Skrjabinocara squamata (Linstow, 1883)

Fig. 76
Hosts: *Phalacrocorax carbo, Ph. auritus* and *Ph. pygmaeus*. Recorded also in *Accipitriformes*.
Localization: proventriculus and gizzard.

Distribution: Europe (England, Rumania, U.S.S.R. — the Volga Region), Asia (U.S.S.R. — Transcaucasus Region, Kazakhstan, republics of Middle Asia). Outside the Palaearctic Region, South-West Asia and Central America.

Description:
Male: Body length 5.95—13.0 mm, maximum body width 0.35 mm, width at level of cloaca 0.12—0.15 mm, vestibule 0.20—0.35 mm × 0.02 mm. Cordons 0.65—1.07 mm long, about 0.04 mm wide. Cervical papillae 0.03 to

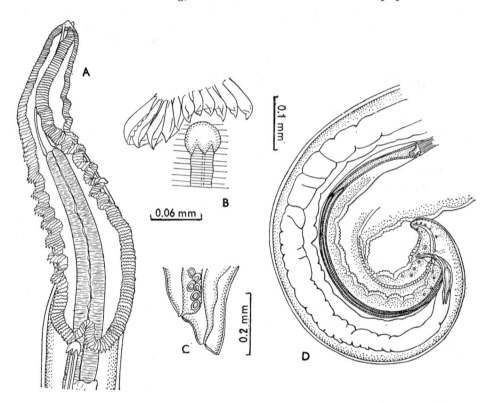

Fig. 76. *Skrjabinocara squamata* (Linstow, 1883). A — anterior end (lateral view); B — cuticular cordons and lateral band; C — posterior end of female (lateral view); D — posterior end of male (lateral view); A, D — After Saidov (1954); B, C — After Sobolev (1949).

0.06 mm long, 0.02—0.06 mm wide. Muscular part of oesophagus 0.22—0.84 mm long, 0.12 mm wide. Tail up to 0.12 mm long, caudal alae wide, 0.59 to 0.99 mm long, 0.045—0.06 mm wide, tapering gradually to posterior end. Ventral surface with characteristic fine granulation. There are nine pairs of pedunculate papillae, including four pairs of preanal, four pairs of postanal and one pair of sessile papillae, which are nearest to the tail tip. Large spicule 0.73—0.99 mm long. Smaller spicule cymbiform, 0.095—0.13 mm long.

Female: Body length 12.0—33.5 mm, maximum body width 0.20—0.90 mm. Vestibule 0.30—0.70 mm long, up to 0.03 mm wide. Length of muscular part of oesophagus 0.60—1.00 mm and glandular part up to 3.0 mm. Cordons 0.6—1.3 mm long, up to 0.07 mm wide. Tail bluntly rounded, up to 0.03 mm long. Vulva 0.25 mm from tail end. Eggs oval, $0.015—0.025 \times 0.023$ to 0.053 mm, shell covered with small protuberances.

The species was first described on the basis of one female recovered from a cormorant in Middle Asia and was assigned to the genus *Filaria*; later it was transferred to other genera. The male described by Saidov (1954) was obtained from a cormorant from the Caspian Sea. The species was placed in the genus *Skrjabinocara* by Kurashvili (1941).

Biology: Some data on the life-cycle of this parasite were included in the paper by Kurochkin (1958) and according to this author, the intermediate host is *Cypris pulera (Ostracoda)*.

Notes: The parasites from cormorants in Australia identified by Johnston & Mawson (1941) as *S. squamata* belong, in our opinion, to the species *S. buckleyi* Ali, 1956, described from cormorants of India.

References: Akhumyan (1966); Babaev (1970); Chiriac (1965); Gushanskaya (1950 a); Kosupko (1963); Kurashvili (1941, 1957); Kurochkin (1958)*; Linstow (1883*, 1886*); Nikolskaya (1939); Saidov (1954)*; Sailov (1962, 1965 a); Shakhtakhtinskaya (1959 b); Skrjabin (1923)*; Sultanov (1963); Turemuratov (1963 a, b); Vaidova (1963, 1965); Vevers (1920); Zhatkanbaeva (1964).

Skrjabinocara parvepapillata Macko, 1962

Fig. 77
Hosts: *Ciconia ciconia* and *C. nigra*.
Localization: under the cuticle of gizzard.
Distribution: Europe (Czechoslovakia). Not reported outside the Palaearctic Region.

Description: Cordons characteristic of *Skrjabinocara*. Lateral cuticular bands present. Cervical papillae small, 0.004—0.005 mm wide, indistinctly bi- or tricuspid.

Male: Body 6.85—7.49 mm long, maximum width in middle part of body 0.27—0.29 mm. Cordons originate between lips, unite posteriorly 0.83 mm from anterior end of body. Vestibule 0.172—0.175 mm long, 0.017—0.018 mm wide. Muscular part of oesophagus 0.41—0.46 mm long, 0.045—0.049 mm wide. Glandular part of oesophagus 1.85—2.10 mm long, maximum width in posterior part 0.192—0.205 mm. Posterior end of body bent ventrally, surrounded by long (0.45 mm) caudal alae. Four pairs of caudal papillae are precloacal and six pairs postcloacal. Left spicule 0.95—0.96 mm long with hook-like process 0.069 mm from distal end. Right spicule distinctly smaller, 0.178 to 0.186 mm long.

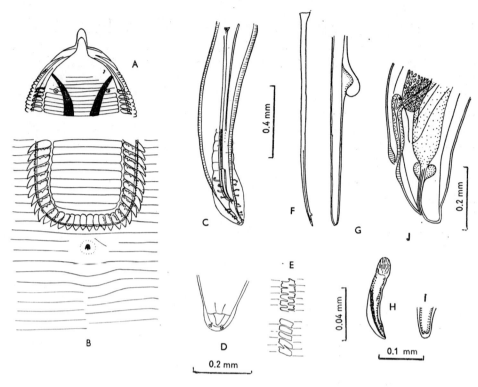

Fig. 77. *Skrjabinocara parvepapillata* Macko, 1962. A — anterior end (lateral view); B — cordons (detail); C — posterior end of male (ventral view); D — posterior end of female; E — structure of cordons; F — long spicule; G — distal end of long spicule; H — small spicule; I — distal end of small spicule; J — posterior end of female (lateral view). After Macko (1962, 1964.)

Female: Body length 10.05—10.56 mm, maximum body width 0.36 to 0.38 mm. Cordons extend up to 0.94—1.19 mm from anterior end of body. Vestibule 0.15—0.16 mm, muscular part of oesophagus measures 0.52 to 0.70 mm and glandular part 2.60—2.87 mm long. Vulva 0.24—0.26 mm, anus 0.061—0.063 mm from posterior end of body. Eggs 0.025—0.030 × × 0.016—0.020 mm.

References: Macko (1962 a*, b*, 1963 a*, b*, 1964 c*); Skrjabin, Sobolev & Ivashkin (1965)*.

Genus *Skrjabinocerca* Shikhobalova, 1930

Cordons straight, not anastomosing and not recurrent. Cervical papillae tricuspid. Lateral alae present. Tail end of female with a number of terminal papillae in form of a rosette. Spicules unequal, dissimilar.

Type and only species *S. prima* Shikhobalova, 1930.

Skrjabinocerca prima Shikhobalova, 1930

Fig. 78
Hosts: *Uria aalge, Lunda cirrhata* and *Halcyon pileata*. Also occurs in many other bird species of different orders *(Galliformes, Charadriiformes, Alciformes, Cuculiformes* and *Passeriformes)*.
Localization: oesophagus.
Distribution: Asia (U.S.S.R. — Far East). In other birds in different areas of the country. Not reported outside the Palaearctic Region.

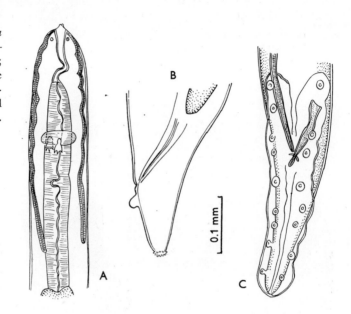

Fig. 78. *Skrjabinocerca prima* Shikhobalova, 1930. A — anterior end (lateral view); B — posterior end of female (lateral view); C — posterior end of male (ventral view). After Daiya (1967).

Description:

Male: Body length 3.4—6.55 mm, maximum body width 0.16—0.23 mm. Cordons 0.33—0.46 mm long. Cervical papillae 0.17—0.30 mm from anterior end. Vestibule 0.09—0.13 mm, muscular part of oesophagus 0.23—0.40 mm and glandular part of oesophagus 1.5—2.47 mm long. Tail 0.16—0.23 mm long. Large spicule 0.30—0.39 mm long, small spicule 0.10—0.15 mm long.
Female: Body length 4.9—6.7 mm, maximum body width 0.25—0.33 mm. Cordons 0.46—0.58 mm long. Cervical papillae 0.23—0.30 mm from anterior end. Vestibule 0.098—0.132 mm, muscular part of oesophagus 0.347 to 0.41 mm, and glandular part 1.5—2.0 mm long. Tail 0.12—0.17 mm long. Vulva 1.0—1.46 mm from tail tip. Eggs 0.042—0.056 × 0.026—0.030 mm.

The description was based on specimens recovered from *Corvus frugilegus* on Sakhalin Island.

Biology: Tsimbalyuk & Kulikov (1966) found large numbers of larvae of this nematode in *Orchestia ochotensis (Amphipoda)* living in the littoral zone of Bering Island.

References: Alekseev & Smetanina (1968); Shikhobalova (1930)*; Skrjabin, Sobolev & Ivashkin (1965)*; Smetanina (1972); Smetanina & Alekseev (1967); Tsimbalyuk & Belogurov (1964); Tsimbalyuk & Kulikov (1966).

Genus *Syncuaria* Gilbert, 1930

Cuticular cordons anastomozing on lateral sides of body, not recurrent. Surface of cordons complicated. Cervical papillae bicuspid. Vulva near posterior end of body or in middle region.

Type species *S. ciconiae* (Gilbert, 1927).

Usually parasitic in birds of the order *Ciconiiformes*. The genus includes nine species, two of which have been found in fish-eating birds of the Palaearctic Region.

KEY TO THE SPECIES OF THE GENUS *SYNCUARIA*

1 Length of spicules 0.76—0.90 and 0.14—0.16 mm . . . . *S. ciconiae*
— Length of spicules 0.46 and 0.17 mm . . . . *S. contorta*

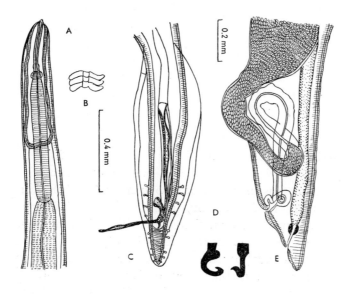

Fig. 79. *Syncuaria ciconiae* (Gilbert, 1927). A — anterior end (lateral view); B — cordons (detail); C — posterior end of male (ventral view); D — distal end of short spicule (different positions); E — posterior end of female (lateral view). After Macko (1964).

Syncuaria ciconiae (Gilbert, 1927)

Fig. 79
Hosts: *Podiceps ruficollis*, *Ciconia ciconia* and *C. nigra*.
Localization: under cuticle of gizzard.
Distribution: Europe (Czechoslovakia, Bulgaria, U.S.S.R. — Byelorussia), Asia (U.S.S.R. — Kazakhstan). Not reported outside the Palaearctic Region.

Description:

Male: Body length 6.79—8.13 mm, maximum body width 0.198—0.307 mm. Vestibule 0.204—0.233 × 0.015—0.018 mm, muscular and glandular parts of oesophagus 0.521—0.685 mm and 2.414—2.76 mm long, respectively. Cordons 0.183—0.545 mm long. Four pairs of preanal and five pairs of postanal papillae present. Tail 0.131—0.205 mm long. Spicules 0.755—0.896 mm and 0.144 to 0.164 mm.

Female: Body length 8.961—11.609 mm, maximum body width 0.274 to 0.324 mm. Vestibule 0.195—0.246 mm, muscular part of oesophagus 0.535 to 0.722 mm and glandular part 2.453—2.901 mm long. Cordons 0.434 to 1.539 mm long. Tail 0.049—0.122 mm long. Vulva 0.343—0.391 mm from posterior end. Eggs 0.029—0.033 × 0.019—0.021 mm.

The species was first described from specimens recovered from *C. ciconia* in Byelorussia and was assigned to the genus *Acuaria*, subgenus *Syncuaria* Gilbert, 1927. Later, Gilbert (1930) raised this subgenus to generic rank, which was recognized by many later authors.

References: Gilbert (1927*, 1930); Gushanskaya (1950 c); Macko (1964 c)*; Skrjabin, Sobolev & Ivashkin (1965)*; Zhelyazkova-Paspaleva (1962 b).

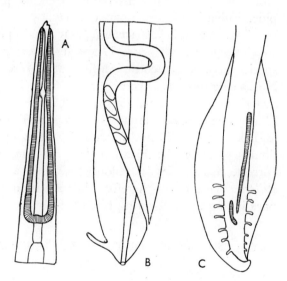

Fig. 80. *Syncuaria contorta* (Molin, 1858). A — anterior end (lateral view); B — posterior end of female (lateral view); C — posterior end of male (ventral view). After Molin (1861).

Syncuaria contorta (Molin, 1858)

Fig. 80

Hosts: *Nycticorax nycticorax*, *Platalea leucorodia* and *Plegadis falcinellus*.
Localization: gizzard.
Distribution: Europe (Austria, Italy, U.S.S.R. — Ukraine, the Volga Region), Asia (U.S.S.R.—Azerbaijan, Kazakhstan).

Description:
Male: Body length 7—8 mm, body width 0.14—0.2 mm. Caudal alae wide, with transverse striations. Seven to eight pairs of caudal papillae, four preanal and three to four postanal. Spicules 0.46 and 0.17 mm long.
Female: Body length 16—19 mm, body width 0.18—0.4 mm. Vulva a little in front of anus.

The species was described from specimens recovered from *P. falcinellus* in Italy and was originally placed in the genus *Dispharagus*. It was transferred to the genus *Syncuaria* by Sobolev (1943).

References: Cram (1927)*; Diesing (1851); Dubinin (1938)*; Dubinin & Dubinina (1940); Gushanskaya (1950c); Feyzullaev (1963a); Linstow (1878); Molin (1858 b*, 1860 b*, 1861 a*); Rudolphi (1819)*; Skrjabin, Sobolev & Ivashkin (1965)*; Smogorzhevskaya (1964); Sobolev (1943 a, b); Statirova (1946).

Genus *Synhimantus* Railliet, Henry & Sisoff, 1912

Cordons recurrent, anastomosing on lateral sides of body, without spines and of the same width throughout. Cervical papillae tricuspid. Females didelphic. Vulva in posterior part of body.
Type species *Synhimantus laticeps* (Rudolphi, 1819).
The genus comprises 17 species, usually parasites of *Falconiformes*. Two species have been recorded in fish-eating birds in the Palaearctic Region.
Notes: Iksanov & Dikambaeva (1962) reported finding *S. laticeps* in *Botaurus stellaris* from Kirghizia. This species is specific for *Falconiformes* and owls and, as far as we know, has never been reported from birds of other orders. Iksanov & Dikambaeva did not describe the nematodes recovered and we have some doubts of their identification. In any case, even if the determination were correct, this species is undoubtedly only an occasional parasite of *Botaurus stellaris*. We do not, therefore, include it in the Key of species of this genus parasitizing the fish-eating birds.

KEY TO THE SPECIES OF THE GENUS *SYNHIMANTUS*

1 Cordons without sinuosities. Parasite of gulls . . . . *S. niloticus*
— Cordons with numerous sinuosities. Parasite of pelicans . . . . *S. sirry*

Synhimantus niloticus Leonov, 1958

Fig. 81
Host: *Gelochelidon nilotica*.
Localization: proventriculus.

Distribution: Europe (U.S.S.R. — coast of the Black Sea). Not reported outside the Palaearctic Region.

Description: Cordons typical of the genus, extending posteriorly from lips, recurrent and anastomosing but without sinuosities.

Male: Body length 4.13 mm, maximum body width 0.2 mm. Cordons 0.248 mm long. Nerve ring 0.186 mm, excretory pore 0.254 mm and cervical papillae 0.31 mm from anterior end. Vestibule 0.155 mm long, muscular part of

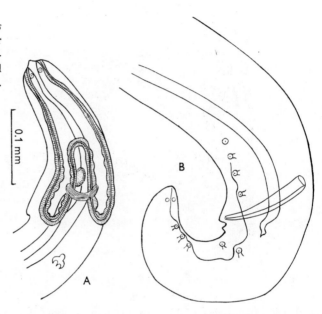

Fig. 81. *Synhimantus niloticus* Leonov, 1958. A — anterior end (lateral view); B — posterior end of male (lateral view). After Leonov (1958).

oesophagus 0.496 mm and glandular part of oesophagus 1.27 mm long. Caudal alae present. Nine pairs of caudal papillae present, four of which are preanal and five postanal. Spicules 0.496 mm and 0.161 mm long.

Female: Body length 4.023 mm long, maximum body width 0.186 mm. Cordons 0.235 mm long. Length of vestibule 0.13 mm, muscular oesophagus 0.415 mm and glandular oesophagus 1.36 mm. Cervical papillae 0.283 mm from anterior end. Vulva 1.55 mm from posterior end.

The description was based on specimens recovered from *G. nilotica* captured on the coast of the Black Sea.

References: Leonov (1958)*; Skrjabin, Sobolev & Ivashkin (1965)*; Smogorzhevskaya (1964).

Synhimantus sirry Khalil, 1931

Fig. 82
Host: *Pelecanus onocrotalus*.
Localization: gizzard.
Distribution: Africa (A.R.E.). Outside the Palaearctic Region in tropical Africa.

Description: Cordons with numerous sinuosities.
Male: Body length 6.8 mm, body width 0.16 mm. Cordons recurrent 0.3 mm

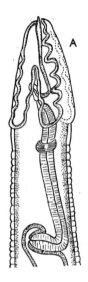
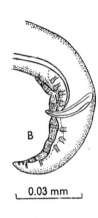

Fig. 82. *Synhimantus sirry* Khalil, 1931. A — anterior end (lateral view); B — posterior end of male (lateral view). After Khalil (1931).

and anastomosing 0.225 mm from anterior end. Vestibule 0.22 mm long, oesophagus 0.4 mm long. Tail 0.16 mm long. Cervical papillae 0.37 mm from anterior end. Nine pairs of caudal papillae present, four preanal and five postanal, and four small additional papillae along the median line of body, near the tail end. Caudal alae present. Right spicule longer and thinner, 0.41 mm long. Left spicule very short, massive, cymbiform.

Female: Body length 9 mm, maximum body width 0.3 mm. Cordons recurrent 0.32 mm, anastomosing 0.237 mm from anterior end. Vestibule 0.32 mm long, oesophagus 0.42 mm long. Tail 0.15 mm long. Cervical papillae 0.40 mm from anterior end. Vulva in posterior part of body, 4/7 of the body length from anterior end. Eggs 0.037 × 0.025 mm.

The description was based on specimens recovered from *P. onocrotalus* from A.R.E.

References: Khalil (1931)*; Skrjabin, Sobolev & Ivashkin (1965)*.

Subfamily *Echinuriinae* Sobolev, 1943

In addition to cordons on anterior end of body, there are longitudinal rows of spines which originate either behind or in front of posterior margin of cordons. Spicules unequal.

The subfamily includes 3 genera. Representatives of all of them have been recorded in fish-eating birds of the Palaearctic Region.

KEY TO THE GENERA OF THE SUBFAMILY *ECHINURIINAE*

1 Longitudinal rows of spines extend to region of cordons. Cuticular collar behind cordons absent . . . . *Echinuria*
— Longitudinal rows of spines do not extend to region of cordons. Cuticular collar behind cordons present 2
2 Cordons with sinuosities . . . . *Skrjabinoclava*
— Cordons without sinuosities . . . . *Cordonema*

Genus *Echinuria* Soloviev, 1912

Cordons rather short, running from base of lips along both sides of body, then anastomosing in pairs, not recurrent. Cervical papillae single, situated approximately at level of nerve ring. Four longitudinal rows of spines present, enlarging gradually from anterior end to middle of body and then diminishing again. Spines reach to region of cordons.

Type species *E. uncinata* (Rudolphi, 1819).

The genus comprises 11 species, usually parasitic in the family *Anatidae* and the order *Charadriiformes*. It is not characteristic of fish-eating birds, in which only three species have been reported, two of them in the Palaearctic Region.

KEY TO THE SPECIES OF THE GENUS *ECHINURIA*

1 Length of spicules 0.24 mm and 0.12 mm. Distal end of large spicule bifurcating . . . . *E. heterobrachiata*
— Length of spicules 0.51—0.87 mm and 0.17—0.24 mm. Distal end of large spicule different in form . . . . *E. uncinata*

Echinuria uncinata (Rudolphi, 1819)

Fig. 83

Hosts: *Podiceps cristatus, P. auritus, P. nigricollis, P. griseigena, P. ruficollis, Mergus albellus, M. serrator* and *Larus minutus*. Recorded also in a large number of other birds, mostly of the family

Anatidae. Only rarely found in fish-eating birds.
Localization: proventriculus.
Distribution: Europe (Poland, U.S.S.R. — central regions, Ukraine), Asia (U.S.S.R. — West Siberia, Tajikistan). Outside the Palaearctic Region, North America and Africa.

Description: Cordons do not extend behind the junction of muscular and glandular part of oesophagus. Longitudinal rows of spines reach nearly to posterior end of body.
Male: Body length 7.0—13 mm, maximum body width 0.22—0.60 mm. Excretory pore 0.52—0.60 mm from anterior end. Length of vestibule 0.13 to 0.152 mm, muscular part of oesophagus 0.48—0.94 mm, and glandular part of

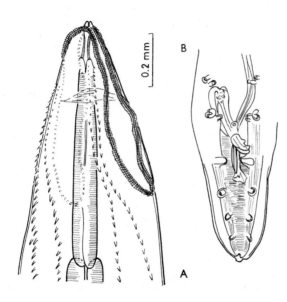

Fig. 83. *Echinuria uncinata* ((Rudolphi, 1819). A — anterior end; B — posterior end of male (ventral view). After Romanova (1946).

oesophagus 1.3—2.7 mm. Caudal alae supported by four pairs of preanal and five pairs of postanal papillae. Single unpaired small papilla present behind fourth pair of preanal papillae. Tail 0.36—0.48 mm long. Right spicule 0.165—0.235 mm long, tubular, bearing at its distal end two sharp processes bent ventrally. Left spicule 0.515—0.87 mm long with proximal part tubular with a bent process and distal end widened and flattened, resembling a frog's foot folded into a spout, with a membrane extending between the slat-like supports.
Female: Body length 9.9—20.5 mm. Excretory pore 0.58—0.63 mm from anterior end. Cordons 0.76—0.94 mm long. Length of vestibule 0.17—0.216 mm, muscular part of oesophagus 0.76—0.92 mm, and glandular part 2.29—2.53 mm. Tail 0.20—0.34 mm long. Vulva 0.66—1.40 mm from tail tip. Eggs 0.028—0.040 × 0.018—0.020 mm.

The species was first described on the basis of specimens recovered from

Anser anser in Europe and was assigned to the genus *Filaria*. Soloviev (1912) transferred it to the genus *Echinuria*.

Biology: According to Skrjabin, Sobolev & Ivashkin (1965), the life-cycle of this species was first observed by Hamann (1891). The most complete studies of the life-cycle of this parasite were carried out by Romanova (1947) but many other authors have published work on it. The main intermediate hosts are different species of water fleas *(Daphnia)*, especially *Daphnia pulex* and *D. magna*. Some other copepods have also been experimentally infected, as for example hog slaters, sand hopper etc. In water fleas, larvae develop to

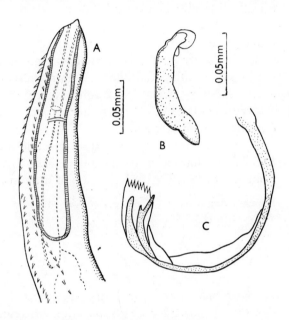

Fig. 84. *Echinuria heterobrachiata* Wehr, 1937. A — anterior end (lateral view); B — small spicule; C — long spicule. After Wehr (1937).

the infective stage at 17—23 °C within 14 days. Development to full maturity in the bird final host (experimentally in ducks) lasts about 50 days.

References: Czaplinski (1962 b)*; Dubinina & Serkova (1951); Hamann (1891); Kambourov & Vasilev (1972); Okorokov & Tkachev (1969); Paskalskaya (1968); Romanova (1947); Rudolphi (1819)*; Shigin (1959); Skrjabin (1923); Skrjabin, Sobolev & Ivashkin (1965)*; Smogorzhevskaya (1962 a, 1964); Soloviev (1912).

Echinuria heterobrachiata Wehr, 1937

Fig. 84
Hosts: *Larus genei* and *L. minutus*.
Localization: proventriculus.
Distribution: Europe (U.S.S.R. — coast of the Black Sea and the Sea of Azov). Outside the Palaearctic Region, North America.

Description: Cordons extend to junction of muscular and glandular parts of oesophagus or posterior to it.

Male: Body length 3—3.4 mm, maximum body width 0.124—0.133 mm. Length of cordons 0.306—0.322 mm, vestibule 0.102—0.115 mm, muscular part of oesophagus 0.175—0.185 mm and glandular part 1.0—1.2 mm. Nerve ring 0.12—0.14 mm from anterior end. Caudal alae present, supported by four pairs of preanal (in two groups of two on each side) and five pairs of postanal papillae, equally spaced. Left spicule 0.24—0.258 mm long, with distal end bifurcating, the inner branch resembling the heterocercal type of fish tail. Sheath distinctly visible at distal end of this spicule and serrated at tip. Right spicule 0.12—0.129 mm long, with distal end tapered and notched at tip and a knob-like formation on ventral surface.

Female: Body length 4—8 mm, maximum body width 0.27 mm. Length of cordons 0.336 mm, vestibule 0.146 mm, muscular part of oesophagus 0.255 mm, and glandular part about 2 mm. Nerve ring 0.175 mm from anterior end. Vulva 0.85—0.90 mm from posterior end. Tail 0.075 mm long with end bent dorsally and bluntly rounded.

The description was based on specimens recovered from *Larus* spp. in the U.S.A. A characteristic morphological feature differentiating this species from all others is the conformation of the distal end of the left spicule.

References: Leonov (1958); Mashtakov (1964)*; Sergeeva (1968 a); Skrjabin, Sobolev & Ivashkin (1965)*; Wehr (1937)*.

Genus *Cordonema* Schmidt & Kuntz, 1972

Cordons arise near bases of lips and extend posteriorly, undulating slightly but never forming sinuosities. Posterior end of cordons anastomoses in pairs on lateral sides of body. Cuticular ridge present posterior to cordons. Two longitudinal rows of spines present on each side of body but not extending to region of cordons. Spicules unequal. Vulva near anus.

Type species *C. venusta* Schmidt & Kuntz, 1972.

This genus includes 3 species: *C. venusta*, described from *Cinclus* spp. from South-East Asia, and *C. solonitzini* and *C. longifuniculata* described from *Charadriiformes* in the U.S.S.R. The last two species were originally assigned to the genus *Skrjabinoclava*. *C. longifuniculata* has been recorded in fish-eating birds of the Palaearctic Region.

Cordonema longifuniculata (Sobolev, 1952)

Fig. 85
Host: *Stercorarius longicaudatus*.

Localization: small intestine.
Distribution: Asia (U.S.S.R. — the lower reaches of the river Yenisei).

Description:

Male: Body length 2.4 mm, maximum body width 0.12 mm. Cordons 0.16 mm long. Length of vestibule 0.012 mm, muscular part of oesophagus 0.36 mm, glandular part of oesophagus 0.80 mm. Left spicule 0.243 mm long, thin, bent, with transparent cuticular alae. Right spicule massive, 0.122 mm

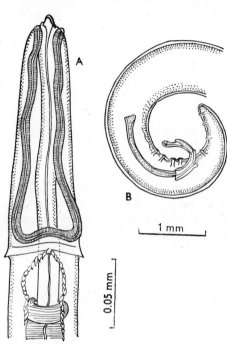

Fig. 85. *Cordonema longifuniculata* (Sobolev, 1952). A — anterior end (lateral view); B — posterior end of male (lateral view). After Sobolev (1952).

long. Tail 0.12 mm long. Caudal alae not visible. Nine pairs of caudal papillae present, four preanal, one adanal and four postanal.

Female: Body length 3.468 mm, maximum body width 0.18 mm. Vestibule 0.20 mm long. Tail 0.07 mm long. Vulva prominent, 0.24 mm from posterior end. Eggs 0.036—0.040 × 0.020—0.024 mm.

The description was based on specimens recovered from *Charadrius* sp. from Far East (U.S.S.R.).

Notes: The specimens from gulls from the lower reaches of the river Yenisei, originally identified by Sergeeva (1969) as *Skrjabinoclava horrida*, also belong to this species. This correction was made by the same author as a result of a re-examination of the material.

References: Sergeeva (1969); Skrjabin, Sobolev & Ivashkin (1965)*; Sobolev (1952)*.

Genus *Skrjabinoclava* Sobolev, 1943

Cordons forming complicated sinuosities and with ends uniting on lateral surfaces of head region. In some, a cuticular ridge marks off the anterior part of body in region of cordons. Double rows of well developed spines originate posterior to cordons on each side of body and extend to posterior end. These spines may cover body at point of origin, but never extend to the region of cordons. Spicules unequal and dissimilar. Caudal alae poorly developed. Vulva in posterior part of body.

Type species *S. decorata* (Solonitsin, 1928).

The genus includes ten species, usually parasitic in birds of the order *Charadriiformes*. Three of them have been recorded in fish-eating birds of the Palaearctic Region.

KEY TO THE SPECIES OF THE GENUS *SKRJABINOCLAVA*

1 Cordons in males less than 0.085 mm long and in females 0.10 mm long. Left spicule four to five times as long as right 2
— Cordons in males more than 0.085 mm long, in females more than 0.15 mm long. Left spicule not more than 3.5 times as long as right . . . . *S. halcyoni*
2 Longitudinal rows of spines bifurcate, surround anterior end of body and then rejoin, extending to posterior end . . . . *S. decorata*
— Longitudinal rows of spines run directly to posterior end . . . . *S. horrida*

Skrjabinoclava decorata (Solonitsin, 1928)

Fig. 86

Hosts: *Podiceps cristatus*, *P. auritus* and *Mergus albellus*. Normally parasitic in *Charadriiformes*; only once (Okorokov & Tkachev 1969) reported in fish-eating birds of South Ural. Also recorded in *Passeriformes*.
Localization: proventriculus.
Distribution: Asia (U.S.S.R. — South Ural). Outside the Palaearctic Region, South-East Asia.

Description:

Male: Body length 2.5—4.0 mm, maximum body width 0.11—0.15 mm. Cordons 0.045—0.081 mm long. Length of vestibule 0.145—0.162 mm, of muscular part of oesophagus 0.254—0.266 mm, and of glandular part 0.982—1.387 mm. Cuticular collar 0.07 mm from head end. Cervical papillae 0.09 mm and nerve ring 0.14 mm from anterior end. Left spicule 0.459 to 0.462 mm long, right spicule 0.084—0.115 mm long. Tail 0.122—0.19 mm long. There are four pairs of preanal and four pairs of postanal papillae.
Female: Body length 3.57—5.0 mm, maximum body width 0.119—0.173

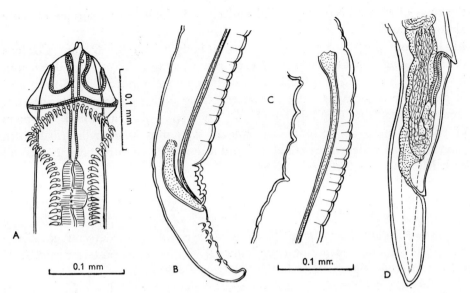

Fig. 86. *Skrjabinoclava decorata* (Solonitsin, 1928). A — anterior end (lateral view); B — posterior end of male (lateral view); C — proximal end of long spicule; D — posterior end of female (lateral view). After Daiya (original).

mm. Cordons 0.053—0.092 mm long. Length of vestibule 0.135—0.15 mm, muscular oesophagus 0.211—0.289 mm, and glandular oesophagus 1.257 to 1.73 mm. Tail 0.124—0.15 mm long. Vulva 0.311—0.369 mm from posterior end. Eggs 0.039—0.040 × 0.018—0.021 mm.

The species was described from specimens recovered from *Tringa* sp. in the Volga Region and was originally placed in the genus *Echinuria*. Sobolev (1943 a) transferred it to the genus *Skrjabinoclava*.

References: Okorokov & Tkachev (1969); Skrjabin, Sobolev & Ivashkin (1965)*; Sobolev (1943 a)*; Solonitsin (1928 a*, b*).

Skrjabinoclava halcyoni Ryzhikov & Khokhlova, 1964

Fig. 87
Hosts: *Halcyon pileata* and *H. smyrnensis*.
Localization: oesophagus.
Distribution: Asia (U.S.S.R. — Far East, China). Outside the Palaearctic Region, South-East Asia.

Description: Longitudinal rows of spines at first run parallel with lower border of cordons to ventral and dorsal surfaces of body, then turn backwards to meet one another. At the level of beginning of muscular part of oesophagus, the rows of spines become parallel and extend along lateral surface of body almost to tail end.

Male: Body length 5.4 mm, maximum body width 0.176 mm. Cordons 0.09 mm long. Transverse part of each row of spines (five to six) 0.046 mm long. Length of vestibule 0.218 mm, of muscular oesophagus 0.36 mm, and of glandular oesophagus 1.92 mm. Nerve ring 0.25 mm from anterior end. Tail 0.15 mm long. Caudal alae absent. Four pairs of preanal and four pairs of postanal papillae present. Left spicule 0.36 mm long, cylindrical, with funnel-shaped widened proximal end and attenuated, slightly bifurcated distal end. Right spicule 0.109 mm long.

Female: Body length 8.32 mm, maximum body width 0.288 mm. Cordons 0.165 mm long. Transverse part of row of spines (six to seven) 0.066 mm long. Length of vestibule 0.29 mm, muscular part of oesophagus 0.56 mm. The length of the glandular part of the oesophagus could not be determined. Nerve ring 0.32 mm from anterior end. Vulva 0.40 mm from posterior end. Tail 0.132 mm long. Eggs 0.043—0.046 × 0.026—0.030 mm.

The description was based on specimens recovered from *H. pileata* in Vietnam.

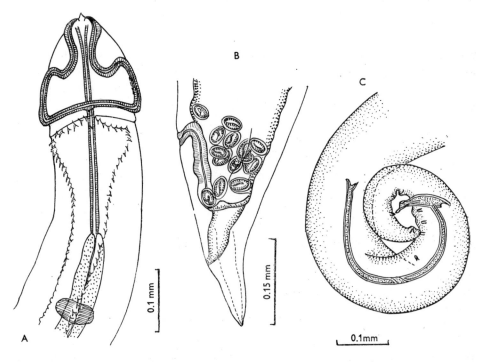

Fig. 87. *Skrjabinoclava halcyoni* Ryzhikov & Khokhlova, 1964. A — anterior end (lateral view); B — posterior end of female (lateral view); C — posterior end of male (lateral view). After Ryzhikov & Khokhlova (1964).

Notes: The nematodes found by Hsü (1957) in *Halcyon smyrnensis* in China and identified by him as *S. cincli* are, in the opinion of Ryzhikov & Khokhlova (1964), *S. halcyoni*, as also are specimens from the same definitive host from India, identified by Singh (1948) as *Echinuria horrida*.

References: Alekseev & Smetanina (1968); Hsü, W. N. (1957)*; Ryzhikov & Khokhlova (1964 a)*; Singh (1948 b)*; Smetanina & Alekseev (1967).

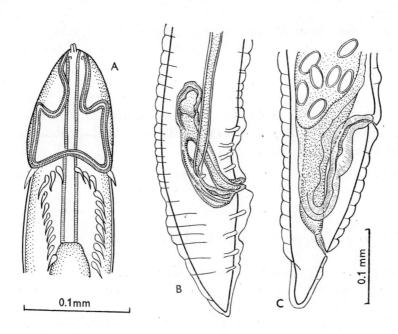

Fig. 88. *Skrjabinoclava horrida* (Rudolphi, 1809). A — anterior end (lateral view); B — posterior end of male (lateral view); C — posterior end of female (lateral view). After Daiya (original).

Skrjabinoclava horrida (Rudolphi, 1809)

Fig. 88
Host: *Larus fuscus* and *Gelochelidon nilotica*. Common parasite of snipe, not characteristic of fish-eating birds.
Localization: proventriculus.
Distribution: Europe (U.S.S.R. — coast of the Black Sea). In snipe in several areas of Europe, also in India and Cuba.

Description:
Male: Body length 2.1—2.5 mm, maximum body width 0.085—0.102 mm. Length of vestibule 0.18—0.189 mm, of muscular oesophagus 0.27—0.275 mm and glandular oesophagus 0.725—0.810 mm. Cordons 0.042—0.045 mm

long. Left spicule 0.33—0.36 mm, right spicule 0.09 mm long. Tail 0.09 to 0.12 mm long. Caudal alae feebly developed, with four pairs of preanal and four pairs of postanal papillae.

Female: Body length 3.05—4.3 mm, maximum body width 0.086—0.102 mm. Length of vestibule 0.16—0.18 mm, of muscular oesophagus 0.449—0.629 mm, and glandular oesophagus 0.85—0.935 mm. Cordons 0.06—0.072 mm long. Vulva 0.24—0.32 mm from posterior end of body. Eggs 0.042 × 0.02 to 0.024 mm.

The species was described as *Strongylus horridus* from specimens recovered from *Charadrius hiaticula* from Europe (precise locality not given). Sobolev (1943 a, b) transferred it to the genus *Skrjabinoclava*.

Notes: The species is normally parasitic in snipe and has only rarely been reported from fish-eating birds of the Palaearctic Region (Leonov 1958; Shigin 1961). Sergeeva's (1969) report of *S. horrida* in *Larus argentatus* proved to be erroneous, more detailed studies showing that the nematodes concerned were *Cordonema longifuniculata*.

References: Leonov (1958); Sergeeva (1969); Shigin (1961); Skrjabin, Sobolev & Ivashkin (1965)*; Smogorzhevskaya (1964); Sobolev (1943 a*, b*).

Family *Desmidocercidae* Cram, 1927

Head end with two small lips and four to eight papillae. Oesophagus divided into two parts. Spicules unequal. Vulva in anterior or posterior part of body. Parasites of fish-eating birds.

This family includes three genera; representatives of two of these have been found in fish-eating birds of the Palaearctic Region.

KEY TO THE GENERA OF THE FAMILY *DESMIDOCERCIDAE*

1 Head end with four submedian papillae. Oesophagus long, about 2/3 of body length. Tail end of both sexes with group of filiform papillae. Genital papillae of males absent
 . . . . *Desmidocerca*

— Head end with eight to twelve papillae. Oesophagus short, about 1/8 of body length. Tail end without filiform papillae in either sex. Genital papillae of males present . . .
 . . . . *Desmidocercella*

Genus *Desmidocerca* Skrjabin, 1915

Characteristic features of this genus are given in the Key. Type and only species *D. areophila* Skrjabin, 1919.

Desmidocerca aerophila Skrjabin, 1915

Fig. 89
Hosts: *Phalacrocorax carbo, Ph. pygmaeus, Ardea cinerea* and *Larus argentatus*.
Localization: air sacs.
Distribution: Europe and Asia (U.S.S.R. — Ukraine, republics of Transcaucasus region, Kazakhstan).

Description:
Male: Body length 3.9 mm, body width 0.26 mm. Mouth opening leading into cylindrical vestibule, which is followed by a very long oesophagus, 2/3 of

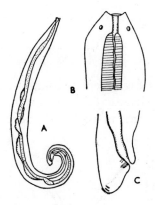

Fig. 89. *Desmidocerca aerophila* Skrjabin, 1915. A — male (overall view); B — anterior end; C — posterior end. After Skrjabin (1915).

body length. Spicules 0.27 mm and 0.66 mm long. Preanal and postanal papillae absent.
Female: Body length 4.7 mm, body width 0.30 mm. Vulva in posterior part of body. Eggs 0.03 × 0.02 mm.

The description was based on specimens recovered from *A. cinerea* and *Ph. carbo* in Kazakhstan.

References: Feyzullaev (1963 a,b); Kulachkova (1950); Skrjabin (1915 a*, 1923); Skrjabin, Sobolev & Ivashkin (1967 a)*; Smogorzhevskaya (1964); Vaidova (1963).

Genus *Desmidocercella* Yorke & Maplestone, 1926

The main characters of representatives of this genus are given in the Key. Type species *D. numidica* (Seurat, 1920).

Opinions differ on the species composition of this genus (Dubinin 1949; Gushanskaya 1953; Chabaud 1957; and others). In agreement with Chabaud (1957) we recognize four species in this genus. Two of them have been recorded in fish-eating birds of the Palaearctic Region. The other two were found in birds of Australia and India.

KEY TO THE SPECIES OF THE GENUS *DESMIDOCERCELLA*

1　Vulva in anterior part of body . . . . *D. numidica*
— 　Vulva in posterior part of body . . . . *D. incognita*

Desmidocercella numidica (Seurat, 1920)

Fig. 90
Hosts: *Phalacrocorax carbo, Ardea cinerea, A. purpurea, Ardeola ralloides, Egretta alba, E. garzetta, Nycticorax nycticorax, Ixobrychus eurythmus, I. minutus, Larus genei* and *L. ridibundus*.
Localization: air sacs.
Distribution: Europe (France, The Netherlands, Czechoslovakia, Poland, U.S.S.R. — the Baltic Sea region, central regions, Ukraine), Asia (U.S.S.R.—Transcaucasus region, republics of Middle Asia, Kazakhstan, West and East Siberia, Far East), Africa (Algeria). Outside the Palaearctic Region, North America, India and the Madagascar.

Description: Eight cephalic papillae and two amphids present.
Male: Body length 2.78—5.3 mm. Body width in region of oesophagus 0.039—0.108 mm, in region of cloaca 0.10 mm. Vestibule 0.020—0.026 ×

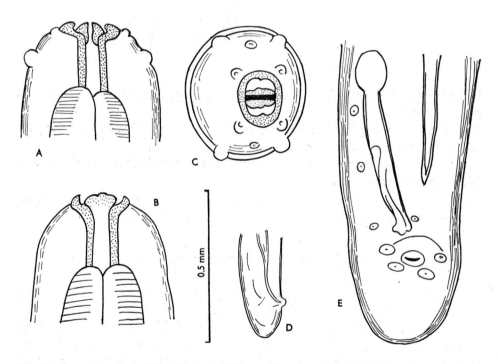

Fig. 90. *Desmidocercella numidica* (Seurat, 1920). A, B — head end (different positions); C — anterior end (apical view); D — posterior end of female (lateral view); E — posterior end of male (ventral view). After Chabaud & Choquet (1953).

× 0.10 mm. Length of muscular part of oesophagus 0.09—0.165 mm and glandular part 0.378—0.58 mm. Nerve ring 0.09—0.11 mm from anterior end. Tail 0.04—0.064 mm long. Spicules 0.225—0.48 mm and 0.11—0.155 mm long. There are four preanal papillae arranged asymmetrically and four postanal papillae arranged symmetrically.

Female: Body length 3.175—8.3 mm, maximum body width 0.13—0.27 mm. Vestibule 0.02—0.03 × 0.01 mm. Length of muscular part of oesophagus 0.1—0.165 mm and glandular part 0.38—0.48 mm. Nerve ring 0.10—0.12 mm and excretory pore 0.18 mm from anterior end. Tail 0.04—0.08 mm long. Vulva 0.54—0.77 mm from anterior end. Eggs 0.045—0.05 × 0.03 mm.

The species was described from specimens recovered from *A. cinerea* in Algeria and was originally placed in the genus *Desmidocerca*. Yorke & Maplestone (1926) transferred it to the genus *Desmidocercella*.

Biology: The larvae parasitize the vitreous body of the eye of different fish species (Dogel & Bykhovsky 1939; Dubinin 1948). Dubinin (1949) experimentally infected herons with larvae from fishes and obtained adult parasites. Skrjabin, Sobolev & Ivashkin (1967) consider fish to be paratenic hosts of this nematode. In their opinion, the intermediate hosts are likely to be aquatic invertebrates.

Notes: The systematic position of *D. numidica* has been revised several times, because specimens of this species were described by some authors under different names and were assigned to diffeent genera. One of the recent papers, in which data on this species are analysed in detail, is that by Chabaud (1957). In agreement with his opinion we regard as *D. numidica* the specimens described under the following names: *Filaria ardeae* Nawrotzky, 1914 (in part); *Filaria marcinowskyi* Skrjabin, 1923; *Desmidtercella leiperi* Singh, 1948; *Pharyngosetaria butoridi* Oshmarin & Belous, 1951; *Doesmidocercella lubimovi* Gushanskaya 1953 and *Lemdana urbaini* Campana, 1949

References: Ablasov & Chibichenko (1961, 1962); Bashkirova (1960); Belopolskaya (1959, 1963); Bezubik (1956); Broek & Jansen (1971); Campana (1949)*; Chabaud (1957); Chabaud & Choquet (1953); Daiya (1971); Dollfus et al. (1961); Dubinin (1949*, 1952); Dubinina & Serkova (1951); Feyzullaev (1963 a, b); Gushanskaya (1950 c); Kasimov & Feyzullaev (1965); Kosupko (1963); Kurashvili (1953, 1956, 1957*); Leonov (1960 a); Lyubimov (1937)*; Macko (1961 a, 1964 a); Michelson (1968); Nikolskaya (1939); Oshmarin (1963, 1965); Oshmarin & Belous (1951); Oshmarin & Parukhin (1963); Sailov (1962, 1965 b); Serkova (1948); Seurat (1920)*; Shakhtakhtinskaya (1959); Skrjabin, Sobolev & Ivashkin (1967 a); Smetanina (1972); Smogorzhevskaya (1962 a, b, 1964, 1967); Spasskaya (1949); Sultanov, Ryzhikov & Kozlov (1960); Turemuratov (1962 a); Yorke & Maplestone (1926); Zhatkanbaeva (1964).

Desmidocercella incognita Solonitsin, 1932

Fig. 91
Hosts: *Phalacrocorax carbo, Ph. capillatus, Ph. pelagicus, Ph. pygmaeus, Ph. urile, Botaurus stellaris* and *Larus argentatus*.
Localization: air sacs, trachea, lungs, heart, liver, gall-bladder etc.
Distribution: Europe and Asia (U.S.S.R. — Ukraine, Volga Delta, Transcaucasus region, republics of Middle Asia and Kazakhstan, West Siberia, Far East.) Not recorded outside the Palaearctic Region.

Description:

Male: Body length 3.56—4.25 mm, maximum body width 0.22 mm. Vestibule 0.02—0.03 mm long. Oesophagus 0.43—0.46 mm long. Spicules 0.47—0.54

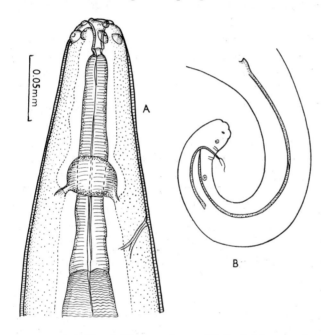

Fig. 91. *Desmidocercella incognita* Solonitsin, 1932. A — anterior end; B — posterior end of male. A — After Gushanskaya (1949); B — After Solonitsin (1932).

mm and 0.21—0.23 mm long. Tail 0.06 mm long. Seven pairs of caudal papillae present, of which three pairs are preanal and four postanal. Ventral surface of body, from the level of mid-region of large spicule to cloaca, covered with small spines.

Female: Body length 4.0—5.0 mm, maximum body width 0.27 mm. Vestibule 0.02—0.03 mm long. Oesophagus 0.48 mm long. Excretory pore 0.18 mm from anterior end. Vulva 1.8—2.8 mm from posterior end. Tail 0.27 mm long. Eggs 0.04—0.05 × 0.03 mm.

The description was based on specimens recovered from *Ph. carbo* in Armenia.

As does Chabaud (1957), we regard *D. skrjabini*, described by Gushanskaya in Skrjabin, Shikhobalova & Sobolev (1949) from cormorants of Transcaucasus region, as synonymous with *D. incognita*.

References: Ablasov & Chibichenko (1961,1962); Alekseev & Smetanina (1968); Chabaud (1957); Dubinina & Serkova (1951); Gushanskaya (1950 c, 1953*); Gushanskaya, In: Skrjabin, Shikhobalova & Sobolev (1949); Kosupko (1963); Leonov (1958); Leonov & Shvetsova (1970); Oshmarin & Parukhin (1963); Petrov & Chertkova (1950); Sailov (1965 a, b); Shakhtakhtinskaya (1959 a, b); Singh (1948 a); Skrjabin, Sobolev & Ivashkin (1967 a)*; Smetanina (1972); Smetanina & Alekseev (1967); Solonitsin (1932)*; Sultanov (1959 a, b); Tsimbalyuk & Belogurov (1964); Turemuratov (1963 a, 1966); Vaidova (1963, 1965, 1969).

Family *Physalopteridae* Leiper, 1908

Mouth with two lips (= pseudolabia) armed with teeth on their inner surfaces. Interlabia absent. Papillae of transverse row fused, situated on pseudolabia. Stoma greatly reduced. Males with or without wide caudal alae, which unite anteriorly across ventral surface of body. Vulva anterior to middle of body. Females with two, four or numerous uteri. Parasites of all classes of vertebrates.

According to Skrjabin & Sobolev (1964) and Skrjabin, Sobolev & Ivashkin (1967 b) the family includes 12 genera and more than 200 species.

The representatives of this family are divided into two subfamilies, namely *Physalopterinae* and *Proleptinae*.

According to available data, only one species of the former subfamily has been recorded in fish-eating birds of the Palaearctic Region.

Genus *Physaloptera* Rudolphi, 1819

Two lateral lips armed with teeth, each lip bearing a terminal and a tripartite inner tooth. Stoma short, vestibule absent. Oesophagus divided into muscular and glandular parts. Four pairs of pedunculate papillae usually present supporting alae. Spicules equal or unequal. Oviparous, eggs contain larva at deposition.

Parasites of digestive tract of vertebrates. The genus includes about 100 species.

Type species *Physaloptera clausa* Rudolphi, 1819.

Physaloptera alata Rudolphi, 1819

Fig. 92

Hosts: *Pandion haliaetus*. Usual hosts are birds of the order *Accipitriformes*.
Localization: gizzard and intestine.
Distribution: ? Europe. Cosmopolitan species, also found in Brazil in fish-eating birds.

Description: Mouth with two lateral lips, each bearing a triangular external and a tripartite internal tooth, and three papillae.

Male: Body length 17—20 mm. Length of tail 0.65 mm. Region around cloaca covered with small protuberances. Caudal alae elongated. Five pairs of long pedunculate papillae present, of which two are preanal, one adanal and two postanal, plus three sessile precloacal papillae and four pairs of postanal papillae. Spicules equal or subequal, measuring 0.28—0.50 mm in length.

Female: Body length 19—27 mm. Vulva 3—7 mm from anterior end. Eggs 0.048—0.055 × 0.025—0.027 mm.

The species was described from specimens recovered from *Accipiter nisus* in Central Europe.

Notes: There are, in the literature, several reports of *Ph. alata* in osprey but

Fig. 92. *Physaloptera alata* Rudolphi, 1819 — posterior end of male (ventral view). After Hartwich (1966).

the exact locality was not given. This species was probably reported from osprey in South America only, as was *Ph. acuticauda* Molin, 1860. As to the species *Ph. (= Spiroptera) tenuicollis* (Rudolphi, 1819), which was also found to parasitize osprey, this is regarded by Skrjabin, Shikhobalova & Sobolev (1949) as *species inquirenda*.

References: Diesing (1851); Oshmarin & Parukhin (1960); Rudolphi (1819)*; Skrjabin, Shikhobalova & Sobolev (1949); Skrjabin & Sobolev (1964).

Family *Schistorophidae* Skrjabin, 1941

Nematodes with cephalic appendages in form of hood-like covering, or various cuticular formations. Cervical cordons and pseudochitinous armament of cuticle absent. Spicules commonly very unequal, dissimilar; more than four pairs of precloacal papillae present in most species.

This familly comprises 13 genera, all species of which are parasites of birds. Seven genera occur in fish-eating birds, of which only *Schistorophus*, *Sciadiocara*, *Viktorocara*, *Stellocaronema* and *Sobolevicephalus* are known from the Palaearctic Region.

KEY TO THE GENERA OF THE FAMILY *SCHISTOROPHIDAE*

1 Cephalic formations in the form of four thorn-like appendages . . . . . *Schistorophus*
— Cephalic formations in the form of wings 2
2 Cephalic wings rounded . . . . *Sciadiocara*
— Cephalic wings not rounded 3
3 Outer margin of each wing pointed . . . . *Viktorocara*
— Outer margin of wings of other form 4
4 Each wing divided into three to four lobes with many irregular serrations on their entire margin . . . . *Sobolevicephalus*
— Cephalic wings divided into four finger-like lobes, without teeth on margins . . . . .
. . . . *Stellocaronema*

Genus *Schistorophus* Railliet, 1916

Mouth with two lateral pseudolabia, each with one pair of submedian papillae and an amphid. Head with four (two dorsal and two ventral) thorn-like wings, directed outwards and backwards, more or less united at their origin, arranged in form of hood or roof surrounding head just behind pseudolabia. Vestibule narrow. Oesophagus consisting of two parts. Mail tail blunt, caudal alae narrow. Precloacal papillae variable in number, postcloacal conical. Vulva in posterior or middle region of body. Oviparous, sometimes viviparous. Parasitic in gizzard of birds.

Type species *S. longicornis* (Hemprich & Ehrenberg, in Schneider 1866).

Only four of ten species belonging to this genus parasitize fish-eating birds in the Palaearctic Region.

KEY TO THE SPECIES OF THE GENUS *SCHISTOROPHUS*

1 Cephalic wings in males more than 0.1 mm long; 14 pairs or more of precloacal papillae 2
— Cephalic wings in males less than 0.05 mm long (0.024—0.040 mm); six pairs of precloacal papillae . . . . *S. bihamatus*
2 Cephalic wings in males ranging 0.10 to 0.14 mm long; shorter spicule longer than 0.10 mm . . . . *S. laciniatus*
— Cephalic wings in males more than 0.20 mm long; shorter spicule about 0.09 mm long
. . . . *S. skrjabini*

Note: The species *S. acanthocephalicus* has not been included in the Key, because only females have been found, but is described below.

Schistorophus acanthocephalicus (Molin, 1860)

Host: *Sterna hirundo* and *Hydroprogne tschegrava*.
Localization: oesophagus, and under cuticle of gizzard.
Distribution: Europe (Italy, G.D.R., U.S.S.R. — Central regions).

Description: Anterior part of body not distinctly separated. Mouth with two small conical lips, each provided with two posteriorly directed processes. Body filiform, with dense transverse striations, tapering at both ends. Tail of female conical, end blunt and bent. Anus situated near posterior extremity. Vulva in posterior part of body, with salient margins. Length of female 15—19 mm, width 0.2 mm.

Molin (1860) described this species as *Spiroptera acanthocephalica* from *Sterna hirundo* from Italy, on the basis of females only. *Spiroptera capillaris* Molin, 1860, is a synonym of *S. acanthocephalicus*. This species was transferred to the genus *Schistorophus* by Railliet (1916).

Notes: According to the original description, this taxon can be placed in the genus *Schistorophus* but without any knowledge of the males the position of the species, in relation to others in the genus, is not clear and there is also the possibility that *Schistorophus skrjabini* (of which no drawings have been published) is a synonym of *S. acanthocephalicus*.

References: Cram (1927)*; Diesing (1861); Drasche (1882—1883); Golovin (1964); Molin (1860 a)*; Railliet (1916); Skrjabin (1923); Skrjabin, Sobolev & Ivashkin (1965); Yorke & Maplestone (1926).

Schistorophus bihamatus (Mueller, 1897)

Fig. 93
Host: *Gelochelidon nilotica*.

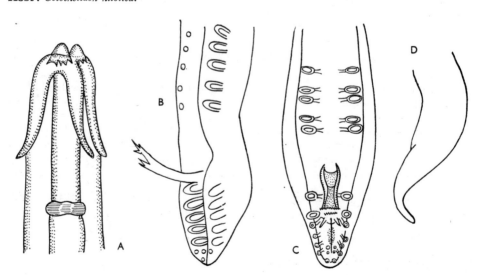

Fig. 93. *Schistorophus bihamatus* (Mueller, 1897). A — anterior end (lateral view); B, C — posterior end of male (lateral and ventral view); D — posterior end of female. After Mueller (1897).

Localization: wall of gizzard.
Distribution: Europe (G.D.R.).

Description: Body filiform, cuticle with fine transverse striations. Anterior part of body, both in males and females, gradually tapering behind the cephalic end to $1/5$ of the mean thickness of body. Head with two lateral pseudolabia and four thorn-like processes, 0.024—0.040 mm long, directed posteriorly. Eight small protuberances (or papillae) arranged in a transverse row at base of pseudolabia. Vestibule 0.05 mm long, muscular part of oesophagus 0.7 to 1.4 mm and glandular part 0.9—1.0 mm long.
Male: Body length 3.5—4.0 mm, body width 0.17—0.18 mm. Posterior end of body spirally coiled, with long, narrow caudal alae. Six pairs of precloacal, five pairs of postcloacal papillae (all pedunculate), plus one pair of small sessile papillae, situated between the last three pairs of postcloacal papillae, and in front of and behind these are two other pairs of small processes (which were not named as papillae by Mueller). Left spicule 0.34 mm long, distal end divided into three sharp points. Right spicule 0.1 mm long, acting as a gubernaculum.
Female: Body length 5.0—6.0 mm, body width 0.2 mm. Vulva situated behind the middle of body, dividing it in the ratio of 4 : 3. Lower part of vagina thick-walled, and shaped like an elongated bell. Posterior part of body conical, tapering to tail 0.1 mm long. Eggs bluntly oval, 0.040 × 0.033 mm, with shell 0.01 mm thick and containing a coiled larva.

This species was described from the host *Sterna risoria* (= *G. nilotica*) in the G.D.R.
Notes: This species was placed by Mueller (1897) in the genus *Ancyracanthus* Schneider, 1866; Cram (1927) transferred it to the genus *Ancyracanthopsis* Diesing, 1861. According to Skrjabin, Sobolev & Ivashkin (1965) it belongs to the genus *Schistorophus* Railliet, 1916.

References: Cram (1927)*; Mueller (1897)*; Skrjabin, Sobolev & Ivashkin (1965)*.

Schistorophus laciniatus (Molin, 1860)

Fig. 94
Hosts: *Xema sabini* and other *Charadriiformes*.
Localization: under the cuticle of gizzard.
Distribution: Asia (U.S.S.R. — Far East, Wrangel Island); outside the Palaearctic Region, in India and Brazil.

Description:
Male: Body length 6.35—7.80 mm, body width 0.14—0.19 mm. Cephalic wings relatively long (0.101—0.140 mm). Nerve ring 0.179—0.241 mm from anterior end. Anterior muscular part of oesophagus 0.638—0.812 mm long,

posterior glandular part 0.112 mm long. Spicules unequal, left 0.392 to 0.588 mm long, right 0.112 mm long. Tail 0.112—0.123 mm long. Caudal alae supported by numerous caudal papillae. Six pairs of postcloacal papillae always present, precloacal papillae arranged asymmetrically; 14 to 18 papillae on the right side and 16 to 20 on the left.

Female: Body length 12.8—13.6 mm, maximum body width 0.31 mm. Cuticle with fine transverse striations, except caudal portion which is smooth. Vestibule 0.05 mm long by 0.033 mm wide anteriorly and 0.02 mm wide

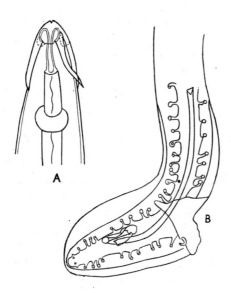

Fig. 94. *Schistorophus laciniatus* (Molin, 1860). A — anterior end (dorso-ventral view); B — posterior end of male (ventral view). After Leonov & Shvetsova (1970).

posteriorly. Muscular portion of oesophagus 0.84 mm long, glandular portion shorter. Nerve ring surrounding muscular oesophagus at level of first third of length. Anus 0.08 mm from tail end. Vulva not salient, in posterior half of body. Eggs 0.048 × 0.027 mm, containing coiled larva.

This species was described by Molin (1860) from a female specimen recovered from *Rallus cayennensis* from Brazil and was placed in the genus *Histiocephalus*. It was transferred to the genus *Schistorophus* by Railliet (1916), **Notes**: A more detailed redescription, based on female specimens from India. was published by Clapham (1945). The first data on the morphology of the males are mentioned in the paper by Leonov & Shvetsova (1970). According to their redescription this form is very similar to *Sch. skrjabini* and synonymy cannot be excluded.

References: Clapham (1945)*; Leonov & Shvetsova (1970)*; Molin (1860 b)*; Railliet (1916); Skrjabin, Sobolev & Ivashkin (1965)*.

Schistorophus skrjabini (Vasilkova, 1926)

Fig. 95
Hosts: *Larus canus, L. argentatus, Rissa brevirostris, R. tridactyla, Stercorarius longicaudatus* and other Charadriiformes.
Localization: oesophagus and under the cuticle of gizzard.
Distribution: Asia (U.S.S.R. — Kazakhstan, West Siberia, Sakhalin Island, Isle of Seals).

Description: Cephalic end with two lateral pseudolabia, each provided with two cuticular wings arising on the dorsal and ventral sides, attached only by

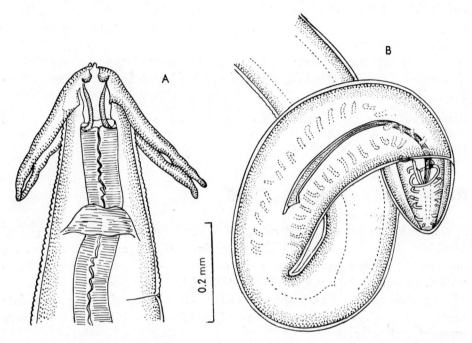

Fig. 95. *Schistorophus skrjabini* (Vasilkova, 1926). A — anterior end (lateral view); B — posterior end of male (dorso-ventral view). After Krotov (1952).

their bases, the greater part of the wings being free and the distal ends surrounding the head in the form of a hood, separated from the rest of the body by a groove which is conspicuous particularly on the ventral side. Vestibule wide, thick-walled, oesophagus consisting of anterior muscular and posterior glandular portions, each of approximately the same length.
Male: Body length 8.27—8.46 mm, width at level of termination of oesophagus 0.26 mm. Length of cephalic cuticular wings 0.228 mm, width at proximal end 0.021 mm and at distal end 0.007 mm. Vestibule 0.046—0.052 mm long and 0.028 mm wide including walls. Muscular portion of oesophagus 0.653 to 0.684 mm and glandular portion 0.833 mm long. Spicules unequal, dissimilar,

the right 0.572 mm long, narrow, with a triangular process in front of the distal end and the left short (0.087 mm long) and wide (0.042 mm at proximal end and 0.021 mm at distal end) with deep indentation on ventral side, functioning as gubernaculum. Caudal papillae pedunculate, except two pairs at end of tail, 22 pairs precloacal and six pairs postclocal (four pairs pedunculate, two pairs sessile). Posterior end surrounded by relatively narrow caudal alae.
Female: Body length 14.63—16.36 mm. Cephalic cuticular wings 0.210 to 0.26 mm long. Vestibule 0.07 mm long. Vulva slightly salient, especially its lower margin, situated in posterior half of body, 6.722—7.486 mm from tail end. Width of body at vulva 0.280—0.357 mm. Eggs 0.042—0.045 × 0.028 mm, with very thick walls and containing larvae when mature. Posterior end of body bluntly rounded. Tail 0.342—0.620 mm long.

The species was recovered from *L. canus* in Kazakhstan and described as *Anenteronema skrjabini*. It was transferred to the genus *Schistorophus* by Gushanskaya (1950 b).

References: Gushanskaya (1950 b); Gushanskaya & Krotov (1952)*; Krotov (1952*, 1959*); Krotov & Delyamure (1952); Sergeeva (1969); Skrjabin, Sobolev & Ivashkin (1965)*; Vasilkova (1926*, 1927*).

Genus *Sciadiocara* Skrjabin, 1916

Mouth with two small, lateral, conical pseudolabia, each bearing a lateral amphid and two submedian papillae. Posterior to the lips are four semicircular, membranous wings arranged in pairs on each side. Vestibule short, thick-walled. Oesophagus long, cylindrical, divided into two parts.
Male: Posterior extremity spirally coiled, alate; six pairs of pedunculate precloacal papillae and several pairs of postcloacal papillae. Spicules very unequal, dissimilar.
Female: Posterior extremity blunt, conical. Vulva in middle or anterior region of body. Oviparous. Eggs oval, thick-shelled, embryonated when deposited. Parasitic in gizzard of birds.

Type species *S. umbellifera* (Molin, 1860).

This genus comprises five species, of which only *S. umbellifera* has been found in fish-eating birds. Other species are known from *Charadriiformes* and *Anseriormes* from Nearctic and Neotropical regions.

Sciadiocara umbellifera (Molin, 1860)

Fig. 96
Hosts: *Larus genei, L. ridibundus, Chlidonias nigra, Ch. hybrida, Sterna hirundo, S. albifrons, Gelochelidon nilotica* and other hosts of the order *Charadriiformes*.

Localization: under cuticle of gizzard.

Distribution: nearly cosmopolitan, in fish-eating birds in Europe (U.S.S.R., Rostov region, the Volga Region, Ukraine) and Asia (U.S.S.R.—Uzbekistan, Kazakhstan, Kirghizia), outside the Palaearctic Region, in Central and South America.

Description: Mouth with two conical pseudolabia and four rounded cephalic wings (two on each side). Outer margin of wings smooth.
Male: Body length 5.97—7.23 mm, maximum body width 0.102—0.164 mm. Vestibule 0.030—0.034 mm long, 0.010—0.016 mm wide. Nerve ring 0.133 to

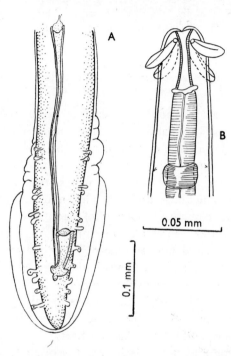

Fig. 96. *Sciadiocara umbellifera* (Molin, 1860). A — posterior end of male (ventral view); B — anterior end (dorso-ventral view). After Daiya (original).

0.145 mm, deirids 0.157—0.174 mm from anterior end of body. Posterior margin of wings 0.027—0.033 mm from anterior end of head. Total length of oesophagus 2.74—3.00 mm. Tail 0.133—0.145 mm long. Right spicule 0.218—0.236 mm long, left spicule 0.503—0.532 mm long. Caudal alae 0.349—0.514 mm long. Six to seven pairs of precloacal and five pairs of postcloacal papillae.
Female: Body length 9.13—12.32 mm, maximum body width 0.189—0.238 mm. Vestibule 0.028—0.042 mm long and 0.012—0.013 mm wide. Nerve ring 0.151—0.196 mm, deirids 0.230—0.234 mm, excretory pore 0.230—0.284 mm, and vulva 5.60—7.52 mm from anterior end of body. Posterior margin of wings 0.034—0.042 mm from anterior end of head. Total length of oesophagus 4.02 mm. Eggs 0.048—0.054 × 0.023—0.028 mm, containing coiled larva. Tail 0.121—0.133 mm long.

The species was described from *Ibis rubra* from Brazil as *Spiroptera umbellifera*. It was transferrred to the genus *Sciadiocara* by Skrjabin (1916 a).
Notes: It should be noted that the species *S. umbellifera* and *S. denticulata* Gibson, 1972 are morphologically very similar. They were separated as late as 1972 by Gibson. Adams & Gibson (1969) regard *S. legrendei* Petter, 1967 as a synonym of *S. umbellifera*.

References: Ablasov & Chibichenko (1962); Adams & Gibson (1969); Gibson (1972)*; Gvozdev & Kasymzhanova (1965); Krotov (1952); Kurochkin & Zablotsky (1961); Leonov (1958); Molin (1860 c)*; Sergeeva (1969); Skrjabin (1916 a); Skrjabin, Sobolev & Ivashkin (1965)*; Smogorzhevskaya (1964); Vasilkova (1926, 1927).

Genus *Sobolevicephalus* Parukhin, 1964

Lateral pseudolabia, each with a median amphid, two submedian papillae and a terminal dentiform process which extends inwards as a square tooth. Dorsal and ventral corners of mouth with processes resembling interlabia but each split longitudinally. These may be called "pseudointerlabia". Head bearing four cuticular wings, two lateroventral and two laterodorsal, separated medially and laterally, and subdivided distally into three or four major lobes; entire margin of each wing with many irregular serrations. Cervical and lateral alae absent.
Male: Caudal alae symmetrical, well developed, with pedunculate preanal and postanal papillae. Left spicule long, slender, with complex tip; right spicule short, stout. Gubernaculum absent.
Female: Vulva inconspicuous, near middle of body. Uterus didelphic. Eggs embryonated when laid; shell of eggs simple. Tail short. Parasites of the gizzards of kingfishers *(Coraciiformes)*.
Type and only species *Sobolevicephalus halcyonis* Parukhin, 1964.

Sobolevicephalus halcyonis Parukhin, 1964

Fig. 97
Host: *Halcyon pileata*.
Localization: under cuticle of gizzard.
Distribution: Asia (U.S.S.R. — Rimsky Korsakov Islands), outside the Palaearctic Region, Vietnam, China — Taiwan.

Description:
Male: Body length 5.0 mm, maximum body width near middle of body 0.16 mm. Vestibule 0.036 mm long. Cephalic wings about 0.040 mm long. Deirids 0.12 mm from anterior end. Muscular oesophagus 1.38 mm, glandular

oesophagus 0.852 mm long. Nerve ring 0.114 mm, excretory pore 0.18 mm from anterior end. Tail 0.12 mm long. Left spicule 0.76 mm long, with complex tip; right spicule 0.18 mm long. Six pairs of preanal and four pairs of postanal papillae. Phasmidial pores near tip of tail.
Female: Body length 7.0—8.0 mm (9.6 mm according to Smetanina & Alekseev 1968), maximum width near middle of body 0.22—0.24 mm. Vestibule 0.038—0.040 mm long. Cephalic wings about 0.040 mm long. Deirids 0.110 to 0.126 mm from anterior end. Muscular oesophagus 1.8—2.07 mm long, glandular oesophagus 1.73—1.76 mm long. Nerve ring 0.118—0.130 mm,

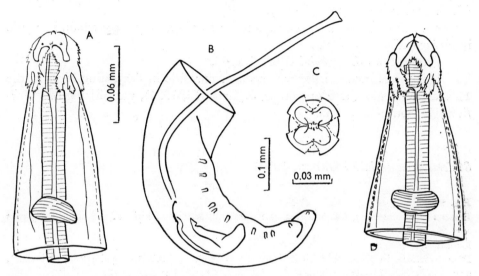

Fig. 97. *Sobolevicephalus halcyonis* Parukhin, 1964. A — anterior end (lateral view); B — posterior end of male (lateral view); C — anterior end (apical view); D — anterior end (dorsoventral view). After Schmidt & Kuntz (1972).

excretory pore 0.16—0.19 mm from anterior end. Vulva about 4.0 mm from anterior end. Tail 0.072—0.096 mm long. Anus inconspicuous. Eggs 0.038 to 0.044 × 0.024—0.028 mm.

This species was described from *H. pileata* from Vietnam.

Notes: Smetanina & Alekseev (1968) recovered this species from *H. pileata* in the Far East and described it under the name *Skrjabinobronema pileati* n. sp., in the family *Schistorophidae*. Smetanina (1972) considered *S. pileati* a synonym of *S. halcyonis*. Schmidt & Kuntz (1972) created for *S. pileati* a new genus *Smetaleksenema* but this is a synonym of the genus *Sobolevicephalus* Parukhin, 1964 described earlier.

References: Parukhin (1964 a)*; Schmidt & Kuntz (1972)*; Smetanina (1972); Smetanina & Alekseev (1968)*.

Genus *Stellocaronema* Gilbert, 1930

Body small, mouth with two lateral interlabia extended as wing-like expansions, which look like a star in apical view, each wing divided latero-ventrally and dorso-ventrally into four digitiform lobes, with a deep notch laterally. Of the four lobes, the two intermediate are smaller than the others. Cervical papillae at level of excretory pore. Vestibule short, cylindrical, without annular markings. Oesophagus divided into two portions.
Male: Posterior extremity spirally coiled, tail short, bluntly pointed. Spicules very unequal. Caudal alae well developed, with five pairs of sessile papillae (three preanal and two postanal).
Female: Tail rounded. Vulva in anterior third of body. Eggs small. Parasitic in gizzard of aquatic birds.

Type species *S. skrjabini* Gilbert, 1930.

This genus includes the type species, parasitic in fish-eating birds in the Palaearctic Region, and one other, *S. fausti* (Li, 1934), parasitizing *Charadriiformes* in South-East Asia.

Stellocaronema skrjabini Gilbert, 1930

Fig. 98
Hosts: *Chlidonias nigra*, *Ch. leucoptera* and numerous species of birds of the order *Ralliformes* and *Charadriiformes*.
Localization: under cuticle of gizzard.
Distribution: in fish-eating birds only in Europe (U.S.S.R.—Byelorussia, Ukraine), and Asia (U.S.S.R.—Azerbaijan, Kirghizia, West Siberia, Far East).

Description:
Male: Body length 6.03—6.33 mm, body width 0.11 mm. Diameter of epaulette-like formations (measured from apical position) 0.04 mm. Vestibule 0.02 mm long and 0.008 mm wide. Deirids and excretory pore 0.22—0.28 mm, nerve ring 0.192—0.264 mm from anterior end of body. Oesophagus 1.268 mm long with muscular part 0.36 mm, glandular 0.908 mm long. Short spicule 0.551 mm long and 0.014 mm wide, long spicule 2.01—2.183 mm long and 0.008 mm wide. Three pairs of precloacal and two pairs of postcloacal papillae. Caudal alae 0.28 mm long and 0.08 mm wide. Tail 0.114 mm long.
Female: Body length 12.96—16.80 mm, body width 0.114—0.122 mm. Oesophagus 1.489 mm long. Vulva 3.06—3.19 mm from anterior end of body. Vestibule 0.028 mm long and 0.010 mm wide. Deirids 0.250 mm from anterior end of body. Tail 0.175 mm long. Eggs 0.030 × 0.016 mm.

The species described was recovered from *Ch. nigra* from the U.S.S.R. (Byelorussia).

Note: Adams & Gibson (1969) consider *S. glareolae* Mawson, 1968 a synonym of *S. skrjabini* Gilbert, 1930.

References: Ablasov & Chibichenko (1962); Adams & Gibson (1969); Gilbert (1930)*; Leonov (1958); Oshmarin (1963); Sailov (1962, 1965 b); Semenov (1927); Serkova (1948); Skrjabin, Sobolev & Ivashkin (1965)*; Smogorzhevskaya (1964).

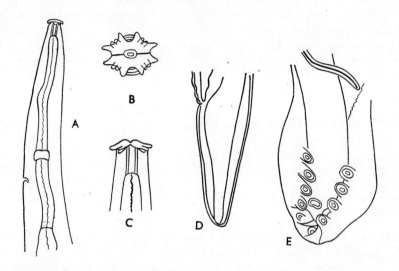

Fig. 98. *Stellocaronema skrjabini* Gilbert, 1930. A — anterior end (lateral view); B — anterior end (apical view); C — anterior end (dorsal view); D — posterior end of female (lateral view); E — posterior end of male (ventral view). After Gilbert (1930).

Genus *Viktorocara* Gushanskaya, 1950

Thin delicate worms. Mouth with two small projecting pseudolabia, at the base of which are two minute papillae. From the lips arise four digitiform or leaf-like, sharply pointed, cuticular wings, each of which is divided by a transverse band above which is another papilla, making eight head papillae in all. The proximal portion of the wings is attached to the body but the distal portion hangs over it freely and a thin delicate membrane extends from the tip of the wings to the lateral fields. Cervical papillae large, nearly level with nerve ring. Vestibulum very long, pseudochitinized, with ring-like striations. Parasites of birds.

Male: Spicules unequal, dissimilar, the short spicule functioning as a gubernaculum. Six to 14 pairs of precloacal and five to six pairs of postcloacal papillae.

Female: Tail digitiform, vulva in posterior part of body.

Type species *V. schejkini* Gushanskaya, 1950.

Of the eight known species belonging to this genus, only three parasitize fish-eating birds in the Palaearctic Region.

KEY TO THE SPECIES OF THE GENUS *VIKTOROCARA*

1	Six pairs of precloacal papillae	2
—	11—14 pairs of precloacal papillae	*V. tenuis*
2	Spicules 0.32—0.35 mm and 0.083—0.087 mm long	*V. schejkini*
—	Spicules 0.49 mm and 0.124 mm long	*V. guschanscoi*

Viktorocara schejkini Gushanskaya, 1950

Fig. 99
Host: *Larus schistisagus*. Known from birds of the order *Charadriiformes* in the U.S.S.R., U.S.A. and Cuba.
Localization: under cuticle of gizzard.
Distribution: Asia (U.S.S.R.—Kamchatka).

Description: Mouth with two pseudolabia bearing cuticular wings with free distal ends. Cervical papillae relatively large.
Male: Body length 5.94—6.00 mm, body width at anterior end 0.026 mm. Length of wings from base of pseudolabia 0.017 mm. Vestibule 0.104 to 0.109 mm long. Nerve ring and cervical papillae 0.144—0.152 mm, and ex-

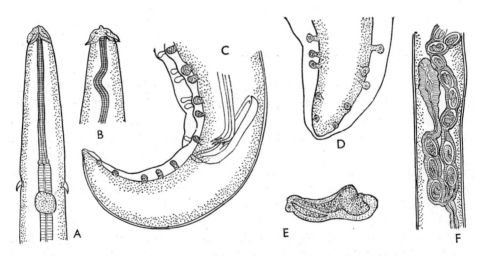

Fig. 99. *Viktorocara schejkini* Gushanskaya, 1950. A — anterior end (dorsal view); B — anterior end (lateral view); C — posterior end of male (lateral view); D — tail of male (ventral view); E — small spicule; D — region of vulva. After Baruš & Lorenzo Hernández (1971)

cretory pore 0.213 mm from anterior end of body. Muscular part of oesophagus 0.470—0.483 mm, glandular part 0.88—0.98 mm long. Posterior end of body coiled, with narrow caudal alae. Six pairs of precloacal and five pairs of postcloacal papillae present. Spicules unequal and dissimilar. Larger spicule 0.326—0.348 mm long, its distal end divided into sharp projections. Smaller spicule massive, 0.083—0.087 mm long. Cloaca situated 0.083 mm from tail end.

Female: Body length 8.12—11.25 mm, width at anterior end 0.026 mm, in region of vulva 0.084 mm. Length of wings from base of pseudolabia 0.017 mm. Vestibule 0.109 mm long. Cervical papillae 0.144 mm and excretory pore 0.209 mm from anterior end. Length of muscular part of oesophagus 0.52 mm, and glandular part 1.01 mm. Vulva 3.75—4.40 mm from posterior end. Margins of vulva not salient. Length of vagina 0.048 mm. Eggs 0.043×0.026 mm, thick-shelled, containing coiled larva. Tip of tail rounded. Anus 0.104 to 0.126 mm from tip of tail.

The species was described from specimens recovered from *Terekia cinerea* in the U.S.S.R. (the Komi Autonomous Soviet Socialist Republic).

References: Gushanskaya (1950 d)*; Leonov & Belogurov (1963).

Viktorocara guschanscoi Leonov, 1958

Fig. 100 A—C
Hosts: *Larus crassirostris, Sterna sandvicensis, S. hirundo* and *Hydroprogne tschegrava*.
Localization: under cuticle of gizzard.
Distribution: Europe (Czechoslovakia, U.S.S.R. — Ukraine), Asia (U.S.S.R. — Far East).

Description: From each pseudolabium arise two cuticular wings with free distal ends. Cervical papillae small, inconspicuous.

Male: Body length 6.24 mm, body width 0.093 mm. Cephalic alae 0.022 mm long. Vestibule 0.093 mm long. Muscular portion of oesophagus 0.49 mm long and 0.018 mm wide, glandular portion 1.73 mm long and 0.055 mm wide. Excretory pore 0.17 mm from anterior end. Posterior end spirally coiled, surrounded by narrow caudal alae. Six pairs of precloacal, five pairs of postcloacal papillae. Spicules unequal, dissimilar. Longer spicule filiform, with widened distal end, 0.496 mm long. Short spicule stout, bifid, 0.124 mm long. Tail 0.155 mm long.

Female: Body length 13.4—14.0 mm, body width 0.155 mm. Cephalic alae 0.027—0.030 mm long. Vestibule 0.086 mm, muscular portion of oesophagus 0.372 mm and glandular portion 1.372 mm long. Vulva 5.6—6.1 mm from posterior end. Eggs numerous, not fully formed. Anus 0.170—0.184 mm from tail end.

This species was recovered and described from *Sterna sandvicensis* from the Ukraine.

References: Leonov (1958)*; Macko (1964 a,d); Macko & Baruš (1973)*; Skrjabin, Sobolev & Ivashkin (1965)*; Smetanina (1972); Smogorzhevskaya (1964).

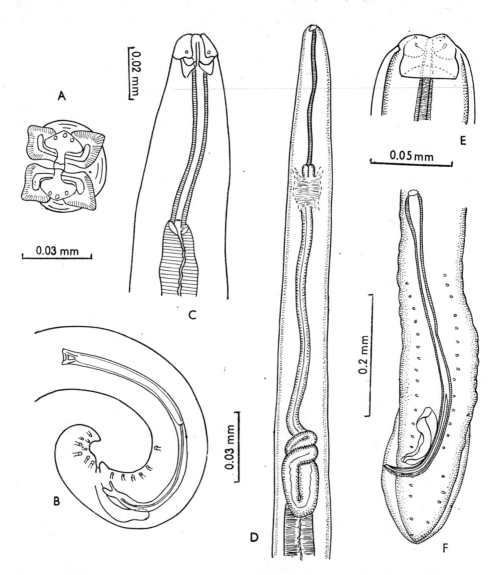

Fig. 100. *Viktorocara guschanscoi* Leonov, 1958 — A, B, C; *V. tenuis* (Maplestone, 1932) — D, E, F. A — anterior end (apical view); B — posterior end of male (lateral view); C — anterior end (dorsal view); D — anterior end (overall view); E — anterior end (dorsal view); F — posterior end of male (ventral view); A, B, C — After Leonov (1958); D, E — After Ryzhikov & Khokhlova (1964); F — After Parukhin in Skrjabin et al. (1965).

Viktorocara tenuis (Maplestone, 1932)

Fig. 100 D—F
Host: *Halcyon pileata*.
Localization: under cuticle of gizzard.
Distribution: Asia (U.S.S.R. — Far East), outside the Palaearctic Region, South-East Asia.

Description: Mouth with two lateral triangular pseudolabia, at the base of which are two small papillae. From each pseudolabium arise two short cuticular alae, directed posteriorly and connected with dorsal and ventral sides by a fine cuticular membrane. Vestibule very long; two small thorn-like cervical papillae present at level of its posterior end. Nerve ring in anterior part of muscular oesophagus.
Male: Body length 7.0—7.5 mm, maximum body width 0.10—0.12 mm. Vestibule 0.10 mm long, 0.006 mm wide. Muscular oesophagus 0.36 mm, glandular 0.80 mm long. Nerve ring 0.10 mm from anterior end. Cloaca 0.088 mm from posterior end. Caudal alae present. Eleven to fourteen pairs of precloacal, five pairs of postcloacal papillae. Spicules unequal, dissimilar. Longer spicule 0.36 mm long, maximum width 0.016 mm. Shorter spicule 0.08 mm long, maximum width 0.020 mm.
Female: Body length 7.5—20.5 mm, maximum body width 0.076—0.12 mm. Cephalic alae 0.013—0.018 mm long. Cervical papillae 0.10—0.13 mm from anterior end. Vestibule 0.083—0.116 mm long. Muscular oesophagus 0.22 to 0.48 mm, glandular 0.64—1.28 mm long. Vulva in posterior part of body. Anus 0.096—0.19 mm from tail end. Mature eggs, 0.043—0.048 × 0.026 to 0.030 mm, containing coiled larva.

The species described was recovered from *H. smyrnensis* in India.

Notes: Ryzhikov & Khokhlova (1964) described *Viktorocara halcyoni*. This species is identical both in morphology and measurements with *V. tenuis* and there are no differences of specific value. *V. halcyoni*, therefore, is a synonym of *V. tenuis*.

References: Maplestone (1932b)*; Parukhin (1964 b)*; Ryzhikov & Khokhlova (1964 b, c*, 1967*); Smetanina (1972); Smetanina & Alekseev (1967); Skrjabin, Sobolev & Ivashkin (1965)*.

Family *Streptocaridae* Skrjabin, Sobolev & Ivashkin, 1965

Cephalic ornamentation in form of collarette, completely or partially surrounding base of lateral lips. Posterior margin of collarette bearing denticles. Cervical papillae large, with three or more denticles. Cuticle smooth or with longitudinal rows of distinct spines.

This family was erected as a result of uniting two subfamilies, *Strepto-*

carinae and *Seuratinae*, which were earlier placed in the family *Acuariidae*. The following composition of the family *Streptocaridae* was proposed: subfamily *Streptocarinae* (genera *Streptocara, Koriakinema, Proyseria, Rusguniella* and *Stegophorus*) and subfamily *Seuratinae* (genus *Seuratia*).

Later, many changes in the structure of the family were made. These changes, which we consider to be valid, concerned the following: 1) Sergeeva (1968) excluded the genus *Rusguniella* and transferred it to the family *Acuariidae*; 2) Gibson (1964) gave evidence that the species *Koriakinema gusi* Oshmarin, 1950 (the only species of the genus) is identical with *Streptocara californica* (Gedoelst, 1919) and Ryzhikov (1966) confirmed this opinion; 3) Gibson (1968) erected a new genus *Ingliseria*, with the type species *I. cirrohamata* (Linstow, 1888) which was earlier assigned to the genus *Streptocara*.

We are including this genus in the family *Streptocaridae*. The composition of the family is, therefore, now as follows: subfamily *Streptocarinae* (genera: *Streptocara, Proyseria, Ingliseria* and *Stegophorus*) and subfamily *Seuratinae* (genus *Seuratia*).

Fish-eating birds of the Palaearctic Region are parasitized by representatives of all genera belonging to this family.

KEY TO THE GENERA OF THE FAMILY *STREPTOCARIDAE*

1 Surface of cuticle without chitinized spines 2
— Surface of cuticle with a double row of chitinized spines . . . . *Seuratia*
2 Cervical papillae with many denticles . . . . *Streptocara*
— Cervical papillae with three denticles 3
3 Cephalic ornamentation in form of cuticular helmet-like formations 4
— Cephalic ornamentation of other form . . . . *Ingliseria*
4 Denticles on helmet-shaped formation distributed along the whole free margin . . . .
   . . . . *Proyseria*
— Denticles on helmet-shaped formation not distributed on the whole free margin, absent on dorsal and ventral slits . . . . *Stegophorus*

Genus *Streptocara* Railliet, Henry & Sisoff, 1912

Mouth with lateral lips, behind which is collar composed of two crescent-shaped outgrowths joining near margins of lips. External margin of collar with small denticles. Each lip bears two submedian papillae. Cervical papillae with many denticles. Lateral alae absent. Spicules unequal, dissimilar. Vulva behind middle of body. Tail end of female rounded. Parasites of various orders of birds.

Type species *S. crassicauda* (Creplin, 1829).

As to the species composition of the genus, different opinions have been expressed in the literature. Skrjabin, Sobolev & Ivashkin (1965) distinguished seven species, although in Ryzhikov's (1966) opinion two of these are not valid and should be synonymized (*S. pectinifera* synonym of *S. crassicauda*; *S. dogieli* synonym of *S. californica*) and two species should be excluded from this genus (*S. transcaucasica* is transferred to the genus *Paracuaria* and *S. penihamata* to the group of poorly known species of *Spirurata*). According to Ryzhikov (1966) the genus thus includes only three species (*S. crassicauda*, *S. californica* and *S. recta*).

The structure of the genus as proposed by Ryzhikov (1966) has since been subjected to a new revision, concerned mainly with the species *S. recta*. Since this species is associated with fish-eating birds, its history is briefly described below.

S. recta (Linstow, 1879) was described on the basis of specimens from *Podiceps cristatus* and it was placed originally in the genus *Filaria*. Mueller (1897) redescribed this species from his own material and pointed out the presence of an inconspicuous dentate collarette, but did not give the measurements of the parasites studied. On the basis of Mueller's description, Railliet, Henry & Sisoff (1912) transferred the species to the genus *Streptocara*.

Daiya (1967) studied the nematodes from *Podiceps* spp. and concluded that *S. recta* does not occur in the territory of the U.S.S.R. Having examined some material from *Podiceps* spp., we obtained the same results and, therefore, we consider all nematodes from the U.S.S.R. described under the name *S. recta* to belong to the species *S. crassicauda*. In our opinion, the species *S. recta* should be regarded as *species inquirenda*.

Gibson (1968) also noted that in all subsequent papers, in which some data on the morphology of *S. recta* were given, the nematodes observed did not correspond to Linstow's description of this species, because their collarette and spicules were smaller.

Having studied the nematodes of the genus *Streptocara* from *P. auritus* from Canada, he found small morphological differences between them and nematodes of that genus from other water birds and regarded the parasites as *S. recta*. However, the nematodes found by Gibson differ considerably in their morphology from *S. recta* sensu Linstow. If studies of a large number of nematodes reveal the constancy and authenticity of the differences between *Streptocara* from grebes and *Streptocara* from other water birds, then it will be necessary to erect a new species. However, we cannot agree with Gibson that the nematodes found by him belong to the species *S. recta*.

In any case, considering all the facts mentioned above, there is no reason to include the species *S. recta* in the list of helminth parasites of fish-eating birds of the Palaearctic Region.

Kreis (1958) described a new species of this genus, namely, *S. rissae* from *Rissa tridactyla* from Iceland. Daiya (1967) proved this species to be a synonym of *Paracuaria tridentata*. Of the two valid species of this genus *(S. crassicauda* and *S. californica)* only the former has been reported from fish-eating birds in the Palaearctic Region.

Streptocara crassicauda (Creplin, 1829)

Fig. 101
Hosts: *Gavia arctica, G. stellata, Podiceps cristatus, P. auritus, P. nigricollis, P. griseigena, P. ruficollis, Puffinus puffinus, Phalacrocorax carbo, Ph. pygmaeus, Botaurus stellaris, Mergus merganser, M. albellus, M. serrator, Rissa tridactyla, Larus canus, L. argentatus, L. genei, L. ichthyaetus, L. minutus, L. ridi-*

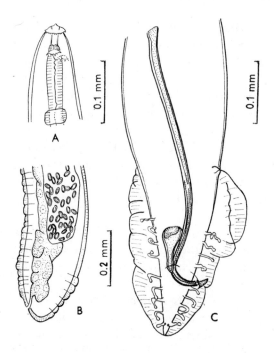

Fig. 101. *Streptocara crassicauda* (Creplin, 1829). A — anterior end (lateral view); B — posterior end of female (lateral view); C — posterior end of male (ventral view). After Daiya (1967).

bundus, L. schistisagus, Hydroprogne tschegrava, Gelochelidon nilotica, Sterna hirundo, S. sandvicensis, Alca torda, Uria aalge, U. lomvia and *Fratercula corniculata*; often encountered also in birds of the order *Anseriformes*, less often in *Charadriiformes* and *Galliformes*.
Localization: under cuticle of gizzard.
Distribution: Europe (Iceland, England, France, Poland, U.S.S.R. — Ukraine, central and north regions, the Volga Region), Asia (U.S.S.R. — West and East Siberia, Transcaucasus region, republics of Middle Asia, Kazakhstan, Far East; Japan). Outside the Palaearctic Region, North America and Australia.

Description: Cuticle with fine transverse striations. Cervical papillae large, with four to seven denticles.
Male: Body 5.5—6.1 mm, maximum body width 0.13—0.17 mm. Length of

vestibule 0.023—0.03 mm, of muscular part of oesophagus 0.36—0.57 mm, and of glandular part 1.8—2.5 mm. Cervical papillae 0.03—0.04 mm and nerve ring 0.09—0.15 mm from head end. Tail end coiled in spiral, with alae measuring 0.23 mm. Four pairs of preanal and five pairs of postanal pedunculate papillae present. Left spicule 0.32—0.41 mm long, bent and thin, with an uncinate process at its distal end. Right spicule short and wide, 0.07—0.08 mm long. Tail 0.06—0.08 mm long.

Female: Body length 6.0—11.7 mm, maximum body width 0.16—0.20 mm. Muscular part of oesophagus 0.5—0.7 mm long, glandular part 1.45—2.0 mm. Cervical papillae 0.02—0.03 mm and nerve ring 0.12—0.20 mm from head end. Vulva situated in posterior half of body, opening 1.7—5.8 mm from posterior end. Tail 0.02—0.05 mm long with rounded end. Eggs 0.01 to 0.02 × 0.026—0.036 mm.

The species was described from specimens recovered from *Gavia stellata* and was originally placed in the genus *Spiroptera*. Molin (1860) transferred it to the genus *Dispharagus* and Railliet, Henry & Sisoff (1912) to the genus *Streptocara*.

Biology: The most complete study of the life-cycle of this species was made by Garkavi (1949 a, 1950, 1953). Data on its biology are also included in papers by Kovalenko (1960, 1963), Vasilev (1968) and other authors. The main intermediate hosts are crustaceans *(Gammarus* spp.).

Some fish species *(Carassius carassius, Phoxinus perenurus* and others) may serve as paratenic hosts.

Notes: We regard the following species as synonyms of *S. crassicauda*: 1) *S. pectinifera* (Neumann, 1900) — this synonymy was proved experimentally by Vasilev (1968); 2) all nematodes identified in the U.S.S.R. as *S. recta* (Kontrimavichus & Bakhmeteva 1960; Shigin 1957; Michelson 1968; Sailov 1966; Golovin 1964 and others); 3) nematodes recovered from *Podiceps ruficollis* in Japan and identified by Yamaguti (1935) as *S. recta*.

References: Ablasov (1957); Ablasov & Chibichenko (1961, 1962); Alekseev (1970); Babaev (1970); Baird (1853), Bakke (1972); Belogurov (1965); Belopolskaya (1952); Cram (1927)*; Creplin (1829*, 1845—1846); Czaplinski (1962 b); Daiya (1967 a*, b, 1971); Diesing (1851); Dujardin (1845); Garkavi (1949 a, 1950*, 1953); Ginetsinskaya (1952); Golovin (1964); Gubanov & Daiya (1967); Gurlt (1845); Gushanskaya (1950 c); Gvozdev & Kasymzhanova (1965); Iksanov & Dikambaeva (1962); Kibakin et al. (1963); Kontrimavichus & Bakhmeteva (1960); Karokhin (1935); Kosupko (1963); Kovalenko (1960, 1963); Kulachkova & Kochetova (1964); Kurashvili (1961); Leonov (1958, 1960 b); Leonov & Belogurov (1963); Leonov & Shvetsova (1970); Linstow (1894); Maksimova (1966); Mashtakov (1964); Michelson (1968); Oshmarin (1950, 1963, 1965); Oshmarin & Parukhin (1963); Paskalskaya (1968); Petrov & Chertkova (1950); Ryzhikov (1963 b, 1966); Ryzhikov & Daiya (1967); Ryzhikov & Kozlov (1959); Sailov (1966); Sergeeva (1968 a, 1969); Shakhtakhtinskaya (1959 a, b); Shigin (1957, 1959); Skrjabin (1915 a, 1916 a*, 1923); Skrjabin, Sobolev & Ivashkin (1965)*; Smetanina (1972); Smogorzhevskaya (1962 a, 1964, 1967); Sonin & Larchenko (1974); Sultanov (1959 a, b); Tolkacheva (1967); Tsimbalyku

(1965); Tsimbalyuk & Belogurov (1964); Turemuratov (1965 a); Vaidova (1965); Vasilev (1968); Yamaguti (1935)*.

Genus *Ingliseria* Gibson, 1968

Head with paired pseudolabia, each bearing an anteriorly directed dentiform process with a pair of tooth-like structures at its base, and two submedian papillae. Cephalic ornamentations appearing as four crescentic, scalloped cordons which arise from oral opening and terminate on lateral surface of head without anastomosing. Each cordon bears about 15 blunt lappet-like teeth and is bordered posteriorly by a deep groove. Vestibule moderately long. Deirid moderately large and tricuspid. Narrow lateral alae present.

Male: Caudal end alate, with four pairs of preanal and five pairs of postanal pedunculate papillae and two pairs of sessile or subsessile papillae near apex. Left spicule long and slender, right spicule short and stout.

Female: Posterior end conical, vulva somewhat posterior to middle of body. Parasitic in stomach of cormorants and other fish-eating birds.

Type and only species *I. cirrohamata* (Linstow, 1888).

Ingliseria cirrohamata (Linstow, 1888)

Fig. 102

Hosts: *Gavia arctica, G. stellata, Mergus merganser, M. serrator* and *Larus argentatus*.
Localization: proventriculus and gizzard.
Distribution: Europe and Asia (U.S.S.R. — Ukraine, East Siberia, Far East). Outside the Palaearctic Region, in Australia.

Description:

Male: Body length 7.58—11.0 mm. Maximum body width 0.25—0.35 mm. Tail 0.24—0.38 mm long. Caudal alae wide, 0.41 mm long. Four pairs of preanal and five pairs of postanal papillae present. Left spicule 0.4—0.6 mm long, with a small process at its end, right spicule 0.116—0.15 mm long.

Female: Body length 9.72—16 mm, maximum body width 0.32—0.35 mm. Length of vestibule 0.14—0.18 mm, of muscular part of oesophagus 0.84 to 0.9 mm and of glandular part 2.8—2.9 mm. Cervical papillae 0.14—0.18 mm and vulva 7.1—8 mm from head end. Cervical papillae 0.025 × 0.022 mm. Tail 0.12—0.17 mm long. Eggs 0.035—0.039 × 0.019 mm.

The species was described from specimens recovered from *Phalacrocorax verrucosus* from Kerguelen Islands. It was originally assigned to the genus *Filaria*, but Skrjabin (1916) transferred it to the genus *Streptocara*. The species was reported from fish-eating birds in the U.S.S.R., but the authors did not give either figures or descriptions of the parasites recovered. Gibson (1968) transferred this species to another genus, *Ingliseria*, and included in his paper the description and figures of the type specimen made by Inglis.

References: Gibson (1968)*; Linstow (1888)*; Oshmarin (1950, 1965); Skrjabin (1916 a)*; Smogorzhevskaya (1967).

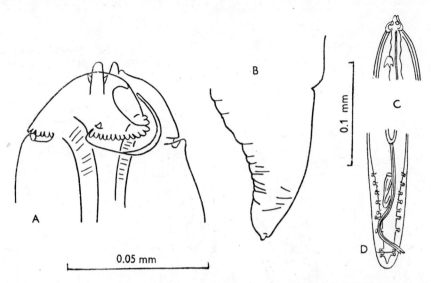

Fig. 102. *Ingliseria cirrohamata* (Linstow, 1888). A, C — anterior part of body; B — posterior part of female; D — posterior part of male (ventral view). A, B — After Inglis in Gibson (1968); C, D — After Linstow (1888).

Genus *Proyseria* Petter, 1959

Head ornamentation in form of two cuticular membranes covering anterior end of body. Thickened margins of membranes have appearance of cordons originating on dorsal and ventral side of mouth opening and uniting laterally. The margins of these cordons are serrated along whole length. Cervical papillae tricuspid. Cuticle at head end swollen in form of head vesicle. Spicules unequal and dissimilar. Caudal alae well developed.

Type and only species *P. decora* (Dujardin, 1845).

Proyseria decora (Dujardin, 1845)

Fig. 103
Host: *Alcedo atthis*.
Localization: under cuticle of gizzard and oesophagus.

Distribution: Europe (France) and Asia (U.S.S.R. — Primorye Territory; Iran). Outside the Palaearctic Region, Indo-China (the Democratic Republic of Vietnam).

Description: Small nematodes, with thick cuticle with fine transverse striations. Two conical lateral lips, each with two papillae at base, present at head end. Two ear-shaped cuticular plates on each side at level of papillae on dorsal and ventral surfaces of lips. From these originate helmet-shaped,

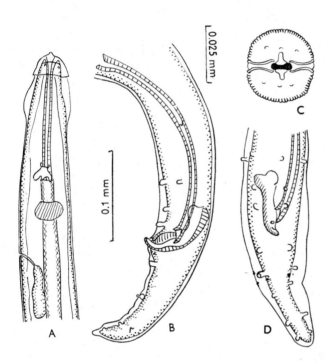

Fig. 103. *Proyseria decora* (Dujardin, 1845). A — anterior end (lateral view); B — posterior end of male (lateral view); C — anterior end (apical view); D — posterior end (ventral view). After Petter (1959).

cuticular structures with denticulate margins and a narrow cuticular ridge running at base of denticles. Vestibule wide, thick-walled. Cervical papillae at level of transition of pharynx to oesophagus.

Male: Body length 2.48—4.7 mm, maximum body width 0.09—0.12 mm. Cephalic cuticular helmet 0.026—0.03 mm long. Cervical papillae 0.076 to 0.13 mm from head end. Length of vestibule 0.04—0.14 mm, of muscular oesophagus 0.40—0.60 mm, and of glandular oesophagus 0.96—1.4 mm. Nerve ring at 0.180 mm from anterior end of body. Tail end with cuticular alae, their maximum width 0.066 mm. Large spicule 0.237—0.281 mm long, with distal end terminating in two pointed processes, one directed lengthwise to the distal end and the other angled. Processes united by a thin cuticular membrane. Smaller spicule 0.068—0.085 mm long. Nine pairs of caudal papillae present, of which four pairs are preanal and five pairs postanal. Tail 0.08—0.13 mm long.

Female: Body length 6.4—8.0 mm, maximum body width 0.20—0.224 mm. Cuticular helmet 0.05 mm long. Cervical papillae 0.132 mm and nerve ring 0.215 mm from head end. Length of vestibule 0.165 mm, of muscular oesophagus 0.64 mm, and glandular oesophagus 1.54 mm. Vulva 0.40—0.50 mm and anus 0.08—0.13 mm from posterior end. Mature eggs 0.039 mm long.

The species was described from specimens recovered from *A. atthis* in France and was originally placed in the genus *Dispharagus*. Later, it was transferred to other genera by several authors and, eventually, Petter (1959) erected for it a new genus *Proyseria*.

References: Chabaud (1953)*; Diesing (1851); Dujardin (1845)*; Gedoelst (1919); Gendre (1920); Gurlt (1845); Petter (1959)*; Ryzhikov & Khokhlova (1965)*; Skrjabin (1916 a*, 1923); Skrjabin, Sobolev & Ivashkin (1965)*.

Genus *Stegophorus* Wehr, 1934

Mouth opening in form of dorso-ventral slit, surrounded by two lips. Each lip bears two submedian papillae and a small amphid. Head end with helmet-like structure composed of two lateral cuticular membranes. Anterior part of membranes joined with body, posterior part forming descending collarettes with dentate margin. Caudal alae well developed. Spicules unequal and dissimilar. Vulva postequatorial. (In accord with other authors we regard the genus *Paryseria* Johnston, 1938 as a synonym of *Stegophorus*.)

Type species *S. stellaepolaris* (Parona, 1901).

The genus includes eight species, all of which parasitize fish-eating birds. Two species have been reported from the Palaearctic Region.

KEY TO THE SPECIES OF THE GENUS *STEGOPHORUS*

1   54—56 denticles on helmet-like structure; large spicule striated in its posterior third . . . . *S. stellaepolaris*
—   24—30 denticles on helmet-like structure; large spicule whithout striations . . . . *S. stercorarii*

Stegophorus stellaepolaris (Parona, 1901)

Fig. 104

Hosts: *Gavia stellata, Fulmarus glacialis, Oceanodroma monorchis, Hydrobates pelagicus, Stercorarius parasiticus, S. longicaudatus, S. pomarinus, Larus argentatus, L. crassirostris, L. glaucescens, Uria aalge, U. lomvia, Cepphus carbo, Synthliboramphus antiguus, Aethia pygmaea, Cyclorhynchus psittacula* and *Lunda cirrhata*.

Localization: under cuticle of gizzard.

Distribution: Europe and Asia (northern regions of the Atlantic, the Arctic and the Pacific Oceans, to the Netherlands in the west and to Primorye Territory in the east).

Description: Helmet-like structure with 54—56 denticles on margin.
M a l e : Body length 5—7.8 mm, maximum body width 0.10—0.145 mm. Length of helmet-like structure 0.075—0.09 mm, maximum width 0.08 to

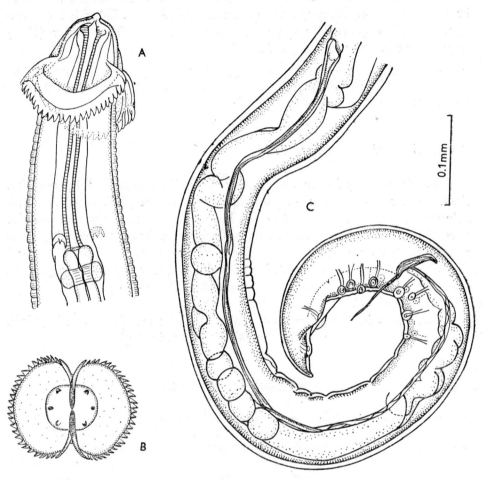

Fig. 104. *Stegophorus stellaepolaris* (Parona, 1901). A — anterior end; B — anterior end (apical view); C — posterior end of male (lateral view). After Ryzhikov (1966).

0.09 mm. Cervical papillae 0.15—0.19 mm and nerve ring 0.17—0.23 mm from head end. Length of vestibule 0.16—0.19 mm, of muscular oesophagus 0.40—0.55 mm, and of glandular oesophagus 1.28—1.9 mm. Tail end coiled in spiral, caudal alae narrow. Nine pairs of pedunculate caudal papillae present (four pairs preanal and five pairs postanal). Large spicule long (0.82

to 2.43 mm) and thin, with proximal end widened and distal end pointed with a small uncinate process; posterior third of spicule with characteristic striations, giving the appearance of a lash or twisted cord. Small spicule 0.045 to 0.10 mm long, in form of a thin plate with bent longitudinal alae. Tail 0.16 mm long.

Female: Body length 10.40—16.00 mm, maximum body width 0.12—0.19 mm. Length of helmet-like formation 0.07—0.10 mm, maximum width 0.09—0.11 mm. Cervical papillae 0.14—0.22 mm and nerve ring 0.17 to 0.27 mm from anterior end. Length of vestibule 0.155—0.20 mm, of muscular oesophagus 0.48—0.56 mm and of glandular oesophagus 1.28—2.1 mm. Vulva 4.5—7.8 mm from anterior end. Tail conoid, 0.15 mm long. Eggs 0.04—0.05 × 0.02—0.03 mm.

The species was described from specimens recovered from *F. glacialis* in Novaya Zemlya and was originally placed in the genus *Histiocephalus*. Wehr (1934) transferred it to the genus *Stegophorus*.

References: Alekseev & Smetanina (1968); Baer (1956)*; Bakke & Baruš (1976)*; Baylis (1928); Broek (1968); Broek & Jansen (1971); Gubanov & Sergeeva (1971); Jansen & Broek (1966); Kulachkova & Kochetova (1964); Leonov & Belogurov (1963); Leonov & Shvetsova (1970); Parona (1901*, 1903); Ryzhikov (1965)*; Sergeeva (1968 a, 1969); Skrjabin, Sobolev & Ivashkin (1965)*; Smetanina (1972); Smetanina & Alekseev (1967); Tsimbalyuk & Belogurov (1964); Wehr (1934)*.

Stegophorus stercorarii Leonov, Sergeeva & Tsimbalyuk, 1966

Fig. 105
Hosts: *Fulmarus glacialis, Stercorarius parasiticus, S. longicaudatus, S. pomarinus, Larus argentatus, Sterna hirundo, Aethia cristatella, A. pusilla, Fratercula corniculata, Lunda cirrhata, Uria lomvia* and *Cepphus grylle*.
Localization: under cuticle of gizzard.
Distribution: Asia (U.S.S.R. — coasts of the Arctic Ocean, of the Bering Sea, the Sea of Japan and the Sea of Okhotsk).

Description: Helmet-like structure with 24—30 denticles on margin.
Male: Body length 4.5—5.3 mm, maximum body width 0.10—0.17 mm. Length of helmet-like structure 0.03—0.04 mm. Cervical papillae 0.07—0.14 mm and nerve ring 0.14—0.16 mm from head end. Length of vestibule 0.13—0.14 mm, of muscular oesophagus 0.72—0.84 mm and of glandular oesophagus 1.10—1.60 mm. Caudal alae 0.39—0.65 mm long. Four pairs of preanal and five pairs of postanal papillae present, the second, fourth and fifth pairs being largest. Right spicule long (0.56—0.73 mm), thin, its distal end with uncinate process. Left spicule short (0.08—0.10 mm), thick, cymbiform.

Female: Body length 6.30—11.90 mm, maximum body width 0.11—0.25 mm. Length of helmet-like formation 0.03—0.06 mm. Cervical papillae 0.09 to 0.17 mm and nerve ring 0.16—0.19 mm from head end. Length of vestibule 0.13—0.15 mm, of muscular oesophagus 0.63—0.65 mm and of glandular oesophagus 1.10—2.00 mm. Vulva 2.8—3.3 mm from posterior end. Eggs 0.04—0.05 × 0.02—0.03 mm.

The description was based on specimens recovered from *Laridae* and *Alcidae* in the lower reaches of the Yenisei River and Chukotka.

References: Alekseev & Smetanina (1968); Belogurov, Leonov & Zueva (1968); Daiya (1967 a, b); Gubanov & Sergeeva (1971); Leonov, Sergeeva & Tsimbalyuk (1966)*; Leonov & Shvetsova (1970); Smetanina (1972); Smetanina & Alekseev (1967).

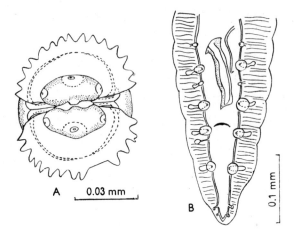

Fig. 105. *Stegophorus stercorarii* Leonov, Sergeeva & Tsimbalyuk, 1966. A — anterior end (apical view); B — posterior end of male (ventral view). After Leonov et al. (1966).

Genus *Seuratia* Skrjabin, 1916

There are four longitudinal rows of spines originating as two parallel rows at anterior end and gradually drawing nearer, extending to middle of body length. Anterior end of body bears a bilobed cap divided by a dorso-ventral fissure. Each lobe has a median incision and numerous denticles on its external margin. Cervical papillae tricuspid. Vestibule long, with sclerotized walls. Oesophagus long, subdivided into portions. Spicules unequal and dissimilar. Caudal end bears small alae. Vulva in middle part of body.

Type species *S. shipleyi* (Stossich, 1900).

The genus includes four species, one of which parasitizes fish-eating birds in the Palaearctic Region.

Seuratia shipleyi (Stossich, 1900)

Fig. 106
Hosts: *Diomedea exulans, Procellaria diomedea, Puffinus griseus, P. tenuirostris, Oceanodroma monorchis, Stercorarius parasiticus, S. longicaudatus, Larus canus, L. argentatus, L. glaucescens* and *Fratercula corniculata.*
Localization: crop, oesophagus, proventriculus and gizzard.
Distribution: the Atlantic, Pacific and Arctic Oceans.

Description: Mouth surrounded by two trilobed lips, each of which bears three small papillae. Cuticle of anterior end of body in form of collarette composed of two wide cordon-like discs with 52 to 100 sharp denticles on external margin.
Male: Body length 7.5—15.0 mm. Cervical pappillae 0.085—0.162 mm and nerve ring 0.11—0.216 mm from anterior end. Vestibule 0.118—0.215 × × 0.018—0.045 mm. Length of muscular oesophagus 0.476—0.83 mm and

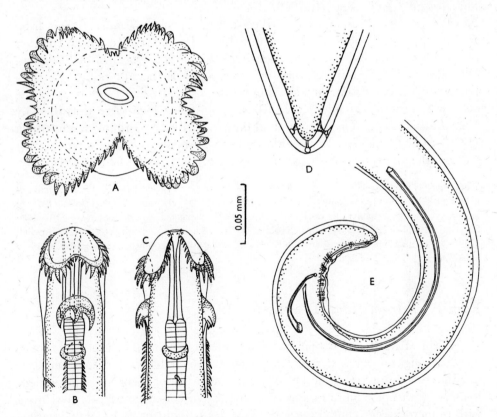

Fig. 106. *Seuratia shipleyi* (Stossich, 1900). A — anterior end (apical view); B, C — anterior end (lateral and dorsal view); D — posterior end of female; E — posterior end of male (lateral view). After Mendonça & Rodrigues (1968).

of glandular oesophagus 3.114—4.58 mm. Tail end rounded, alate, 0.179 to 0.28 mm long. Four pairs of pedunculate preanal and four to five pairs of postanal papillae present plus, sometimes, three small papillae at end of tail. Left spicule long (0.65—1.20 mm), thin, with sharp end. Right spicule 0.196 to 0.23 mm long.

Female: Body length 10.2—35 mm. Cervical papillae 0.08—0.28 mm and nerve ring 0.21—0.264 mm from anterior end. Vestibule measures 0.10 to 0.215 × 0.02—0.045 mm. Length of muscular oesophagus 0.71—0.985 mm and of glandular oesophagus 4.15—10.2 mm. Vulva in mid-body, 5.3—14.6 mm from anterior end. Tail 0.183 mm long, sometimes with three small papillae at tip. Eggs 0.025—0.046 × 0.012—0.021 mm.

The species was described from specimens recovered from *Diomedea exulans* caught in eastern areas of the Pacific Ocean and was originally assigned to the genus *Gnathostoma*. Skrjabin (1916) included it in the genus *Seuratia* as type species and synonymized with it the species *Rictularia paradoxa* Linstow, 1903 and *Acuaria pelagica* Seurat, 1916.

Notes: In 1941 Yamaguti described a new species, *Seuratia puffini*, from *Puffinus griseus*. Mendonça & Rodrigues (1968), having studied a large number of specimens, concluded that the species *S. puffini* is a synonym of *S. shipleyi*. We agree with their opinion, although some other authors do not, e. g. Vassiliades (1970). We also regard as synonyms of *S. shipleyi* those specimens of the species *S. yamaguti* Mendonça & Rodrigues, 1968 which were described by Yamaguti (1941) as *S. procellariae*, and those described by Leonov & Shvetsova (1970) as *S. procellariae*. In accordance with Mendonça & Rodrigues (1968) we consider the species *S. procellariae* (Diesing, 1851) as *species inquirenda*.

References: Bakke (1972); Daiya (1967 a,b); Gubanov & Sergeeva (1971); Leonov & Belogurov (1963); Leonov, Belogurov & Tsimbalyuk (1964)*; Leonov & Shvetsova (1970)*; Linstow (1903)*; Mendonça & Rodrigues (1968)*; Sergeeva (1969); Seurat (1916 c)*; Skrjabin (1916 c, 1923)*; Skrjabin, Sobolev & Ivashkin (1965)*; Sprehn (1962); Stossich (1900)*; Tsimbalyuk & Belogurov (1964); Tsimbalyuk, Leonov & Belogurov (1964); Vassiliades (1970); Yamaguti (1941)*.

Family *Tetrameridae* Travassos, 1914

Form of head end rather variable; lateral pseudolabia usually present. There are eight cephalic papillae arranged in two circles, each of four papillae, but those of the inner circle are often rudimentary or completely atrophied. Females with body more or less fusiform, vulva coiled posteriorly and uteri with numerous eggs containing larvae. Males without caudal alae. Gubernaculum absent. Females located either in glands of proventriculus, or in cysts in gizzard.

The family comprises two subfamilies, namely, *Tetramerinae* and *Geopetitiinae;* representatives of the former have been reported from fish-eating birds. This subfamily contains two genera and members of both have been found in fish-eating birds in the Palaearctic Region.

Notes: In many papers published recently this family was named *Tropisuridae*. This name was proposed by Yamaguti (1961), who assumed the type genus of the family *Tetrameres* Creplin, 1846 to be a synonym of the genus *Tropisurus* Diesing, 1831. Skrjabin & Sobolev (1963) did not recognize the name proposed by Yamaguti and we use the names of the family and of the taxa included in it as set out in their monograph.

KEY TO THE GENERA OF THE FAMILY *TETRAMERIDAE*

1 Male body surface usually with longitudinal rows of cuticular spines (except in species of the subgenus *Gynaecophila*). Female body saccular, globular or oval, with four longitudinal fissures . . . . *Tetrameres*
— Male body surface without cuticular spines. Female body spirally twisted, without longitudinal fissures . . . . *Microtetrameres*

Genus *Tetrameres Creplin*, 1846

Lips feebly developed; buccal capsule (vestibule), muscular and glandular oesophagus present. Males thin, colourless, usually with spines arranged medially and laterally. Spicules very unequal, left spicule sometimes reaching $^2/_3$ of body length. Females fusiform, with transverse striations, more distinct in anterior part of body, and with four longitudinal fissures corresponding to median and lateral lines. Vulva in posterior part of body, uterus well developed.

Type species *T. paradoxa* (Diesing, 1835).

The genus includes 50 species divided into three subgeneric groups: *Tetrameres, Petrowimeres* Chertkova, 1953 and *Gynaecophila* Gubanov, 1950. (The main characters of subgenera are given in the Key.)

Nine species of all three subgenera parasitize fish-eating birds of the Palaearctic Region. Their assignment to individual subgenera is given in the Key. Only females are known of one of these species *(T. coccinea)* and it is, therefore, not included in the Key although a description is given on p. 211.

KEY TO THE SPECIES OF THE GENUS *TETRAMERES*

1 Male body surface with longitudinal rows of cuticular spines 4
— Male body surface without cuticular spines (subgenus *Gynaecophila*) 2
2 Male without spicules . . . . *T. (G.) gynaecophila*
— Male with spicules 3

3	Spicules 1.42—1.53 mm and 0.11 mm long	*T. (G.) ardea*
—	Spicules 1.6 mm and 0.05—0.06 mm long	*T. (G.) schigini*
4	Head end with lateral cuticular appendages (subgenus *Petrowimeres*)	5
—	Head end without lateral cuticular appendages (subgenus *Tetrameres*)	7
5	Praecervical papillae present	*T. (P.) pavonis*
—	Praecervical papillae absent	6
6	Larger spicule 0.3—0.5 mm long	*T. (P.) fissispina*
—	Larger spicule 0.2—0.4 mm long	*T. (P.) crami*
7	Spicule length ratio 1 : 30; larger spicule 4 mm long, small spicule 0.13 mm long	*T. (T.) gubanovi*
—	Spicule length ratio 1 : 15; larger spicule 1.5 mm long, small spicule 0.1 mm long	*T. (T.) skrjabini*

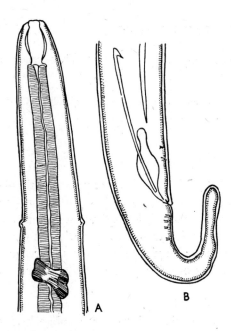

Fig. 107. *Tetrameres ardea* Shigin, 1953. A — anterior end of male (dorsoventral view); B — posterior end of male (lateral view). After Shigin (1953).

Note: It should be pointed out that the species *T. ardea* and *T. schigini* on the one hand and *T. fissispina*, *T. pavonis* and *T. crami* on the other are closely related in their morphology. A more detailed study may reveal synonymy in these two cases.

Tetrameres ardea Shigin, 1953

Fig. 107

Hosts: *Ardea cinerea*, *A. purpurea*, *Ardeola ralloides* and *Egretta alba*.
Localization: oesophagus and proventriculus.
Distribution: Europe and Asia (U.S.S.R. — Rybinsk water reservoir, Ukraine, Azerbaijan, Far East). Outside the Palaearctic Region, India.

Description:
Male: Body length 2.97 mm, maximum body width 0.072 mm. Cuticular spines absent. Buccal capsule urceolate, 0.01 mm deep and 0.01 mm in diameter. Length of muscular oesophagus 0.29 mm and of glandular oesophagus 0.53 mm. Tail 0.10 mm long. Cervical papillae 0.13 mm and nerve ring 0.17—0.18 mm from head end. Six pairs of caudal papillae present of which two pairs are preanal and four pairs postanal. Spicules 1.42—1.53 mm and 0.11 mm long. Gubernaculum present, 0.036 mm long and 0.14 mm wide. Female: Body length 2.64 mm, maximum body width 1.35 mm. Buccal capsule urceolate, 0.03 mm × 0.01 mm. Nerve ring 0.16 mm from head end. Vulva 0.22 mm from tail end.

The desctiption was based on specimens recovered from *A. cinerea* in the European part of the U.S.S.R.

References: Feyzullaev (1963 a); Shigin (1953*, 1957); Skrjabin & Sobolev (1963)*; Smetanina (1972); Smogorzhevskaya (1967).

Tetrameres coccinea (Seurat, 1914)

Fig. 108
Hosts: *Bubulcus ibis* and *Platalea leucorodia*. Reported also from flamingo.
Localization: proventriculus.

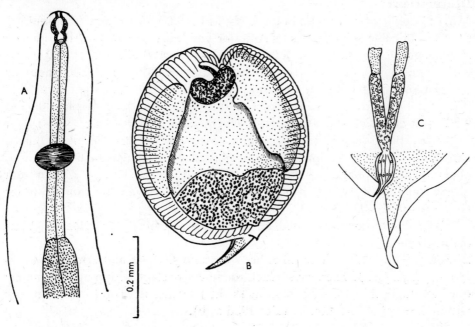

Fig. 108. *Tetrameres coccinea* (Seurat, 1914). A — anterior end of female; B — female (overall view); C — posterior end of female (detail). After Seurat (1914).

Distribution: Europe and Asia (U.S.S.R. — Volga Delta, Turkestan), Africa (Algeria, Tunis). Not reported outside the Palaearctic Region.

Description: Only females have been found. Body length 2.2 mm, body width 2.5 mm. Buccal capsule 0.02 mm long. Muscular oesophagus 0.35 mm long. Cervical papillae 0.12 and 0.18 mm from anterior end. Glandular oesophagus 1.4 mm long. Vulva 0.48 mm from anus. Eggs 0.026—0.030 × 0.015 to 0.018 mm.

The species was described from specimens recovered from *B. ibis* in Algeria as *Tropidocerca coccinea*. Travassos (1914) assigned it to the genus *Tetrameres*.

References: Dubinin & Dubinina (1940); Seurat (1914 a)*; Skrjabin (1915 a, 1923); Skrjabin & Sobolev (1963)*; Travassos (1914).

Tetrameres crami Swales, 1933

Fig. 109 A,B
Hosts: *Mergus albellus* and *M. serrator*. Parasite of different species of ducks.
Localization: proventriculus.
Distribution: Asia (U.S.S.R. — West and East Siberia, Far East). Recorded also in North America.

Description:
Male: Body length 1.9—4.58 mm, body width 0.07—0.13 mm. Longitudinal rows of cuticular spines present. Anterior end with pair of lateral cuticular appendages 0.05—0.11 mm long. Buccal capsule 0.015—0.02 × 0.005 to 0.015 mm. Length of muscular oesophagus 0.27—0.039 mm and of glandular oesophagus 0.64—0.88 mm. Cervical papillae 0.12—0.145 mm and nerve ring 0.15—0.20 mm from anterior end. Tail 0.15—0.25 mm long. Spicules 0.09 to 0.185 mm and 0.24—0.350 mm long, length ratio 1 : 1.9—3.0.
Female: Body length 1.5—3.25 mm, body width 1.2—2.2 mm. Buccal capsule 0.014 × 0.012 mm. Length of muscular oesophagus 0.16—0.25 mm, of glandular oesophagus 0.92—1.3 mm. Cervical papillae 0.21—0.24 mm from anterior end. Tail 0.11—0.16 mm long. Vulva 0.32—0.35 mm from posterior end. Eggs 0.04—0.057 × 0.026—0.034 mm.

The description was based on specimens recovered from *Anas platyrhyncha* in Canada.

Biology: The intermediate hosts are *Gammarus fasciatus* and *Hyalella knickerbockeri (Amphipoda)*. The larvae reach the infective stage within approximately one month. Within the organs of ducks the parasites develop to full maturity within 53 days of infection (Swales 1936 a, b).

References: Bélogurov (1965); Daiya (1967 a,b); Ryzhikov & Daiya (1967); Skrjabin & Sobolev (1963)*; Swales (1933*, 1936 a*, b); Tolkacheva (1967).

Tetrameres fissispina (Diesing, 1861)

Fig. 109 C—F
Hosts: *Podiceps cristatus, P. nigricollis, P. griseigena, P. ruficollis, Ardea cinerea, A. purpurea, Ardeola ralloides, Egretta alba, Nycticorax nycticorax, Botaurus stellaris, Plegadis falcinellus, Mergus merganser, M. albellus, M. serrator, Larus canus, L. argentatus, L. fuscus, L. ridibundus, Chlidonias leucoptera* and *Ch. nigra*. Reported also from ducks, pigeons, *Galliformes* and rails.
Localization: proventriculus.
Distributions: Europe (England, Sweden, Norway, Poland, Czechoslovakia, U.S.S.R.

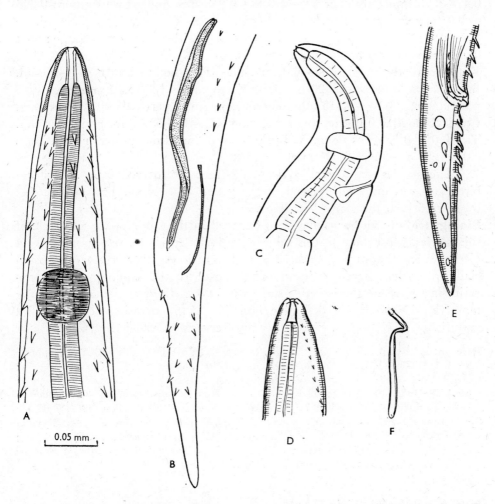

Fig. 109. *Tetrameres crami* Swales, 1933 — A, B; *T. fissispina* (Diesing, 1861) — C, D, E, F. A — anterior end of male; B — posterior end of male (ventral view); C — anterior end of female (lateral view); D — anterior end of male (dorsal view); E — posterior end of male (lateral view); F — small spicule (ventral view). A, B — After Tolkacheva (original) and C, D, E, F — After Czaplinski (1962).

north and central regions, Ukraine, the Volga Region), Asia (U.S.S.R.—Transcaucasus Region, republics of Middle Asia, Kazakhstan, West and East Siberia, Far East).

Description:
Male: Body length 3—6 mm, body width 0.14—0.2 mm. Longitudinal rows of cuticular spines present. Anterior end with lateral cuticular appendages. Buccal capsule 0.02—0.03 mm long and 0.009—0.013 mm in diameter. Cervical papillae asymmetrical, 0.14—0.16 mm from anterior end. Oesophagus 0.79—1.44 mm long with muscular part 0.32—0.45 mm and glandular 0.47—0.76 mm long. Nerve ring 0.18—0.245 mm and excretory pore 0.225 to 0.27 mm from anterior end. Long spicule 0.28—0.49 mm long, short spicule 0.09—0.2 mm long. Tail 0.21—0.265 mm long with conical appendage about 0.01 mm long and provided with spines.
Female: Body length 1.58—6.0 mm, maximum body width 0.96—3.6 mm. Buccal capsule 0.018—0.023 mm long. Oesophagus 1.2—1.49 mm long, with muscular part 0.21—0.315 mm long and glandular part 0.97—1.23 mm long. Cervical papillae 0.11 mm and nerve ring 0.13—0.185 mm from anterior end. Tail 0.07—0.175 mm long. Vulva 0.1—0.41 mm from posterior end. Eggs 0.03—0.06 × 0.025—0.035 mm.

The species was described from specimens recovered from anseriform birds and was originally placed in the genus *Tropidocerca*. Travassos (1914) transferred it to the genus *Tetrameres*.

Biology: Many authors have been engaged in the study of the life-cycle of this species. Most data on this subject are included in the paper by Garkavi (1949 b). Sandhoppers *(Amphipoda)*, and sometimes other crustaceans, especially water fleas, serve as intermediate hosts. In the sandhoppers the parasite larvae reach the infective stage within 8 to 18 days, depending on the temperature, and in the definitive host (duck) the parasites develop to full maturity within 18 days. Fishes may also serve as paratenic hosts.

References: Bakke (1970, 1972); Bakke & Baruš (1976); Belogurov (1965); Czaplinski (1962 b)*; Diesing (1861)*; Dubinin & Dubinina (1940); Dubinina (1937); Garkavi (1949 b); Ginetsinskaya (1952); Gubsky (1960); Guildal (1964, 1966, 1968); Kibakin et al. (1963); Kurashvili (1956, 1957*); Okorokov & Tkachev (1969); Paskalskaya (1968); Pemberton (1963); Ryzhikov & Daiya (1967); Ryzhikov & Kozlov (1959); Sailov (1962, 1965 b); Sergienko (1971, 1972); Shakhtakhtinskaya (1959 b); Shigin (1959); Škarda (1964); Skrjabin (1923); Skrjabin & Sobolev (1963)*; Smetanina (1972); Smogorzhevskaya (1962 a, 1964); Sonin & Larchenko (1974); Sultanov (1958, 1959 a, b); Threlfall (1965 b, 1967); Tkachev (1971); Travassos (1914); Zhatkanbaeva (1964).

Tetrameres gubanovi Shigin, 1957

Fig. 110 A,B
Hosts: *Podiceps cristatus, P. nigricollis* and *P. griseigena*.
Localization: oesophagus, proventriculus.

Distribution: Europe (Czechoslovakia, U.S.S.R. — central regions, Ukraine), Asia (Far East, Kazakhstan, Turkmenia).

Description:

Male: Body length 6.67 mm, maximum body width 0.13 mm. Longitudinal

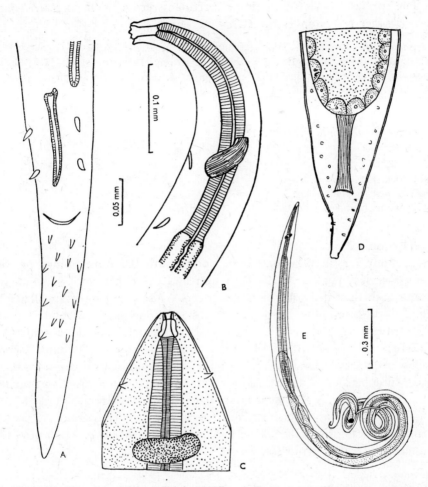

Fig. 110. *Tetrameres gubanovi* Shigin, 1957 — A, B; *T. gynaecophila* (Molin, 1858) — C, D; *T. schigini* Oshmarin, 1956 — E. A — posterior end of male (ventral view); B — anterior end; C — anterior end (dorso-ventral view); D — posterior end of male (ventral view); E — male (overall view). A, B — After Shigin (1957); C, D — After Seurat (1915); E — After Oshmarin (1956).

rows of cuticular spines present. Buccal capsule 0.03 mm deep and 0.013 mm in diameter. Length of muscular oesophagus 0.31 mm and of glandular oesophagus 0.98 mm; cervical papillae 0.26 mm and nerve ring 0.22 mm from head end. Excretory pore at level of nerve ring. Tail 0.33 mm long. Four pairs

of conical papillae plus, on margin of tail, three pairs of hardly visible pedunculate papillae. Spicules 0.13 and 4.00 mm long.
Female: Body length 6.2 mm, maximum body width 3.2 mm. Eggs 0.063 × × 0.026 mm.

The species was described from specimens recovered from *P. cristatus* in Rybinsk water reservoir (U.S.S.R.).

References: Alekseev & Smetanina (1968); Baruš & Zajíček (1967)*; Golovin (1964); Gvozdev & Kasymzhanova (1965); Ryzhikov & Kozlov (1959); Shigin (1957)*; Skrjabin & Sobolev (1963)*; Smetanina (1972); Smetanina & Alekseev (1967); Smogorzhevskaya (1967); Turemuratov (1965 a).

Tetrameres gynaecophila (Molin, 1858)

Fig. 110 C, D
Host: *Nycticorax nycticorax*.
Localization: proventriculus.
Distribution: Europe (Italy, U.S.S.R. — Volga Delta). Not known outside the Palaearctic Region.

Description:
Male: Body length 6.4—10.0 mm, body width 0.6 mm. Spines on body absent. Cervical papillae 0.12—0.13 mm and excretory pore 0.28 mm from anterior end. Buccal capsule 0.02 mm deep. Muscular oesophagus 0.265 mm long. Tail 0.12 mm long. Two pairs of postanal, one pair of adanal and seven sessile preanal papillae on the right and six on the left. Spicules absent.
Female: Body length 8.0—13.0 mm, body width 7.0—13.0 mm. Cervical papillae near nerve ring, 0.085 mm from anterior end. Buccal capsule 0.03 mm long. Muscular oesophagus 0.31 mm, glandular oesophagus 2.34 mm long. Tail 0.14 mm long. Vulva near anus. Eggs 0.05 × 0.02 mm.

The species was described from specimens recovered from *N. nycticorax* in Italy and was originally placed in the genus *Tropidocerca*. Travassos (1914) transferred it to the genus *Tetrameres*.

References: Cram (1927)*; Dubinin & Dubinina (1940); Dubinina (1937); Gubanov (1950); Molin (1858 b)*; Parona (1894); Seurat (1912, 1915); Skrjabin (1923); Skrjabin & Sobolev (1963)*; Travassos (1914).

Tetrameres pavonis Chertkova, 1953

Fig. 111
Host: *Larus ridibundus*. This species is not characteristic of fish-eating birds.
Localization: proventriculus.
Distribution: Asia (U.S.S.R. — Turkmenistan); outside the Palaearctic Region reported from *Pavo cristatus* in India (Ali 1970).

Description:

Male: Body length 4.7—5.0 mm, maximum body width 0.16—0.18 mm. Buccal capsule 0.019—0.035 mm deep, 0.015—0.017 mm wide. Oesophagus 1.12—1.4 mm long. Two lateral cuticular processes, 0.077—0.105 mm long, arise at base of lips and at the end of these the cuticular spines begin. According to the original description, the spines are irregularly distributed in the anterior part of the body but according to Ali (1970) they are arranged only in the lateral fields in a single row on each side. In the middle part of the body the spines are indistinct but become more conspicuous again posteriorly. Spines on tail distributed in four longitudinal rows. Cervical papillae 0.147 mm from anterior end. Two praecervical papillae situated anteriorly to beginning

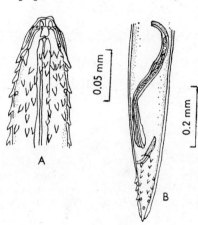

Fig. 111. *Tetrameres pavonis* Chertkova, 1953. A — anterior end; B — posterior end of male (ventral view). After Chertkova (1953).

of spines. Two unequal and dissimilar spicules present, the longer measuring 0.40—0.43 mm and the shorter 0.105—0.14 mm.

Female: Body length 5.0 mm, body width 1.33 mm. Oesophagus 1.45 mm long, buccal capsule about the same as in male. Tail 0.16 mm long, sharp and bent ventrally. Vulva 0.22 mm from tail end. Eggs 0.050—0.052 × 0.028 to 0.030 mm, with polar cap at each end.

The species was described from specimens from *Pavo cristatus* from the Zoological Gardens in Moscow (U.S.S.R.).

References: Ali (1970)*; Babaev (1970); Chertkova (1953)*; Kibakin (1965); Skrjabin & Sobolev (1963)*.

Tetrameres schigini Oshmarin, 1956

Fig. 110 E
Hosts: *Ardea cinerea*, *A. purpurea* and *Egretta alba*.
Localization: stomach.
Distribution: Asia (U.S.S.R. — Far East). Not known outside the Palaearctic Region.

Description:
Male: Body length 3.36—3.87 mm. Cuticle without spines. Buccal capsule 0.019—0.022 mm long and 0.007—0.008 mm wide. Muscular oesophagus 0.255—0.3 mm long and glandular oesophagus 0.735—0.8 mm long. Nerve ring 0.14—0.145 mm from anterior end. Four pairs of caudal papillae. Spicules 1.6 mm and 0.05—0.06 mm long.
Female: Body length 3.6—4.12 mm, maximum body width 0.85—1.44 mm. Buccal capsule urceolate, 0.020 mm long and 0.015 mm wide. Genital opening 0.23 mm from tail which is 0.115 mm long. Eggs 0.046—0.048 × 0.023 to 0.024 mm.

The species was described from specimens recovered from *A. purpurea* in Primorye Territory (U.S.S.R.).

References: Oshmarin (1956*, 1963); Skrjabin & Sobolev (1963)*; Smetanina (1972).

Tetrameres skrjabini Panova, 1926

Fig. 112 A,B
Hosts: *Stercorarius longicaudatus, S. parasiticus, Larus canus, L. argentatus, L. crassirostris, L. genei, L. ichthyaetus, L. minutus, L. ridibundus, L. schistisagus, Chlidonias hybrida, Ch. nigra, Ch. leucoptera, Gelochelidon nilotica, Sterna hirundo* and *St. sandvicensis*.
Localization: proventriculus; found also under cuticle of gizzard.
Distribution: Europe and Asia (Czechoslovakia; U.S.S.R. — northern regions, the Volga Region, North Caucasus, Ukraine, West Siberia, Far East).

Description:
Male: Body whitish in colour, with fine transverse striations, 2.50—3.12 mm long and 0.070—0.096 mm wide. Lateral alae absent. Buccal capsule 0.005 to 0.010 mm deep and 0.008—0.009 mm wide. Muscular oesophagus 0.20 to 0.26 mm long and 0.011 mm wide, glandular oesophagus 0.37—0.70 mm long and 0.034—0.038 mm wide. Nerve ring 0.11—0.14 mm and excretory pore 0.13—0.15 mm from anterior end. Cuticular spines arranged in four longitudinal rows. First spines 0.023—0.026 mm from anterior end of body, last spines 0.073—0.080 mm from tail end. Posterior end of body tapering to a short, finger-like protuberance. Cloaca 0.14—0.16 mm from tail end. Six pairs of cuticular spines: four pairs medial and 2 pairs lateral. Spicules unequal and dissimilar. Long spicule strongly sclerotized, 1.23—1.57 mm long; its proximal end 0.006—0.008 mm wide, distal end 0.004—0.007 mm wide. Short spicule feebly sclerotized, 0.077—0.115 mm long. Its proximal and distal ends 0.003—0.004 mm wide. Distal end of both spicules rounded.
Female: Body length 2.7—5.0 mm and maximum width 1.20—3.5 mm. Body globular, with protruding anterior and posterior part. Buccal capsule 0.017—0.019 mm deep and 0.011—0.019 mm wide. Oesophagus 1.16 mm

long, with muscular part 0.18 mm and glandular part 0.98 mm long. Nerve ring 0.096 mm from anterior end. Posterior part of body terminating in a finger-like tail with rounded tip. Anus 0.065—0.10 mm and vulva 0.21 mm from tail tip. Eggs relatively thick-walled, 0.051—0.060 × 0.026—0.030 mm with somewhat blunted poles, which in fully mature eggs bear filaments.

The description was based on specimens recovered from *L. canus* in North Caucasus.

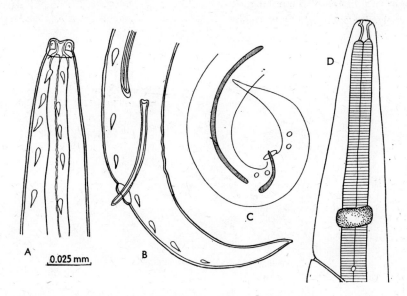

Fig. 112. *Tetrameres skrjabini* Panova, 1926 — A, B; *Microtetrameres spiralis* (Seurat, 1915) — C, D. A — anterior end; B — posterior end of male (lateral view); C — posterior end of male (lateral view); D — anterior end. A, B — After Panova (1926); C, D — After Seurat (1915).

References: Alekseev & Smetanina (1968); Bakke (1972); Belogurov, Leonov & Zueva (1968); Krivonogova (1963); Kulachkova (1950); Kulachkova & Kochetova (1964); Kurochkin & Zablotsky (1961); Leonov (1958); Mozgina (1967, 1969); Panova (1926)*; Serkova (1948); Skrjabin & Sobolev (1963)*; Smetanina (1972); Smetanina & Alekseev (1967); Smogorzhevskaya (1962 a, 1964); Vasilkova (1927*, 1930).

Genus *Microtetrameres* Travassos, 1917

Body of female spirally twisted; longitudinal fissures, which are characteristic of females of the genus *Tetrameres*, are usually absent in this genus. Body of male without spines. Left spicule very long. Parasitic in proventriculus.

Type species *M. cruzi* (Travassos, 1914).

The genus includes more than 30 species usually parasitic in *Falconiformes* and *Passeriformes*. Two species have been recorded in fish-eating birds of the Palaearctic Region.

KEY TO THE SPECIES OF THE GENUS *MICROTETRAMERES*

1 Spicules 0.145 and 2.3 mm long . . . . *M. spiralis*
— Spicules 0.2 and 1.15 mm long . . . . *M. pelecani*

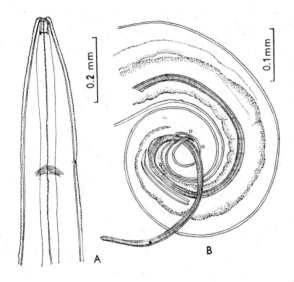

Fig. 113. *Microtetrameres pelecani* Skrjabin, 1949. A — anterior end of male (lateral view); B — posterior end of male (lateral view). After Skrjabin (1949).

Microtetrameres pelecani Skrjabin, 1949

Fig. 113
Host: *Pelecanus onocrotalus*.
Localization: proventriculus and posterior part of oesophagus.
Distribution: Asia (U.S.S.R. — Turkmenia). Not known outside the Palaearctic Region.

Description:

Male: Body length 9.5—11 mm, maximum body width 0.29 mm. Mouth opening surrounded by six papillae, two lateral and four submedian. Buccal capsule 0.06 × 0.03 mm. Nerve ring, excretory pore and cervical papillae situated 0.47 mm, 0.45 mm and 0.575 mm, respectively, from anterior end. Muscular part of oesophagus 0.76 mm long. Tail 0.25 mm long. There are one pair of preanal and two pairs of postanal papillae. Spicules 1.15 mm and 0.2 mm long.

The description was based on male specimen recovered from *P. onocrotalus*.

References: Skrjabin in Skrjabin, Shikhobalova & Sobolev (1949); Skrjabin & Sobolev (1963)*.

Microtetrameres spiralis (Seurat, 1915)

Fig. 112 C, D
Host: *Bubulcus ibis*.
Localization: proventriculus.
Distribution: Asia (Japan), Africa (Egypt, Algeria). Outside the Palaearctic Region, India and Madagascar.

Description:
Male: Body length 4.75 mm, body width 0.13 mm. Buccal capsule 0.03 mm long. Oesophagus measures $1/3$ of body length with muscular part $1/4$ of total length of oesophagus. Cervical papillae 0.3 mm and excretory pore 0.325 mm from anterior end. Four pairs of caudal papillae present, of which two pairs are preanal and two postanal. Spicules 0.145 mm and 2.3 mm long.
Female: Body length 2.5 mm, body width 2.0 mm. Buccal capsule 0.030 mm long. Muscular oesophagus 0.175 mm long. Tail 0.225 mm long. Vulva a short distance in front of anus. Eggs 0.050 × 0.030 mm.

The species was described from specimens recovered from *B. ibis* in Algeria and was originally placed in the genus *Tropidocerca*. Cram (1927) transferred it to the genus *Microtetrameres*.

References: Cram (1927)*; Myers, Kuntz & Wells (1962); Seurat (1915)*; Skrjabin & Sobolev (1963)*.

Family *Thelaziidae* Skrjabin, 1915

Mouth rounded or hexagonal, without lips. Buccal capsule (vestibule) divided or not divided into two parts, armed with teeth on inner surface. Spicules unequal, dissimilar or subequal, gubernaculum present or absent. Vulva either in anterior part of body or short distance from anus. Eggs thick- or thin-shelled, containing fully formed larva.

Parasites of orbital cavity and lachrymal duct of birds and mammals. According to Skrjabin, Sobolev & Ivashkin (1967 a, b) the family comprises six genera and more than 120 species.

The representatives of this family are divided into two subfamilies, namely, *Oxyspirurinae* and *Thelaziinae*. Fish-eating birds from the Palaearctic Region were found to be parasitized by two species of the same genus, *Thelaziella*, belonging to the subfamily *Thelaziinae*.

Genus *Thelaziella* Skrjabin, Sobolev & Ivashkin, 1967

Buccal capsule short, with thick chitinous walls which are concave and smooth from the inner side. Spicules unequal, dissimilar, gubernaculum usually present. Vulva in anterior part of body, eggs thick-shelled. Parasites of orbital region and conjunctival sac of birds.

Type species *Th. stereura* (Rudolphi, 1819).

The genus comprises about 30 species (Skrjabin, Sobolev & Ivashkin 1967, a, b). Fish eating-birds from the Palaearctic Region were reported to be parasitized by two species. One of them, *Th. nyctardeae*, was described only on the basis of female specimens.

Thelaziella aquillina (Baylis, 1934)

Fig. 114 A—F
Hosts: *Pandion haliaetus* and other falconiform birds.
Localization: conjunctival sac, orbit.
Distribution: Asia (U.S.S.R. — Far East). Found also in fish-eating birds *(Haliaetus leucoga-*

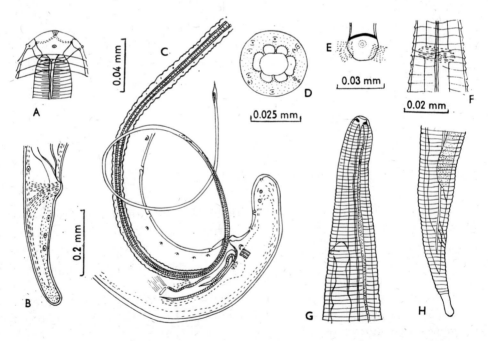

Fig. 114. *Thelaziella aquillina* (Baylis, 1934) — A, B, C, D, E, F and *Th. nyctardeae* (Dubinina, 1937) — G, H. A — anterior end of female (lateral view); B — posterior end of female (lateral view); C — posterior end of male (lateral view); D — anterior end of female (apical view); E — region of vulva; F — striation of cuticle; G — anterior end of body; H — posterior end of female. A to F — After Anderson & Díaz-Ungría (1959); G, H — After Dubinina (1937).

ster) in Australia, and in other birds in the Palaearctic Region, North and South America and Australia.

Description: Cuticular rings clearly distinct, 19 to 25 in number. Tail long tapering, ending bluntly.
Male: Body length 12—18 mm, maximum body width 0.25—0.475 mm. Buccal capsule 0.016—0.029 mm deep, 0.025—0.04 mm wide. Nerve ring 0.31—0.40 mm, cervical papillae 0.49—0.60 mm from head end. Oesophagus 0.72—1.05 mm long. Tail 0.13—0.25 mm long. At posterior end of body eight to ten pairs of large preanal papillae and three to five pairs of postanal papillae. In addition, there is an unpaired median papilla situated on anterior lip of cloaca, and a pair of small terminal phasmids. Right spicule 0.17—0.23 mm, left one 1.7—2.28 mm long. Gubernaculum present.
Female: Body length 16—19 mm, maximum body width 0.33—0.47 mm. Buccal capsule 0.02—0.039 mm deep, 0.032—0.040 mm wide. Nerve ring 0.34—0.04 mm, cervical papillae 0.51—0.60 mm and vulva 0.66—0.79 mm from head end. Oesophagus 0.85—1.20 mm and tail 0.27—0.38 mm long.

The description was based on nematodes from *H. leucogaster* from Australia. The species was placed in the genus *Thelaziella* by Skrjabin, Sobolev & Ivashkin (1967 a).
Notes: Anderson & Díaz-Ungría (1959) revised the genus *Thelazia* and synonymized *Th. chui* Hsü, 1935, *Th. spizaeti* Strachan, 1957 and *Th. platyptera* Hwang & Wehr, 1957 with the species *Th. aquillina*. We agree with the opinion of these authors and additionally synonymize *Th. skrjabilina* Timofeeva, 1964, described from *Pernis ptilorhynchus* from the Far East, with *Th. aquillina*.

References: Anderson & Díaz-Ungría (1959); Baylis (1934)*; Skrjabin, Sobolev & Ivashkin (1967 a)*; Smetanina (1972).

Thelaziella nyctardeae (Dubinina, 1937)

Fig. 114 G, H
Hosts: *Ardea cinerea* and *Nycticorax nycticorax*.
Localization: under eyelids.
Distribution: Europe (U.S.S.R. — Volga Delta), Asia (U.S.S.R. — Far East).

Description: Only female known. Body length 18—23 mm. Cuticle with transverse striations, which are interrupted on lateral fields in middle part of body. Maximum width of body 0.60—0.73 mm, at head end 0.15 mm, at level of anus 0.11—0.14 mm. Buccal capsule 0.037 mm deep, 0.056 mm wide. Vulva 0.85—1.1 mm from head end. Posterior end of body conoid, ending in a small swelling. Tail 0.40—0.45 mm long.

The description was based on specimens recovered from *N. nycticorax* from the Volga Delta and named *Thelazia nyctardeae*. The species was transferred to the genus *Thelaziella* by Skrjabin, Sobolev & Ivashkin (1967 a).

References: Dubinin & Dubinina (1940); Dubinina (1937)*; Krastin (1957)*; Oshmarin (1963); Skrjabin, Sobolev & Ivashkin (1967 a)*.

SUBORDER *FILARIATA* SKRJABIN, 1915

Filiform nematodes, usually with thin slender cuticle. Head end without lips or with indistinct lips. Some forms possess peculiar cuticular formations. Oesophagus relatively short, usually divided into two parts. Spicules equal or unequal, similar or dissimilar. Vulva in anterior part of body. Tail end usually bluntly rounded. Male may possess caudal alae. Most species viviparous.

All representatives of this suborder require an intermediate host during development. For the viviparous forms, in which the larvae which emerge from the females are not expelled into the external environment, the intermediate hosts are blood-sucking arthropods. For the oviparous forms insects act as intermediaries.

The overwhelming majority of species of this suborder parasitize birds and mammals. The fauna of filariids parasitizing birds all over the world includes about 250 species. Fifteen species belonging to eleven genera and four families have been recorded from fish-eating birds of the Palaearctic Region.

KEY TO THE FAMILIES OF THE SUBORDER *FILARIATA*

1 Oviparous; eggs thick-shelled, containing differentiated larva 2
— Viviparous; larvae of microfilariid type, with or without sheath 3
2 Spicules equal or subequal, similar; head end without chitinous armament . . . . .
. . . . *Aproctidae*
— Spicules distinctly unequal; head end with peculiar chitinous armament in the form of tridents or epaulette-like structures with projecting teeth . . . . . . *Diplotriaenidae*
3 Spicules equal or subequal, caudal end of male without alae . . . *Splendidofilariidae*
— Spicules unequal; if subequal, then the caudal end of male with alae . . . .
. . . . *Oswaldofilariidae*

Family *Aproctidae* Skrjabin & Shikhobalova, 1945

Oviparous. Eggs thick-shelled, containing fully formed larva without spines at posterior end of body. Vulva in oesophageal region. Spicules equal or subequal, similar, tapering gradually from proximal to distal end. Parasites of birds.

Representatives of this family are divided into three subfamilies *(Aproctinae, Squamofilariinae* and *Tetracheilonematinae)*.

According to Sonin (1966) this family includes seven genera and 50 species. Fish-eating birds of the Palaearctic Region are parasitized by only one species of the subfamily *Aproctinae*.

Genus *Aprocta* Linstow, 1883

Mouth simple, leading into small slightly chitinized vestibule. Four pairs of submedian papillae and one pair of lateral amphids. Oesophagus short, not distinctly divided into two parts. Vulva in anterior part of body. Tail of male spirally twisted. Spicules similar, short, equal or subequal. Parasitic in orbital, nasal or body cavity of birds.

Type species *A. cylindrica* Linstow, 1883, parasitic in *Passeriformes*.

Other 30 species of this genus are known to parasitize birds of various orders. In fish-eating birds of the Palaearctic Region only one species has been recorded.

Aprocta turgida Stossich, 1902

Fig. 115
Hosts: *Larus argentatus, L. genei, L. ichthyaetus, L. ridibundus, Larus* sp., *Chlidonias leucoptera, Gelochelidon nilotica* and *Sterna hirundo*.
Localization: nasal and orbital cavities.
Distribution: Europe (Spain, France, Italy, U.S.S.R. — Ukraine, Volga Region, North Caucasus), Asia (U.S.S.R. — West and East Siberia, Far East, Azerbaijan, Kazakhstan, Turkmenia; Mongolian People's Republic), Africa (Morocco). Outside the Palaearctic Region in gulls in Canada.

Description: Body wide, rounded at both extremities, with smooth cuticle. Mouth without lips, surrounded by eight papillae (four external, large, lateromedian and four small medio-median). Oesophagus simple and short. Nerve ring near the head end. Excretory pore behind nerve ring.
Male: Body length 18—27 mm, body width 0.8—1.10 mm. Oesophagus 0.78—0.88 mm long. Tail short, 0.19—0.23 mm, rounded, with symmetrically situated subterminal phasmids. Spicules equal or subequal, 0.27—0.40 mm long. One unpaired papilliform process on posterior lip of cloaca, but other papillae not found.
Female: Body length 30—38 mm, body width 1.0—1.2 mm. Oesophagus 0.75 to 0.93 mm long. Nerve ring 0.15—0.20 mm from head end; vulva 0.7—1.03 mm from head end. Anus opens ventrally at end of tail. Eggs thick-

shelled, 0.047—0.058 × 0.022—0.034 mm, containing fully formed larvae.

This species was described from gulls from Italy.

Notes: *A. turgida* resembles the species *A. matronensis* the definitive hosts of which are birds of the family *Corvidae*. It is possible that they are synonyms (Sonin 1966).

In 1958, Leonov described the species *A. milinskii*, also parasitizing gulls. In his opinion this species differs fundamentally from *A. turgida* in the presence

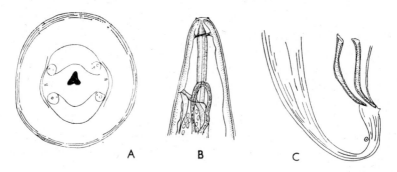

Fig. 115. *Aprocta turgida* Stossich, 1902. A — anterior end (apical view); B — anterior end of female (lateral view); C — posterior end of male (lateral view). After Chabaud & Choquet (1955).

of subequal spicules but Sonin (1966) placed *A. milinskii* in synonymy with *A. turgida*.

References: Babaev (1970); Bakke (1972); Chabaud & Choquet (1955)*; Danzan (1964); Dollfus (1962); Hartwich (1966); Kurochkin & Zablotsky (1961); Leonov (1958*, 1960 b); López-Neyra (1956); Sailov (1970); Sergeeva (1968 a, 1969); Shigin (1961); Skrjabin (1916 b*, 1917, 1923); Skrjabin & Shikhobalova (1948); Smetanina (1972); Smogorzhevskaya (1964); Smogorzhevskaya, Kornyushin, Iskova & Eminov (1964); Sonin (1959*; 1966*); Sonin & Larchenko (1974); Stossich (1902); Threlfall (1966 c); Zhatkanbaeva (1971).

Family *Splendidofilariidae* Sonin, 1962

Viviparous. Microfilariae with or without sheath. Spicules equal or subequal, similar. Proximal part (handle) of spicule widened, between it and the body of spicule there is a characteristic constriction. Parasites of reptiles and birds.

Representatives of this family are divided into two subfamilies *(Eufilariinae* and *Splendidofilariinae)*. According to Sonin (1966) this family comprises 18 genera and about 80 species. In fish-eating birds of the Palaearctic Region three species belonging to the two subfamilies have been reported.

KEY TO THE GENERA OF THE FAMILY *SPLENDIDOFILARIIDAE*

1 Anus subterminal in both males and females, caudal end widened, oesophagus narrow, not divided . . . . *Eufilaria*
— Anus some distance from end of tail, tail finger-shaped 2
2 Oesophagus not divided, massive, club-shaped, parasites of inner parts of eye . . . . *Skrjabinocta*
— Oesophagus distinctly divided into muscular and glandular parts, caudal end of male short, shorter (or somewhat longer) than spicules; parasites of internal organs . . . . *Parornithofilaria*

Genus *Eufilaria* Seurat, 1921

Mouth without lips, oesophagus short, narrow and transparent, not divided. Body straight, with rounded extremities. Cuticle smooth or with transverse striations. Spicules subequal, short, needle-like. Anus subterminal in both sexes.

Type species *E. sergenti* Seurat, 1921, parasitic in *Passeriformes*. Six species of this genus are known.

Eufilaria lari Yamaguti, 1935

Fig. 116

Hosts: *Larus argentatus, L. canus, L. genei, L. glaucescens, L. marinus, L. ridibundus, Rissa tridactyla, Xema sabini, Sterna hirundo, Chlidonias leucoptera, Cepphus grylle, Uria lomvia* and *U. aalge*. Recorded also in *Charadriiformes*.
Localization: subcutaneous tissue in region of oesophagus.
Distribution: Europe (U.S.S.R. — White Sea coast, Volga Region) and Asia (U.S.S.R. — West and East Siberia, Far East; Japan).

Description: Body cylindrical, tapering somewhat at both extremities. Cuticle with fine transverse striations. Small cuticular thickenings in form of two lip-like structures on both sides of mouth opening. Two pairs of indistinct papillae on head end (laterally).
Male: Body length 6.6—7.5 mm, body width 0.44—0.54 mm. Oesophagus 0.38—0.54 mm long, 0.043—0.049 mm wide. Spicules 0.155—0.170 mm long.
Female: Body length 9.8—11.7 mm, body width 0.35—0.50 mm. Oesophagus fine, 0.30—0.33 mm long, posterior end sligthly widened. Terminal portion of gut atrophied. Vulva 0.40—0.65 mm from head end. Length of microfilariae in uterus 0.084 mm. Microfilariae sheathed.

The description of the species was based on female nematodes recovered from *L. canus* in Japan. The males were first described by Belopolskaya (1952), who found *E. lari* in some species of gulls and guillemots on the coasts of the White Sea.

References: Bakke (1972); Belogurov & Smetanina (1965); Belogurov, Leonov & Zueva (1968); Belopolskaya (1952)*; Krivonogova (1963); Kulachkova & Kochetova (1964); Kurochkin & Zablotsky (1961); Leonov, Belogurov, Kazachenko & Zueva (1965); Leonov & Shvetsova (1970); Serkova (1948); Shigin (1961); Skrjabin & Shikhobalova (1948)*; Smetanina (1965, 1972); Sonin (1963, 1966)*; Tsimbalyuk & Belogurov (1964); Yamaguti (1935)*.

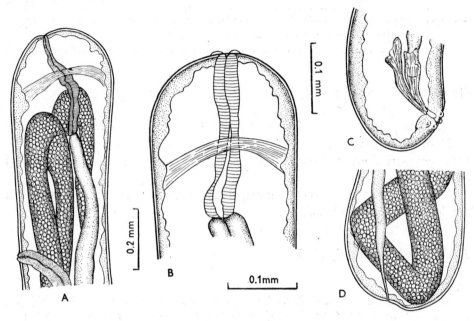

Fig. 116. *Eufilaria lari* Yamaguti, 1935. A — anterior end of female; B — anterior end of male; C — posterior end of male (lateral view); D — posterior end of female (lateral view). After Sonin (1963).

Genus *Parornithofilaria* Sonin, 1965

Cuticle without bosses. Oesophagus divided into two parts. Vulva in oesophageal region, somewhat protruding. Spicules equal or subequal. Tail of male nearly the same length as spicules, not markedly exceeding width of body at level of cloaca.

Type species *P. stantchinskyi* (Gilbert, 1930), parasitic in birds of the family *Corvidae*. Ten species of the genus are known, parasitizing birds of various orders.

Parornithofilaria shaldybini (Gubanov, 1954)

Host: *Phalacrocorax urile*.
Localization: kidneys.
Distribution: Asia (U.S.S.R. — Kuril Islands).

Description: Body filiform, slightly attenuated towards anterior and posterior ends. Cuticle smooth. Mouth surrounded by six small papillae and two pairs of larger papillae. Oesophagus divided into two parts. Nerve ring in posterior third of muscular portion of oesophagus. Excretory opening behind nerve ring.
Male: Body length 17.34 mm, maximum body width 0.16 mm. Muscular portion of oesophagus 0.18 mm long, glandular portion 1.1 mm. Nerve ring 0.16 mm from head end. Tail end spirally twisted. Three pairs of preanal papillae present. Cloaca 0.084 mm from posterior end. Spicules short, divided into two parts — the distal grooved chitinized and cup-shaped, the proximal membranous. Distal part of larger spicule 0.084 mm long, proximal 0.024 mm long.
Female: Body length 38.5 mm, maximum body width 0.26 mm. Length of muscular oesophagus 0.18 mm and glandular 1.15 mm. Nerve ring 0.18 mm from head end. Vulva in anterior part of body, 0.60 mm from head end. Uteri parallel. Eggs oval, 0.020 × 0.032 mm. Mature eggs contain larvae.

The first description of this parasite (based on Gubanov's manuscript) was published by Sonin in 1963. Gubanov placed this species in the genus *Chandlerella;* Anderson & Chabaud (1959) assigned it to *Splendidofilaria* and Sonin (1965, 1966) to *Parornithofilaria*. (No drawings of this species have been published.)

References: Anderson & Chabaud (1959); Gubanov (1954); Sonin (1963*, 1965, 1966*).

Genus *Skrjabinocta* Chertkova, 1946

Mouth simple, without lips. Head with four pairs of submedian papillae arranged in two circles. Oesophagus simple, cylindrical. Spicules longer than the distance from cloaca to tail tip, subequal, similar, tapering at distal ends. Anus obliterated in females. Vulva in region of oesophagus. Viviparous. Parasitic in inner parts of eyes of birds.

Type species *S. petrowi* Chertkova, 1946, parasitic in *Columbiformes*. Other species, *S. ciconiae* Morozov, 1958, parasitic in *Ciconiiformes*.

Skrjabinocta ciconiae Morozov, 1958

Host: *Ciconia ciconia*.
Location: inner parts of eye.
Distribution: Europe (U.S.S.R. — Byelorussia).

Description: Male unknown. Body length 22.2 mm, maximum body width 1.027 mm. Oesophagus 0.837 mm long, maximum width 0.07 mm. Nerve ring 0.426 mm from head end, vulva 0.758 mm. Viviparous. Length of microfilariae in uteri about 0.130 mm, width 0.009—0.010 mm. (No drawings of this species have been published.)

References: Morozov (1958)*; Sonin (1965, 1966*).

Family *Diplotriaenidae* Anderson, 1958

Anterior end of body with cuticular formations in form of lateral epaulettes, tridents or two circumoral odontoid processes. Oesophagus divided into two parts. Spicules often unequal, dissimilar. Oviparous. Eggs thick-shelled, containing fully formed larva. First-stage larvae with several rows of spines, rounded anterior and posterior ends.

Parasites of hypodermic tissue of reptiles and birds, air sacs and body cavity of birds.

Representatives of this family are divided into two subfamilies *(Diplotriaeninae* and *Dicheilonematinae)*, which include (according to Sonin, 1968) 11 genera and more than 100 species. Two species belonging to the subfamily *Dicheilonematinae* occur in fish-eating birds of the Palaearctic Region.

KEY TO THE GENERA OF THE FAMILY *DIPLOTRIAENIDAE*

1 Epaulette-like structures indistinct, four pairs of double cephalic papillae in external circle . . . . *Monopetalonema*
— Eight cephalic papillae arranged in two circles, epaulette-like structures clearly distinct . . . . *Dicheilonema*

Genus *Dicheilonema* Diesing, 1861

Large nematodes, females significantly larger than males. Mouth oval, with pair of chitinous tooth-like elevations on each side. Trilobate epaulette-like structures lie close to the posterior margin of each „tooth". There is a pair of lateral amphids and eight submedian papillae. Papillae of inner circle situated on epaulette-like structures. Tail of male with alae supported by pedunculate papillae. Spicules unequal and dissimilar. Small spicule bent. Vulva near head end. Eggs thick-shelled, containing fully formed larva at deposition.

Two of the seven species belonging to this genus, *D. acutum* (Diesing, 1851) and *D. bilabiatum* (Diesing, 1851), parazitizing fish-eating birds, were described very incompletely. We regard them as *species inquirendae* and do not include them in our account.

Type species — *D. ciconiae* (Schrank, 1788) parasitizing *Ciconiiformes* in the Palaearctic Region.

Dicheilonema ciconiae (Schrank, 1788)

Fig. 117
Hosts: *Ciconia ciconia, C. nigra, Ardea cinerea, A. purpurea, Egretta alba, E. garzetta, Nycticorax nycticorax* and *Phalacrocorax carbo* (?).
Localization: air sacs, trachea, bronchi and under mucous membrane of oral cavity.
Distribution: Europe (France, Italy, F.R.G., G.D.R., Poland, Czechoslovakia, U.S.S.R. — Volga Region), Asia (U.S.S.R. — Transcaucasus, republics of Middle Asia).

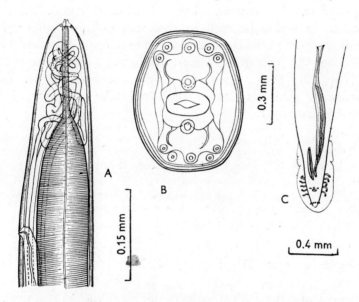

Fig. 117. *Dicheilonema ciconiae* (Schrank, 1788). A — anterior end of female (lateral view); B — anterior end (apical view); C — posterior end of male (ventral view). After Macko (1964.)

Description: Body cylindrical, of nearly the same thickness throughout, tapering somewhat towards tail end. Cuticle with fine transverse striations. A massive chitinized tooth in the form of a flat truncated cone is present on each side of the mouth. Laterally, at base of each tooth are characteristic cuticular epaulette-like structures forming oval plates with three small irregular outgrowths on their outer margin. On each of these structures is a small papilla; larger papillae of outer ring situated in incisions between outgrowths.
Male: Body length 75—138 mm. Oesophagus 6.80—24.00 mm long. Posterior portion of body provided with alae supported by four pairs of preanal

and one pair of postanal pedunculate papillae. One pair of pedunculate adanal papillae present plus eight additional small sessile papillae, one situated above cloaca, the other symmetrically at end of tail. Tail 0.16—0.18 mm long. Length of longer spicule 0.97—1.25 mm, shorter one 0.31—0.35 mm. Spicules provided with small alae.

Female: Body length 454—735.5 mm. Oesophagus 17.47—46.02 mm long. Vulva opens 2.20—3.70 mm from head end. Vagina divided into four or five uteri. Eggs 0.022—0.024 × 0.045—0.054 mm, containing differentiated larvae when mature.

This species was first described as *Filaria ciconiae* from European storks.
Notes: Many authors have described nematodes parasitic in the body cavities of storks under different names. Skrjabin (1915 a) established the genus *Contortospiculum*, in which he included parasites of storks under the name *C. ciconiae* (Gmelin, 1917). He considered the species *F. ardeae nigrae* and *F. labilata* to be synonyms of *C. ciconiae*. Witenberg (1925) placed *C. ciconiae* (Gmelin, 1791) in synonymy with *C. ciconiae* (Schrank, 1788). The genus *Contortospiculum* Skrjabin, 1915 is now considered to be a synonym of the genus *Dicheilonema* Diesing, 1861.

References: Blanchard (1848); Condorelli (1895); Creplin (1825); Diesing (1851, 1861); Dubinin & Dubinina (1940); Dubinina & Serkova (1951); Dujardin (1845); Feyzullaev (1963 a); Gmelin (1790—1791); Gundlach (1969); Gurlt (1845); Macko (1963 b, 1964 b, c*); Molin (1858 a); Mühling (1898); Nathusius (1837); Nikolskaya (1939); Parona (1894); Rudolphi (1819); Schneider (1866); Schrank (1788)*; Siebold (1838); Skrjabin (1915 a, 1923); Skrjabin & Shikhobalova (1948)*; Sonin (1968)*; Stossich (1890 a, 1897); Turemuratov (1962 a); Witenberg (1925)*; Zeder (1803).

Genus *Monopetalonema* Diesing, 1861

Head and rounded, with two pairs of double submedian papillae and one pair of amphids situated laterally. Epaulette-like structures not distinct. Mouth opening with two chitinous lateral tooth-like elevations. Tail end of male rounded and provided with alae which are supported by clearly distinguishable pedunculate papillae. Vulva near head end. Eggs with fully formed larvae. Parasites of body cavity and joint cavities of birds.

Type species — *M. alcedinis* (Rudolphi, 1819). Five species of this genus are known, four of them parasitic in kingfishers.

Monopetalonema alcedinis (Rudolphi, 1819)

Fig. 118
Host: *Ceryle* (= *Alcedo*) *torquata* and outside the Palaearctic Region, in other species of kingfishers.

Localization: body cavity, air sacs.
Distribution: Africa (A.R.E.), outside the Palaearctic Region, North and South America.

Description: Cuticle thick, with flat thickenings. Lateral alae absent or present.

Male: Body length 63—85 mm, maximum body width 0.65—0.67 mm. Tail 0.10 mm long. Caudal alae fuse behind the end of body and in front of opening of cloaca. Four pairs of pedunculate preanal papillae, situated ventro-laterally, one unpaired median papilla immediately above the opening of cloaca, one

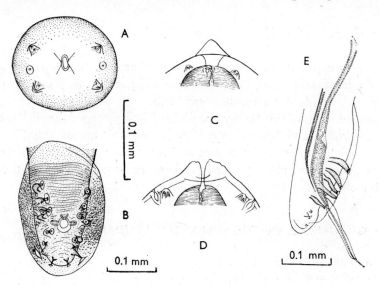

Fig. 118. *Monopetalonema alcedinis* (Rudolphi, 1819). A — anterior end (apical view); B — posterior end of male (ventral view); C — anterior end (lateral view); D — anterior end (dorsal view); E — posterior end of male (lateral view). After Anderson (1959).

pair of adanal papillae and two pairs of pedunculate postanal papillae present. In addition there are two pairs of phasmid-like formations on ventro-lateral surface of tail. Right spicule narrow, pin-like, 0.25 mm long. Left spicule 0.57 mm long, divided into tubular manubrium (about 0.20 mm long) and widened portion running into narrow filiform process provided with alae.

Female: Body length 243—355 mm, maximum body width 0.89—1.02 mm. Vulva 0.70—0.75 mm from anterior end. Tail short, blunt, 0.14—0.15 mm long. Eggs 0.042—0.050 × 0.032—0.040 mm, oval, thick-shelled, containing larvae.

This species was first described as *Filaria alcedinis* from *Alcedo* spp. in South America.

Notes: Diesing (1851) regarded this species as a synonym of *F. physalura* Bremser in Diesing (1851) and placed it in the genus *Monopetalonema* created by him. Freitas & Lent (1936) were justified in considering this synonymy to be ill-founded and renewed the name *M. alcedinis*, placing the previously described species of filariae in synonymy with it. Boulenger (1928) recovered a nematode which he ascribed to this species from the body cavity of *C. torquata* in Egypt.

References: Anderson (1959)*; Boulenger (1928)*; Diesing (1851, 1861); Freitas & Lent (1936); Lent & Freitas (1948); Rudolphi (1819)*; Skrjabin & Shikhobalova (1948)*; Sonin (1968)*.

Family *Oswaldofilariidae* Sonin, 1968

Spicules unequal and dissimilar; circumoral armament absent. Viviparous. Larvae of microfilariid type. Parasites of hypodermic tissue, blood system and body cavity of amphibians, reptiles and birds.

Representatives of this family are divided into three subfamilies *(Oswaldofilariinae, Icosiellinae* and *Lemdaninae)*. Parasites of birds are included in the last subfamily which comprises 18 genera and about 60 species (Sonin 1968). Of these, five genera and nine species have been recorded in birds of the Palaearctic Region.

KEY TO THE GENERA OF THE SUBFAMILY *LEMDANINAE*

1 Spicules subequal . . . . *Pelecitus*
— Spicules dissimilar 2
2 Epaulette-like structures present, but chitinous teeth absent; spicules very distinctly unequal (left spicule more than 30 times as long as right one) 3
— Epaulette-like structures absent, or rudimentary; left spicule not more than three times as long as right one 4
3 Caudal end of male with alae; right spicule with chitinous alae, 1/60 to 1/90 length of left one . . . . *Heterospiculum*
— Caudal end of male and right spicule without alae, left spicule about 35 times as long as right one . . . . *Heterospiculoides*
4 Spicules distinctly unequal, right spicule 1/4 to 1/8 length of left one . . . *Lemdana*
— Spicules unequal, left spicule not more than three times as long as right one . . . . .
. . . . *Paronchocerca*

Genus *Lemdana* Seurat, 1917

Filiform nematodes. Mouth simple, without lips, surrounded by cephalic papillae. Cuticle thick, smooth. Lateral and caudal alae absent. Oesophagus divided into two parts. Tail end short. Spicules unequal. Vulva in region of oesophagus. Viviparous. Microfilaria without sheath.

Type species *L. marthae* Seurat, 1917, parasitic in *Galliformes*.

Five species of this genus are known, three of them having been recorded in fish-eating birds of the Palaearctic Region. Two of these three species were described only from the females and no Key is, therefore, given.

Lemdana behningi Levashov, 1929

Fig. 119 A,B
Host: *Phalacrocorax carbo*.
Localization: wall of intestine.
Distribution: Europe (U.S.S.R. — Astrakhan Region).

Description:
Male: Body length 26 mm, maximum body width 0.39 mm. Oesophagus 16.3 mm long. Left spicule 2.17 mm long with proximal end wavy; right spicule 0.33 mm long. Four pairs of preanal papillae, two pairs of postanal papillae, of which last pair is indistinct. Cloaca 0.05 mm from posterior end.
Female: Body length 69—94.5 mm, maximum body width 0.84—0.92 mm. Vulva 0.76—0.89 mm from head end. Eggs 0.053×0.030 mm, containing differentiated larvae.

This species was described from nematodes recovered from the intestinal wall of cormorant.

Notes: This species differs from typical representatives of the genus *Lemdana* in the presence of thick-shelled eggs containing formed larvae. Having regard to this fact, as well as to the morphological peculiarities of the parasites (according to Levashov's description and figures), Sonin (1968) assumed that Levashov had examined specimens belonging to the genus *Hamatospiculum* which are not normal parasites of cormorants. This may explain their unusual location. Unfortunately, Levashov's original material was lost.

References: Levashov (1929)*; Skrjabin & Shikhobalova (1948)*; Sonin (1968)*.

Lemdana limboonkengi Hoeppli & Hsü, 1929

Fig. 119 C,D
Host: *Ardeola bacchus*.
Localization: external surface of pericardium.
Distribution: Asia (People's Republic of China).

Description: Male unknown. Body length 32—33 mm. Cuticle smooth, anterior end rounded, without lips. Very small submedian cephalic papillae present, their number not given. Nerve ring at level of mid-length of muscular portion of oesophagus. Length of oesophagus 5.0—6.3 mm and of its muscular

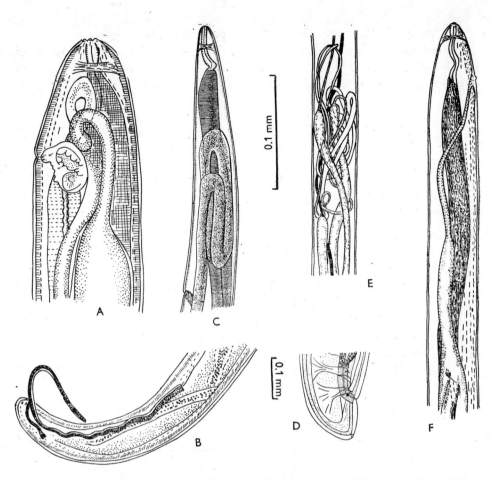

Fig. 119. *Lemdana behningi* Levashov, 1929 — A, B; *L. limboonkengi* Hoeppli & Hsü, 1929 — C, D; *L. lomonti* Desportes, 1947 — E, F. A — anterior end of female (lateral view); B — posterior end of male (lateral view); C — anterior end of female (lateral view); D — posterior end of female (lateral view); E — middle part of body; F — anterior end of female (lateral view). A, B — After Levashov (1929); C, D — After Hoeppli & Hsü (1929); E, F — After Desportes (1947).

portion 0.4 mm. Vulva 2.3 mm from head end. Vagina with thick muscular walls. Tail end bears three small subterminal papillae. None of the females examined contained mature eggs or microfilariae.

The description of this species was based on three sexually immature females and the position of these parasites in the systematics of filariae cannot be determined at present. Therefore, this species remains in the genus *Lemdana* for the time being.

References: Hoeppli & Hsü (1929)*; Skrjabin & Shikhobalova (1948)*; Sonin (1968)*.

Lemdana lomonti Desportes, 1947

Fig. 119 E,F
Host: *Ardea purpurea.*
Localization: femur musculature.
Distribution: Europe (France).

Description: Only one female known. Body length 37 mm, maximum body width 0.50 mm. Cuticle thin, with dense transverse striations. Mouth simple, without lips, surrounded by thin, oval chitinous ring. Four small submedian papillae and one pair of lateral amphids. Oesophagus divided into two parts. Anterior portion of oesophagus 0.55 mm long, glandular portion 6.88 mm long. Intestine narrower than oesophagus, rectum thin and short, anus obliterated. Vulva 0.56 mm from anterior end. Muscular portion of vagina 1.80 mm long, running into an unpaired thin-walled portion of vagina, 1.45 mm long, then dividing into three uteri passing in parallel nearly to posterior end of body.

This species was described from a nematode from a purple heron from France.

Notes: Desportes (1947) reports that this species cannot be with certainty assigned to any of the known filariid genera. He explains the presence of three uteri and three ovaries in *L. lomonti* as teratological changes in the morphology of the parasite. In Sonin's (1968) opinion, *L. lomonti*, in respect of many morphological features and localization, conforms with and is most probably identical to *Heterospiculum sobolevi*. However, a more detailed comparison of their morphology is necessary for confirmation.

References: Desportes (1947)*; Sonin (1968)*.

Genus *Heterospiculoides* Shigin, 1957

Nematodes of medium size, tapering towards both ends. Cuticle with fine transverse striations. Mouth opening spherical, surrounded by cuticular ring with "epaulettes" on both sides. Chitinous tooth-like structures absent. Two pairs of large papillae situated on lateral lobes of "epaulettes" and a pair of smaller papillae on top of medial lobes; two papillae located dorsally and ventrally to the margin of "epaulettes" and, finally, two large papillae, most probably amphids, laterally behind the borders of "epaulettes". Caudal alae absent. Spicules without alar thickening. Parasites of articular capsules of herons.

Type and only species *H. skrjabini* Shigin, 1957.

Heterospiculoides skrjabini Shigin, 1957

Fig. 120
Host: *Ardea cinerea*.
Localization: articular capsule of calcaneal joint.
Distribution: Europe (U.S.S.R. — Volga Region).

Description: Female unknown. Length of male body 24 mm, maximum width 0.394 mm. Length of muscular part of oesophagus 0.408 mm and glandular part 6.67 mm. Nerve ring 0.079 mm from head end. Tail short, 0.030 mm long. 14 large caudal pedunculate papillae, eight of them preanal (three in right and five in left rows) and the rest arranged in two semicircles behind cloaca. Large spicule 3.225 mm long, filiform, with proximal end slightly widened and distal end tapered. Short spicule 0.084 mm long, thicker-walled, proximal end widened and separated from spicule body by distinct isthmus, longitudinal groove-like depression on ventral surface along which the long spicule slides.

This species was described from nematodes from herons in the Volga Region.

References: Shigin (1957)*; Sonin (1968)*.

Genus *Heterospiculum* Shigin, 1951

Nematodes of medium size. Around the mouth opening is an annular formation with eight papillae: the lateral sides of this formation are extended into paired epaulette-like structures, each of which bears a papilla. Male tail

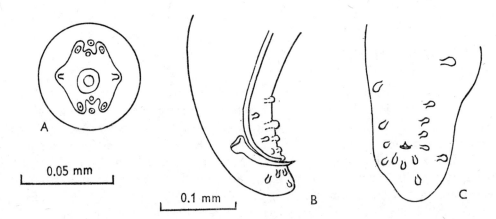

Fig. 120. *Heterospiculoides skrjabini* Shigin, 1957. A — anterior end (apical view); B — posterior end of male (lateral view); C — posterior end of male (ventral view). After Shigin (1957).

alate, with two, large rounded processes at end; caudal alae supported by pair of very large papillae. Preanal papillae pedunculate, globular. Postanal papillae not found. Spicules unequal, dissimilar. Gubernaculum absent. Vulva at level of junction of muscular with glandular portion of oesophagus. Viviparous.

Type and only species *H. sobolevi* Shigin, 1951.

Heterospiculum sobolevi Shigin, 1951

Fig. 121
Host: *Ardea cinerea*.
Localization: intermuscular connective tissue of tibia.
Distribution: Europe and Asia (U.S.S.R. — Ukraine, Volga Region, Transcaucasus Region).

Description: Cuticle with fine transverse striations, covered with numerous small verrucous formations.
Male: Body length 18—49 mm, maximum body width 0.332—0.495 mm. Length of muscular part of oesophagus 0.726 mm and glandular part 4.65 to 7.16 mm. Nerve ring 0.118—0.169 mm from head end. Tail end bears distinct alae not longer than 0.264 mm, maximum width 0.032 mm. Number of caudal papillae variable, ranging from 11 to 14. Preanal papillae, gradually enlarging anteriorly from cloaca. Spicules unequal, dissimilar. Short spicule 0.058 to

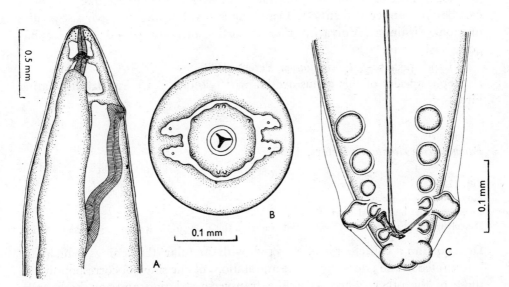

Fig. 121. *Heterospiculum sobolevi* Shigin, 1951. A — anterior end of female (lateral view); B — anterior end (apical view); C — posterior end of male (ventral view). After Shigin (1951).

0.073 mm long, strongly chitinized, its distal end bearing a small hook, proximal end funnel-shaped, widened. Body of short spicule slightly bent, bearing paired alae on considerable part of ventral surface. Longer spicule 4.785 to 7.92 mm long, cylindrical, poorly chitinized, without alae, proximal end spoon-shaped, widened, distal end somewhat attenuated. Tail very short, bluntly rounded, with a pair of large rounded lateral outgrowths. Opening of cloaca 0.032—0.044 mm from posterior end.

Female: Body length 110 mm, maximum body width 0.792 mm. Vulva in anterior portion of body, 0.528—0.825 mm from head end. Anus situated on a small terminal process. Microfilaria 0.159—0.200 mm long, 0.004—0.005 mm wide, with rounded anterior end and pointed posterior end.

This species was described from nematodes from herons in the Volga Region.

References: Feyzullaev (1963 a, 1965); Leonov (1960 a); Mozgina (1969); Shigin (1951*, 1957); Smogorzhevskaya (1964); Sonin (1968)*; Vaidova (1970).

Genus *Paronchocerca* Peters, 1936

Body filiform, gradually tapering towards both ends. Cuticle with fine transverse striations, bearing (in most species) prominent annular thickenings. Mouth simple, without lips, surrounded by four submedian papillae and one pair of lateral amphids. Some species possess cuticular thickenings in quadrate form on head end, sometimes rudiments of trilobate epaulette-like structures can be seen under the cuticle. Oesophagus divided into two parts. Spicules unequal, dissimilar. Vulva in anterior portion of body, pre- or postoesophageal. Viviparous.

Type species — *P. ciconiarum* Peters, 1936.

Four species of this genus are known; two of them parasitize fish-eating birds.

Paronchocerca ciconiarum Peters, 1936

Fig. 122
Host: *Ciconia ciconia*.
Localization: heart and blood vessels.
Distribution: Europe (Czechoslovakia). Outside the Palaearctic Region, in storks of Africa.

Description: Cuticle thick, covered with annular thickenings which are interrupted at the lateral lines, the annulations of one side not corresponding to those of the other. There is a fine transverse striation and, on both ends, longitudinal striation also.

Head flat, with slightly constriction immediately behind it. Apically the head appears rectangular with sides directed dorsally, ventrally and laterally, the appearance caused by the presence of four distinct submedian processes each of which bears a small papilla. Amphids lateral.
Male: Body length 35.6—45.1 mm, maximum body width 0.42—0.492 mm. Length of oesophagus 1.10—1.538 mm, with anterior part 0.16 mm long. Tail end bluntly rounded, 0.121—0.137 mm long. Left spicule 0.330—0.37 mm

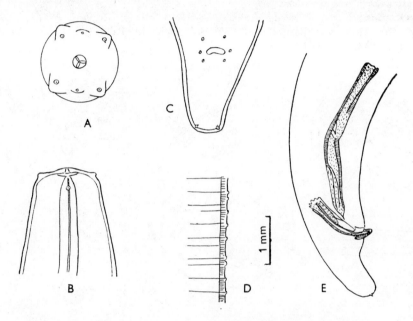

Fig. 122. *Paronchocerca ciconiarum* Peters, 1936. A — anterior end (apical view); B — anterior end; C — posterior end of male (ventral view); D — ornamentation of cuticle; E — posterior end of male (lateral view). After Peters (1936).

long, bent approximately in the middle of its length, width at proximal end 0.035—0.036 mm, distal end pointed. Rigth spicule 0.16—0.17 mm long, its apex rounded. Three pairs of caudal papillae visible ventrally. Phasmids subterminal.
Female: Body length 56.0—70.0 mm, maximum body width 0.678—0.90 mm. Total length of oesophagus 1.349—1.855 mm, length of muscular part 0.192 to 0.24 mm. Vulva post-oesophageal, 3.892—4.70 mm from head end. Tail 0.08—0.11 mm long. Microfilaria 0.092×0.005 mm, short and thick, with pointed tail end, without sheath.

The description of this species was based on specimens recovered from the heart of African storks which died in the London Zoological Gardens.

References: Macko (1963 b)*; Peters (1936)*; Skrjabin & Shikhobalova (1948)*; Sonin (1968)*.

Paronchocerca tonkinensis (Chow, 1939)

Fig. 123
Hosts: *Ardea cinerea, Ardeola bacchus, Egretta garzetta* and *Botaurus stellaris*.
Localization: venous system.
Distribution: Europe (Czechoslovakia, U.S.S.R. — Volga Region), Asia (U.S.S.R. — Azerbaijan, Turkmenia, East Siberia, Far East), outside the Palaearctic Region, Democratic Republic of Vietnam, Taiwan.

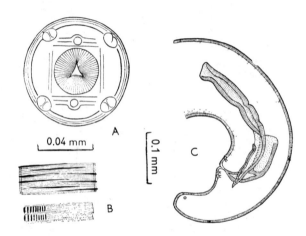

Fig. 123. *Paronchocerca tonkinensis* (Chow, 1939). A — anterior end (apical view); B — ornamentation of cuticle; C — posterior end of male (lateral view). After Macko (1964).

Description: Cuticle with fine transverse striations, provided with annular thickenings. Mouth simple, without lips, followed by a small vestibule. Rudimentary epaulette-like trilobate structures present.
Male: Body length 21—31 mm, maximum body width 0.28—0.35 mm. Length of muscular part of oesophagus 0.33—0.46 mm, glandular part 2.31 to 3.24 mm. Nerve ring 0.14—0.22 mm from anterior end. Tail 0.10—0.13 mm long. Three pairs of adanal papillae distributed on annular thickening around cloaca and two pairs of postanal papillae. Left spicule 0.19—0.25 mm long, its distal end pointed. Right spicule 0.10—0.14 mm long, with rounded end.
Female: Body length 34—47 mm, maximum body width 0.42—0.87 mm. Length of muscular part of oesophagus 0.31—0.46 mm, glandular part 4.73 to 6.38 mm. Nerve ring 0.14—0.19 mm, vulva 1.61—3.25 mm from head end. Two uteri, opisthodelphic. Length of microfilaria in utero 0.16—0.18 mm.

The description of this species was based on nematodes recovered from the heart of *A. bacchus* in the Democratic Republic of Vietnam, which were assigned to the genus *Houdemeres*. It is now placed in the genus *Paronchocerca*

(López-Neyra 1947; Chabaud & Biocca 1951 and others). In the Palaearctic Region this species was first recorded by Oshmarin (1950) in *B. stellaris* in the Far East.

References: Chow (1939)*; Feyzullaev (1963 a,b); López-Neyra (1947 a); Macko (1964 b)*; Oshmarin (1950, 1963); Ryzhikov & Kozlov (1959); Shigin (1957); Skrjabin & Shikhobalova (1948)*; Sonin (1963*, 1968*).

Genus *Pelecitus* Railliet & Henry, 1910

Head end rounded, without lips. Mouth opening spherical surrounded, by four pairs of submedian cephalic papillae and a pair of amphids. Mouth leads into a rudimentary vestibule which runs into a cylindrical oesophagus, not divided into parts. Lateral alae present or absent. Tail end of male with symmetrical or asymmetrical alae. Two relatively short, slightly chitinized spicules nearly similar, equal, or subequal. Two to three pairs of preanal papillae, postanal papillae variable in number in different representatives of the same species. Vulva in region of oesophagus. Viviparous.

Type species *P. helicinus* (Molin, 1858), parasite of birds in South America.

Fifteen species of this genus are known; two of them have been recorded in fish-eating birds of the Palaearctic Region.

Pelecitus fulicaeatrae (Diesing, 1861)

Fig. 124 A—D
Hosts: *Ardea cinerea*, *A. purpurea*, *Egretta alba*, *Botaurus stellaris*, *Larus ridibundus*, *Chlidonias nigra* and *Pandion haliaetus*. Recorded also in rails, *Anatidae*, *Falconiformes* and *Coraciae*.
Localization: synovial cavity of calcaneal and knee joint.
Distribution: fish-eating birds in Asia (U.S.S.R. — East Siberia and Far East), rails and other birds in the Palaearctic Region and Central Africa.

Description: Cuticle with fine transverse striations. Lateral alae absent.
Male: (Based on specimens from fish-eating birds.) Body length 8.5—10 mm, maximum body width 0.43 mm. Oesophagus 0.86—1.04 mm long. Tail 0.05—0.056 mm long. Spicules unequal, left 0.105 mm long, consisting of two parts, right 0.07—0.075 mm long. Number of tail papillae varies but three pairs of preanal and four to five pairs of postanal papillae usually present, plus unpaired papillae.
Female: Body length 22—28 mm, maximum body width 0.91 mm. Oesophagus 1.10—1.60 mm long. Vulva 0.65—0.72 mm from head end.

This species was described from nematodes recovered from coots in England.

Notes: Linstow (1899) described the species *Spiroptera helix* which was found in coots in Germany. Baylis (1944) considered the species described by Linstow and by Diesing to be synonymous and placed this species in the genus *Spirofilaria*. Skrjabin & Shikhobalova (1948), without knowledge of the paper by Baylis, included the species *S. helix* in the genus *Pelecitus*, and López-Neyra (1956) synonymized the genus *Spirofilaria* with the genus *Pelecitus*. In our opinion, the species *P. ardeae* Vuylsteke, 1957, recovered from herons in Central Africa, is a synonym of *P. fulicaeatrae*.

References: Alekseev & Smetanina (1968); Bashkirova (1960); Baylis (1944); Belopolskaya (1959, 1963); Diesing (1861)*; Gubanov & Daiya (1967); Linstow (1899)*; López-Neyra (1956); Oshmarin (1950)*; Oshmarin & Parukhin (1960, 1963); Skrjabin & Shikhobalova (1948)*; Smetanina (1972); Smetanina & Alekseev (1967); Sonin (1968)*.

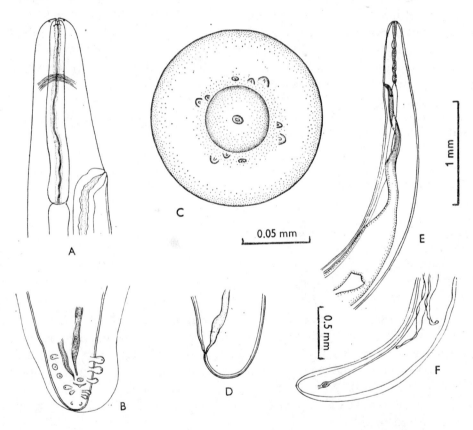

Fig. 124. *Pelecitus fulicaeatrae* (Diesing, 1861) — A, B, C, D; *P. podicipitis* (Yamaguti, 1935) — E, F. A — anterior end of female (lateral view); B — posterior end of male (ventral view); C — anterior end (apical view); D — posterior end of female (lateral view); E — anterior end of female; F — posterior end of female. A, B, D — After Skrjabin & Shikhobalova (1948); C — After Sonin (1968); E, F — After Yamaguti (1935).

Pelecitus podicipitis (Yamaguti, 1935)

Fig. 124 E,F
Host: *Podiceps cristatus* and *P. ruficollis*.
Localization: leg joints.
Distribution: Europe (England, France), Asia (U.S.S.R.-Far East; Japan).

Description: Body tapering at both ends, spirally twisted, provided with distinct cuticular alae along the whole length. Cuticle with fine transverse striations. Mouth without lips, surrounded by two lateral and four submedian papillae. Oesophagus divided into two parts.
Male: Body length (measured in straight line) 7—9.4 mm, maximum width, including alae, 0.35—0.45 mm. Tail end asymmetrical. Left spicule 0.08—0.11 mm long, right spicule 0.06—0.07 mm, both consisting of tubular proximal and membranous distal part. Nine pairs of caudal papillae and three or four additional unpaired papillae; the most anterior pair is situated 0.10 mm from posterior end of body; three pairs of lateral preanal papillae and one unpaired median papilla in front of cloaca. Four small papillae on posterior margin of cloaca and, nearly at the same level, the fourth pair of lateral papillae. Around the tail pulvinate formations bearing three pairs of papillae.
Female: Body length (measured in straight line) 15—22 mm, maximum width (including alae) 0.64—0.80 mm. Vulva salient, situated at level of transition from oesophagus to gut.
    Yamaguti (1935) described the nematodes from grebes as *Spirofilaria podicipitis*.
Notes: Baylis (1944) examined nematodes from the same species of grebe collected in England and assigned them to the species *S. fulicaeatrae*. López-Neyra (1956) placed the genus *Spirofilaria* in synonymy with the genus *Pelecitus*. Yamaguti (1961) considered the species *P. podicipitis* to be independent, because *P. fulicaeatrae* lacks the lateral alae, which are distinct in *P. podicipitis*.

References: Baylis (1944)*; Dollfus et al. (1961); Smetanina (1972); Skrjabin & Shikhobalova (1948)*; Sonin (1968)*; Yamaguti (1935*, 1961).

SUBORDER *CAMALLANATA* CHITWOOD, 1936

Body cylindrical, slender, but may be rather stout. Mouth usually without lips. Buccal cavity present or absent. Oesophagus usually long, more or less divided into two portions, rarely degenerate.
Male: Cloaca terminal or ventral. Caudal alae and papillae present or absent. Spicules equal, similar, occasionally absent.
Female: Anus and vulva atrophied or present. Vulva, when present, situated in anterior, middle or posterior parts of body. Vagina reduced or not. Parasitic

in body cavity, blood sinus, air-bladder, or tissues of vertebrates, with crustaceans as intermediate hosts.

Ivashkin, Sobolev & Khromova (1971) included seven families in this suborder. Representatives of six of them parasitize cold-blooded animals. Consistently with Yamaguti (1961) we place in this suborder also *Robertdollfusidae*, the eighth family, representatives of which parasitize birds.

KEY TO THE FAMILIES OF THE SUBORDER *CAMALLANATA*

1 Head end with cuticular shield, digestive tract well developed .... *Dracunculidae*
— Head end without cuticular shield, digestive tract atrophied (mouth, oesophagus, intestine and anus absent) .... *Robertdollfusidae*

Family *Dracunculidae* Leiper, 1912

Body long and thin, females considerably larger than males. Head end rounded, with cuticular shield. Lips and pseudolabia absent. Oesophagus divided into anterior, muscular and posterior glandular portions or muscular throughout. Anus of adults obliterated or rudimentary, as is vulva. Viviparous. Gubernaculum present or absent.

The representatives of this family are divided into three subfamilies. Parasites of birds belong to the subfamily *Avioserpentinae* Wehr & Chitwood, 1934, which includes the type and only genus *Avioserpens*.

Genus *Avioserpens* Wehr & Chitwood, 1934

Female many times longer than male. Head end with cuticular shield, mouth without circumoral chitinized ring. 14 cephalic papillae arranged in two circles. Oesophagus subdivided into two parts. Anterior part of glandular portion of oesophagus with distinct swelling. Spicules equal or subequal. Gubernaculum present, its distal end covered with spines on the lower side. Cervical papillae projecting slightly above surface of body. Vulva behind middle of body, atrophied in adult females.

Type species *A. taiwana* (Sugimonto, 1919).

Consistently with other authors we regard the genera *Oshimaia* Sugimoto, 1934 and *Petroviprocta* Shakhtakhtinskaya, 1951 as synonyms of this genus. The genus includes three species, two of which are parasitic in fish-eating birds of the Palaearctic Region.

KEY TO THE SPECIES OF THE GENUS *AVIOSERPENS*

1 Spicules more than 0.35 mm long; gubernaculum 0.11—0.15 mm long . . . .
. . . . *A. galliardi*
— Spicules less than 0.20 mm long; gubernaculum 0.08—0.11 mm long . . . . . .
. . . . *A. mosgovoyi*

Avioserpens galliardi Chabaud & Campana, 1949

Fig. 125
Hosts: *Gavia stellata*, *Ardea cinerea*, *A. purpurea*, *Egretta alba*, *E. garzetta*, *Nycticorax nycticorax*, *Botaurus stellaris* and *Mergus merganser*.
Localization: hypodermal tissue, aponeurosis in upper part of oesophagus.
Distribution: Europe (France, U.S.S.R. — Vogla Delta), Asia (mouth of the Ob River, republics of Transcaucasus Region and Middle Asia). Outside the Palaearctic Region, in herons of India and Canada.

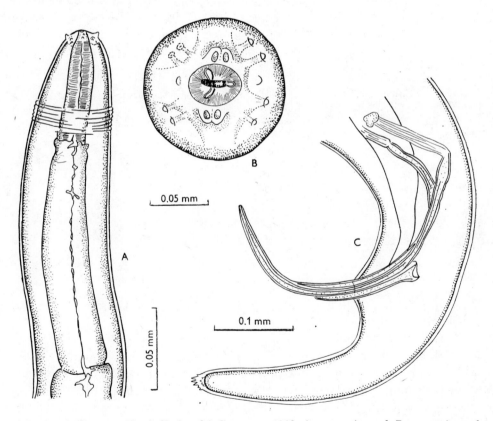

Fig. 125. *Avioserpens galliardi* Chabaud & Campana, 1949. A — anterior end; B — anterior end (apical view); C — posterior end (lateral view). After Chabaud & Campana (1949).

Description: Body filiform, tapering towards both extremities. Cuticle with thin, dense transverse striations. Mouth circular, small, slightly protruding.
Male: Body length 5.35—12.12 mm, maximum body width 0.14—0.19 mm. Length of muscular part of oesophagus 0.10—0.125 mm, glandular part 3.32—3.68 mm. Tail narrow, 0.22—0.30 mm long. Spicules 0.38—0.45 mm long. Gubernaculum 0.11—0.15 mm long.
Female: Body length 105—210 mm, maximum body width 0.60—0.80 mm. Tail 0.68—0.75 mm long, terminating in a small pointed process, about 0.01 mm long. Muscular part of oesophagus 0.17—0.35 mm long. Uterus filled with larvae, occupying nearly the whole body.

The description was based on specimens recovered from *E. garzetta* in France.

Biology: The larvae are ingested by *Cyclops* in which they grow and develop (Chabaud & Campana 1949).

Notes: Chabaud & Campana (1952) synonymized with this species *Petroviprocta vigissi* described by Shakhtakhtinskaya (1951) from *N. nycticorax* from Azerbaijan. Supryaga (1971 a) synonymized *A. multipapillosa* Singh, 1949, described from herons in India and *A. nana* Mawson, 1957 described from herons in Canada with *Avioserpens galliardi*. We agree with the opinion of these authors.

References: Ablasov & Chibichenko (1961, 1962); Chabaud & Campana (1949 a*, 1952); Daiya (1967 a, b); Dollfus et al. (1961); Feyzullaev (1963 a, b); Ivashkin, Sobolev & Khromova (1971)*; Kasimov & Feyzullaev (1969); Kosupko (1963); Shakhtakhtinskaya (1951*, 1959 a, b); Supryaga (1971 a); Turemuratov (1962 a).

Avioserpens mosgovoyi Supryaga, 1965

Fig. 126
Hosts: *Gavia arctica, Podiceps cristatus, P. griseigena* and *P. ruficollis*. From birds other than fish-eating, *Fulica atra* and *Nyroca fuligula*.
Location: hypodermal tissue.
Distribution: Europe (U.S.S.R. — lower reaches of the Kuban River).

Description: Body filiform, widening towards both ends. Cuticle thin, transversely striated. Mouth simple, terminal.
Male: Body length 6.8—14.0 mm, maximum body width 0.14—0.21 mm. Length of muscular part of oesophagus 0.09—0.13 mm, of glandular part 1.40—2.31 mm. Glandular thickening of oesophagus 0.11—0.22 mm long and 0.07—0.15 mm wide. Nerve ring 0.25—0.35 mm from head end. Spicules 0.16—0.19 mm long. Gubernaculum cuneiform, 0.08—0.11 mm long. Tail end of body conical, terminating in three processes. Four pairs of sessile postanal papillae.

Female: Body length 800—1,130 mm, maximum body width 0.8—1.0 mm. Distance from head end to swollen part of oesophagus 0.20—0.29 mm, length of this part 0.47—0.96 mm. Nerve ring 0.74—1.95 mm and cervical papillae 1.14 mm from head end. Body cavity of females completely occupied by sac-like uterus with enormous number of larvae 0.45—0.50 mm long.

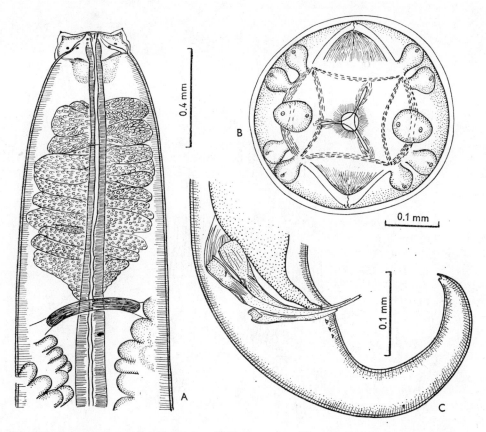

Fig. 126. *Avioserpens mosgovoyi* Supryaga, 1965. A — anterior end; B — anterior end (apical view); C — posterior end (lateral view). After Supryaga (1966, 1967).

Biology: The intermediate hosts of this nematode are different species of crustaceans of the genera *Cyclops* and *Diaptomus*. Paratenic hosts are the fry of fishes and larvae of dragon-flies. In *Cyclops* at 22—29 °C the larvae develop to the infective stage within three to four days. Under optimal conditions the whole life-cycle is completed within 36 to 40 days (Supryaga 1965 b, 1971 b).

References: Ivashkin, Sobolev & Khromova (1971)*; Supryaga (1965 a*, b, 1968, 1969 a, b, 1971 a, b, 1972).

Family *Robertdollfusidae* Chabaud & Campana, 1950

Nematodes with fine, smooth cuticle. Digestive tract atrophied (mouth, oesophagus and anus absent). Amphidelphic, viviparous. Embryos not numerous, large, without digestive canal. Parasitic in tissues of vertebrates.

Type and single genus *Robertdollfusa* Chabaud & Campana, 1950.

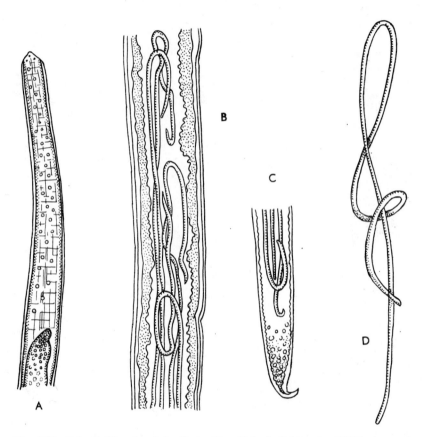

Fig. 127. *Robertdollfusa longimicrofilaria* (Parukhin & Oshmarin, 1960). A — anterior end of female; B — part of body with larvae; C — posterior end of female; D — larva. After Parukhin & Oshmarin (1960).

Genus *Robertdollfusa* Chabaud & Campana, 1950

Cephalic end bluntly pointed. Male unknown.
Female: Posterior end tapering to a transparent digitiform appendage. Genitalia reduced to uterine sac, occupying almost entire body; vulva without sphincter, behind middle of body. Embryos about one sixth the length of

adults; head end rounded, with transparent cuticular cap provided with two refractile points; caudal end digitiform, without cuticular ornamentations. Parasites of birds.

Type species *R. paradoxa* Chabaud & Campana, 1950.

The genus includes two species, one of which parazitizes fish-eating birds.

Robertdollfusa longimicrofilaria (Parukhin & Oshmarin, 1960)

Fig. 127
Host: *Pandion haliaetus*.
Localization: brain tissue.
Distribution: Asia (U.S.S.R. — Far East).

Description: Small nematodes. Length of female body 5.8—6.0 mm, maximum body width 0.120—0.125 mm. Cuticle relatively smooth. Head end with six papillae arranged in two rows, two lateral in anterior row and four submedial in the other row. Mouth opening, oesophagus, intestine and anal opening absent. Anterior end of body somewhat narrowed and regularly cylindrical in form. Anterior end of body terminates in thimble-like process. Cuticular-muscular sac, with the exception of a portion of 0.4 mm from head end, occupied by uterus and ovary. Uterus contains filiform larvae 1.1 mm long and 0.04 mm wide. Genital opening absent. Posterior end of body pointed.

The species was described from specimens recovered from osprey under the name *Encephalonema longimicrofilaria*. Chabaud (personal communication) places this species in the genus *Robertdollfusa*. We are of the same opinion.

References: Chabaud (personal communication); Chabaud & Campana (1950 b); Oshmarin & Parukhin (1960, 1963); Parukhin & Oshmarin (1960)*; Smetanina (1972); Sonin (1968)*.

List of fish-eating birds and their nematodes recorded in the Palaearctic Region

The host species and the nematodes found in them are arranged sytematically. Hosts have sometimes been recorded in the helminthological literature under older names, now regarded as synonyms. We have, on principle, used, recent nomenclature for the birds and in the cases mentioned above we have used the host names valid at present. It should be stressed that the note concerning hosts recorded as free of nematodes refers to the Palaearctic Region only.

Order *Gaviiformes*
Family *Gaviidae*
Genus *Gavia* Forster, 1788
G. adamsii Gray, 1859
Contracaecum rudolphii Hartwich, 1964

G. arctica (L., 1758)

Capillaria carbonis (Rudolphi, 1819)
Eustrongylides tubifex (Nitzsch, 1819)
Contracaecum rudolphii Hartwich, 1964
Cosmocephalus obvelatus (Creplin, 1825)

Paracuaria tridentata (Linstow, 1877)
Streptocara crassicauda (Creplin, 1829)
Ingliseria cirrohamata (Linstow, 1888)
Avioserpens mosgovoyi Supryaga, 1965

G. immer (Brünn., 1764)

Capillaria mergi Madsen, 1945
Eustrongylides tubifex (Nitzsch, 1819)

Contracaecum variegatum (Rudolphi, 1809)

G. pacifica Lawrence, 1858
(No nematodes yet recorded.)

G. stellata (Pontop., 1763)

Capillaria mergi Madsen, 1945
Eustrongylides mergorum (Rudolphi, 1809)
E. tubifex (Nitzsch, 1819)
Syngamus arcticus Ryzhikov, 1952

Contracaecum rudolphii Hartwich, 1964
C. variegatum (Rudolphi, 1809)
Cosmocephalus aduncus (Creplin, 1846)
Paracuaria tridentata (Linstow, 1877)

Streptocara crassicauda (Creplin, 1829)
Ingliseria cirrohamata (Linstow, 1888)
Stegophorus stellaepolaris (Parona, 1901)

Avioserpens galliardi Chabaud & Campana, 1949

Order *Podicipediformes*
Family *Podicipedidae*
Genus *Podiceps* Latham, 1787
P. auritus (L., 1758)

Capillaria podicipitis Yamaguti, 1941
Eustrongylides mergorum (Rudolphi, 1809)
Contracaecum microcephalum (Rudolphi, 1809)
C. ovale (Linstow, 1907)
C. rudolphii Hartwich, 1964
Porrocaecum praelongum (Dujardin, 1845)

Cosmocephalus obvelatus (Creplin, 1825)
C. aduncus (Creplin, 1846)
Echinuria uncinata (Rudolphi, 1819)
Skrjabinoclava decorata (Solonitsin, 1928)
Streptocara crassicauda (Creplin, 1829)

P. cristatus (L., 1758)

Capillaria podicipitis Yamaguti, 1941
Eustrongylides mergorum (Rudolphi, 1809)
E. tubifex (Nitzsch, 1819)
Epomidiostomum uncinatum (Lundahl, 1848)
Contracaecum microcephalum (Rudolphi, 1809)
C. ovale (Linstow, 1907)
C. rudolphii Hartwich, 1964
Cosmocephalus obvelatus (Creplin, 1825)
C. aduncus (Creplin, 1846)
Decorataria decorata (Cram, 1927)

Rusguniella elongata (Rudolphi, 1819)
R. wedli Williams, 1929
Echinuria uncinata (Rudolphi, 1819)
Skrjabinoclava decorata (Solonitsin, 1928)
Streptocara crassicauda (Creplin, 1829)
Tetrameres fissispina (Diesing, 1861)
T. gubanovi Shigin, 1957
Pelecitus podicipitis (Yamaguti, 1935)
Avioserpens mosgovoyi Supryaga, 1965

P. griseigena (Bodd., 1783)

Capillaria carbonis (Rudolphi, 1819)
C. ryjikovi Daiya, 1972
Thominx anatis (Schrank, 1790)
Epomidiostomum uncinatum (Lundahl, 1848)
Contracaecum microcephalum (Rudolphi, 1809)
C. ovale (Linstow, 1907)
C. rudolphii Hartwich, 1964
Porrocaecum crassum (Deslongchamps, 1824)

Cosmocephalus obvelatus (Creplin, 1825)
Decorataria decorata (Cram, 1927)
Paracuaria tridentata (Linstow, 1877)
Rusguniella eloganta (Rudolphi, 1819)
Echinuria uncinata (Rudolphi, 1819)
Streptocara crassicauda (Creplin, 1829)
Tetrameres fissispina (Diesing, 1861)
T. gubanovi Shigin, 1957

P. nigricollis Brehm, 1831

Capillaria carbonis (Rudolphi, 1819)
C. podicipitis Yamaguti, 1941
C. ryjikovi Daiya, 1972
Amidostomum fulicae (Rudolphi, 1819)
Contracaecum ovale (Linstow, 1907)
C. praestriatum Mönnig, 1923
C. rudolphii Hartwich, 1964
Cosmocephalus obvelatus (Creplin, 1825)

C. aduncus (Creplin, 1846)
Decorataria decorata (Cram, 1927)
Rusguniella wedli Williams, 1929
Echinuria uncinata (Rudolphi, 1819)
Streptocara crassicauda (Creplin, 1829)
Tetrameres fissispina (Diesing, 1861)
T. gubanovi Shigin, 1957

P. ruficollis (Pall., 1764)

Capillaria mergi Madsen, 1945
C. podicipitis Yamaguti, 1941
C. ryjikovi Daiya, 1972
Thominx contorta (Creplin, 1839)
Eustrongylides mergorum (Rudolphi, 1809)
E. tubifex (Nitzsch, 1819)
Amidostomum anseris (Zeder, 1800)
A. fulicae (Rudolphi, 1819)
Epomidiostomum uncinatum (Lundahl, 1848)
Contracaecum andersoni Vevers, 1923
C. micropapillatum (Stossich, 1890)
C. ovale (Linstow, 1907)
C. rudolphii Hartwich, 1964
Cosmocephalus aduncus (Creplin, 1846)
Decorataria decorata (Cram, 1927)
Rusguniella wedli Williams, 1929
Syncuaria ciconiae (Gilbert, 1927)
Echinuria uncinata (Rudolphi, 1819)
Streptocara crassicauda (Creplin, 1829)
Tetrameres fissispina (Diesing, 1861)
Pelecitus podicipitis (Yamaguti, 1935)
Avioserpens mosgovoyi Supryaga, 1965

Podiceps sp.

Contracaecum rudolphii Hartwich, 1964
Heterakis gallinarum (Schrank, 1788)
Decorataria decorata (Cram, 1927)

Order *Procellariiformes*
Family *Diomedeidae*
Genus *Diomedea* Linné, 1758
D. albatrus Pall., 1769
(No nematodes yet recorded.)

D. immutabilis Roth., 1839
(No nematodes yet recorded.)

D. nigripes Audubon, 1839
(No nematodes yet recorded.)

Family *Procellariidae*
Genus *Fulmarus* Stephens, 1826
Fulmarus glacialis (L., 1761)

Contracaecum rudolphii Hartwich, 1964
Paracuaria tridentata (Linstow, 1877)
Stegophorus stellaepolaris (Parona, 1901)
S. stercorarii **Leonov, Sergeeva & Tsim-balyuk, 1966**

Genus *Bulweria* Jardine & Selby, 1828
Bulweria bulwerii Jardine & Selby, 1828
(No nematodes yet recorded.)

Genus *Pterodroma* Bonaparte, 1856
P. hypoleuca Salvin, 1888
(No nematodes yet recorded.)

P. inexpectata (Forst, 1844)
(No nematodes yet recorded.)

P. molllis Gould, 1844
(No nematodes yet recorded.)

P. solandri (Gould, 1844)
(No nematodes yet recorded.)

Genus *Procellaria* L., 1758
P. diomedea Scopoli, 1769

Contracaecum bodenheimeri Witenberg, 1929   *Seuratia shipleyi* (Stossich, 1900)

P. leucomelas Temm., 1835
(No nematodes yet recorded.)

Genus *Puffinus* Brisson, 1760

P. assimilis Gould, 1838
(No nematodes yet recorded.)

P. bulleri Salvin, 1888
(No nematodes yet recorded.)

P. griseus (Gm., 1789)
Seuratia shipleyi (Stossich, 1900)

P. pacificus (Gm., 1789)
(No nematodes yet recorded.)

P. puffinus (Brünn., 1764)
Streptocara crassicauda (Creplin, 1829)

P. tenuirostris (Temm., 1835)
Seuratia shipleyi (Stossich, 1900)

Family *Hydrobatidae*
Genus *Hydrobates* Boie, 1822
H. pelagicus (L., 1758)
Stegophorus stellaepolaris (Parona, 1901)

Genus *Oceanodroma* Reichenbach, 1853
O. castro (Harc., 1851)
(No nematodes yet recorded.)

O. furcata (Gm., 1789)
(No nematodes yet recorded.)

O. leucorrhoa (Vieill., 1817)
(No nematodes yet recorded.)

O. matsudairae Kuroda, 1922
(No nematodes yet recorded.)

O. monorchis (Swinh., 1867)

Stegophorus stellaepolaris (Parona, 1901)   Seuratia shipleyi (Stossich, 1900)

O. tristrami Salin, 1896
(No nematodes yet recorded.)

Genus *Pelagodroma* Reichenbach, 1853
P. marina Lath., 1790
(No nematodes yet recorded.)

Order *Pelecaniformes*
Suborder *Phaethontes*
Family *Phaethontidae*
Genus *Phaethon* L., 1758
Ph. aetherus L., 1758
(No nematodes yet recorded.)

Ph. rubricauda Bodd., 1783
(No nematodes yet recorded.)

Suborder *Fregatae*
Family *Fregatidae*
Genus *Fregata* Lacépède, 1799

F. ariel (Gray, 1845)
(No nematodes yet recorded.)

F. magnificiens Math., 1914
(No nematodes yet recorded.)

F. minor (Gm., 1789)
(No nematodes yet recorded.)

Suborder *Pelecani*
Family *Pelecanidae*
Genus *Pelecanus* L., 1758
P. crispus Bruch, 1832

Capillaria carbonis (Rudolphi, 1819)   Eustrongylides africanus Jägerskiöld, 1909

E. mergorum (Rudolphi, 1809)
E. tubifex (Nitzsch, 1819)
Cyathostoma verrucosum (Hovorka & Macko, 1959)

Contracaecum microcephalum (Rudolphi, 1889)
C. micropapillatum (Stossisch, 1890)
C. rudolphii Hartwich, 1964

P. onocrotalus L., 1758

Capillaria carbonis (Rudolphi, 1819)
Eustrongylides africanus Jägerskiöld, 1909
E. mergorum (Rudolphi, 1809)
E. tubifex (Nitzsch, 1819)
Syngamus trachea (Montagu, 1811)
Cyathostoma microspiculum (Skrjabin, 1915)
C. verrucosum (Hovorka & Macko, 1959)

Contracaecum microcephalum (Rudolphi, 1809)
C. rudolphii Hartwich, 1964
Cosmocephalus faridi Khalil, 1931
Desportesius raillieti (Skrjabin, 1924)
Synhimantus sirry Khalil, 1931
Microtetrameres pelecani Skrjabin, 1949

Pelecanus sp.
Contracaecum micropapillatum (Stossisch, 1890)

Family *Sulidae*
Genus *Sula* Brisson, 1760
S. bassana (L., 1758)
(No nematodes yet recorded.)

S. leugocaster Bodd., 1783
(No nematodes yet recorded.)

S. sula L., 1766
(No nematodes yet recorded.)

Family *Anhingidae*
Genus *Anhinga* Brisson, 1760
A. rufa Lacépède & Daudin, 1802
(No nematodes yet recorded.)

Family *Phalacrocoracidae*
Genus *Phalacrocorax* Brisson, 1760
Ph. africanus Gm., 1789
(No nematodes yet recorded.)

Ph. aristotelis (L., 1761)

Contracaecum rudolphii Hartwich, 1964
C. septentrionale Kreis, 1955

Acuaria phalacrocoracis (Smogorzhevskaya, 1961)

Ph. carbo (L., 1758)

Capillaria carbonis (Rudolphi, 1819)
Thominx contorta (Creplin, 1839)
Eustrongylides excisus Jägerskiöld, 1909
E. tubifex (Nitzsch, 1819)
Syngamus trachea (Montagu, 1811)
Cyathostoma microspiculum (Skrjabin, 1915)
Contracaecum microcephalum (Rudolphi, 1809)
C. micropapillatum (Stossich, 1890)
C. rudolphii Hartwich, 1964
C. travassosi Gutierrez, 1943

Cosmocephalus aduncus (Creplin, 1846)
C. jaenschi Johnston & Mawson, 1941
Skrjabinocara squamata (Linstow, 1884)
Desmidocerca aerophila Skrjabin, 1915
Desmidocercella numidica (Seurat, 1920)
D. incognita Solonitsin, 1932
Streptocara crassicauda (Creplin, 1829)
Dicheilonema ciconiae (Schrank, 1788)
Lemdana behningi Levashov, 1929

Ph. capillatus (Temm. & Schlegel, 1850)

Contracaecum rudolphii Hartwich, 1964
Porrocaecum phalacrocoracis Yamaguti, 1941

Desmidocercella incognita Solonitsin, 1932

Ph. nigrogularis Olilvie-Grant & Forbes, 1899
(No nematodes yet recorded.)

Ph. pelagicus Pall., 1811

Contracaecum microcephalum (Rudolphi, 1809)
C. rudolphii Hartwich, 1964

Paracuaria tridentata (Linstow, 1877)
Desmidocercella incognita Solonitsin, 1932

Ph. perspicillatus Pall., 1811
(No nematodes yet recorded.)

Ph. pygmaeus Pall., 1773

Capillaria carbonis (Rudolphi, 1819)
Eustrongylides excisus Jägerskiöld, 1909
E. tubifex (Nitzsch, 1819)
Cyathostoma microspiculum (Skrjabin, 1915)
Contracaecum andersoni Vevers, 1923
C. micropapillatum (Stossich, 1890)
C. microcephalum (Rudolphi, 1809)

C. rudolphii Hartwich, 1964
Cosmocephalus obvelatus (Creplin, 1825)
C. jaenschi Johnston & Mawson, 1941
Skrjabinocara squamata (Linstow, 1884)
Streptocara crassicauda (Creplin, 1829)
Desmidocerca aerophila Skrjabin, 1915
Desmidocercella incognita Solonitsin, 1932

Ph. urile (Gm., 1789)

Contracaecum rudolphii Hartwich, 1964
Desmidocercella incognita Solonitsin, 1932

Parornithofilaria shaldybini (Gubanov, 1954)

Phalacrocorax sp.

Cyathostoma microspiculum (Skrjabin, 1915)

Contracaecum rudolphii Hartwich, 1964

Order *Ciconiiformes*
Suborder *Ardeae*
Family *Ardeidae*
Genus *Botaurus* Stephens, 1819
B. stellaris L., 1758

Thominx contorta (Creplin, 1839)
Contracaecum microcephalum (Rudolphi, 1809)
C. rudolphii Hartwich, 1964
Desportesius brevicaudatus (Dujardin, 1845)
D. orientalis (Wu, 1933)
Paracuaria tridentata (Linstow, 1877)
Streptocara crassicauda (Creplin, 1829)

Desmidocercella incognita Solonitsin, 1932
Tetrameres fissispina (Diesing, 1861)
Paronchocerca tonkinensis (Chow, 1939)
Pelecitus fulicaeatrae (Diesing, 1861)
Avioserpens galliardi Chabaud & Campana, 1949

Genus *Ixobrychus* Billberg, 1828
I. cinnamomeus (Gm., 1789)
(No nematodes yet recorded.)

I. eurythmus (Swinh., 1873)
Desmidocercella numidica (Seurat, 1920)

I. minutus (L., 1766)

Contracaecum microcephalum (Rudolphi, 1809)
Desportesius brevicaudatus (Dujardin, 1845)

Desmidocercella numidica (Seurat, 1920)

I. sinensis (Gm., 1789)
Rusguniella elongata (Rudolphi, 1819)

Genus *Nycticorax* Forster, 1817
N. caledonicus (Gm., 1789)
(No nematodes yet recorded.)
N. nycticorax (L., 1758)

Capillaria carbonis (Rudolphi, 1819)
Eustrongylides mergorum (Rudolphi, 1809)
Contracaecum microcephalum (Rudolphi, 1809)
C. rudolphii Hartwich, 1964
Porrocaecum ardeae (Frölich, 1802)
P. reticulatum (Linstow, 1899)
Heterakis pavonis Maplestone, 1931
Desportesius groffi (Li, 1934)
D. sagittatus (Rudolphi, 1809)

Syncuaria contorta (Molin, 1858)
Desmidocercella numidica (Seurat, 1920)
Tetrameres fissispina (Diesing, 1861)
T. gynaecophila (Molin, 1858)
Thelaziella nyctardeae (Dubinina, 1937)
Dicheilonema ciconiae (Schrank, 1788)
Avioserpens galliardi Chabaud & Campana, 1949

Genus *Butorides* Blyth, 1849
B. striatus (L., 1758)

Contracaecum microcephalum (Rudolphi, 1809)
Porrocaecum ardeae (Fröhlich, 1802)

Genus *Gorsachius* Bonaparte, 1855

G. goisagi (Temm., 1835)
(No nematodes yet recorded.)

Genus *Ardeola* Boie, 1822

A. bacchus (Bonap., 1855)

Porrocaecum microcephalum (Rudolphi, 1809)
P. reticulatum (Linstow, 1899)
Lemdana limboonkengi Hoeppli & Hsü, 1929
Paronchocerca tonkinensis (Chow, 1939)

A. grayii Sykes, 1832
(No nematodes yet recorded.)

A. ralloides (Scop., 1769)

Contracaecum microcephalum (Rudolphi, 1809)
Porrocaecum reticulatum (Linstow, 1899)
Desmidocercella numidica (Seurat, 1920)
Tetrameres ardea Shigin, 1953
T. fissipina (Diesing, 1861)

Genus *Bubulcus* Bonaparte, 1855

B. ibis (L., 1758)

Contracaecum microcephalum (Rudolphi, 1809)
Porrocaecum reticulatum (Linstow, 1899)
Cyrnea ficheuri (Seurat, 1916)
C. monoptera (Gendre, 1922)
Desportesius invaginatus (Linstow, 1901)
Tetrameres coccinea (Seurat, 1914)
Microtetrameres spiralis (Seurat, 1915)

Genus *Egretta* Forster, 1817

E. alba (L., 1758)

Capillaria carbonis (Rudolphi, 1819)
Eustrongylides mergorum (Rudolphi, 1809)
Contracaecum microcephalum (Rudolphi, 1809)
C. micropapillatum (Stossich, 1890)
C. ovale (Linstow, 1907)
C. rudolphii Hartwich, 1964
Porrocaecum ardeae (Frölich, 1802)
P. reticulatum (Linstow, 1899)
Desportesius invaginatus (Linstow, 1901)
D. brevicaudatus (Dujardin, 1845)
Desmidocercella incognita Solonitsin, 1932
D. numidica (Seurat, 1920)
Tetrameres ardea Shigin, 1953
T. fissispina (Diesing, 1861)
T. schigini Oshmarin, 1956
Dicheilonema ciconiae (Schrank, 1788)
Pelecitus fulicaeatrae (Diesing, 1861)
Avioserpens galliardi Chabaud & Campana, 1949

E. eulophotes (Swinh., 1860)
(No nematodes yet recorded.)

E. garzetta (L., 1766)

Capillaria carbonis (Rudolphi, 1819)
Eustrongylides mergorum (Rudolphi, 1809)
E. sinicus Wu & Liu, 1943
Hystrichis tricolor Dujardin, 1845

Contracaecum microcephalum (Rudolphi, 1809)
Porrocaecum ardeae (Frölich, 1802)
P. reticulatum (Linstow, 1899)
Cosmocephalus obvelatus (Creplin, 1825)
Desportesius invaginatus (Linstow, 1901)
D. brevicaudatus (Dujardin, 1845)
D. equispiculatus (Wu & Liu, 1943)

D. spinulatus (Chabaud & Campana, 1954)
Desmidocercella numidica (Seurat, 1920)
Dicheilonema ciconiae (Schrank, 1788)
Paronchocerca tonkinensis (Chow, 1939)
Avioserpens galliardi Chabaud & Campana, 1949

E. gularis (Bosc., 1792)
(No nematodes yet recorded.)

E. intermedia (Wagler, 1829)

Porrocaecum angusticolle (Molin, 1860)   *P. reticulatum* (Linstow, 1899)

E. sacra (Gm., 1789)
(No nematodes yet recorded.)

Genus *Ardea* L., 1758
A. cinerea L., 1758

Capilaria carbonis (Rudolphi, 1819)
C. herodiae Boyd, 1966
C. mergi Madsen, 1945
Eustrongylides mergorum (Rudolphi, 1809)
E. sinicus Wu & Liu, 1943
Cyathostoma lari Blanchard, 1849
Contracaecum microcephalum (Rudolphi, 1809)
C. rudolphii Hartwich, 1964
Porrocaecum ardeae (Frölich, 1802)
P. reticulatum (Linstow, 1899)
Desportesius invaginatus (Linstow, 1901)
D. brevicaudatus (Dujardin, 1845)
Rusguniella elongata (Rudolphi, 1819)

Desmidocerca aerophila Skrjabin, 1915
Desmidocercella numidica (Seurat, 1920)
Tetrameres ardea Shigin, 1953
T. fissispina (Diesing, 1861)
T. schigini Oshmarin, 1956
Thelaziella nyctardeae (Dubinina, 1937)
Dicheilonema ciconiae (Schrank, 1788)
Heterospiculoides skrjabini Shigin, 1957
Heterospiculum sobolevi Shigin, 1951
Paronchocerca tonkinensis (Chow, 1939)
Pelecitus fulicaeatrae (Diesing, 1861)
Avioserpens galliardi Chaubaud & Campana, 1949

A. goliath Cret., 1826
(No nematodes yet recorded.)

A. purpurea L., 1766

Capillaria carbonis (Rudolphi, 1819)
Eustrongylides mergorum (Rudolphi, 1809)
Amidostomum anseris (Zeder, 1800)
Contracaecum microcephalum (Rudolphi, 1809)
C. micropapillatum (Stossich, 1890)
C. ovale (Linstow, 1907)
C. rudolphii Hartwich, 1964
Porrocaecum ardeae (Frölich, 1802)
P. reticulatum (Linstow, 1899)
Desportesius invaginatus (Linstow, 1901)
D. sagittatus (Rudolphi, 1809)

Rusguinella elongata (Rudolphi, 1819)
Desmidocercella numidica (Seurat, 1920)
Tetrameres ardea Shigin, 1953
T. fissispina (Diesing, 1861)
T. schigini Oshmarin, 1956
Dicheilonema ciconiae (Schrank, 1788))
Lemdana lomonti Desportes, 1947
Pelecitus fulicaeatrae (Diesing, 1861)
Avioserpens galliardi Chabaud & Campana, 1949

Suborder *Ciconiae*
Family *Threskiornithidae*
Genus *Platalea* L., 1758
P. leucorodia L., 1758

Thominx spirale (Molin, 1858)
Eustrongylides africanus Jägerskiöld, 1909
Contracaecum microcephalum (Rudolphi, 1809)
C. ovale (Linstow, 1907)

Cosmocephalus obvelatus (Creplin, 1825)
Syncuaria contorta (Molin, 1858)
Tetrameres coccinea (Seurat, 1914)

P. minor Temm. & Schlegel, 1849
(No nematodes yet recorded.)

Genus *Plegadis* Kaup, 1829
P. falcinellus (L., 1766)

Capillaria carbonis (Rudolphi, 1819)
Thominx spirale (Molin, 1858)
Eustrongylides mergorum (Rudolphi, 1809)
Hystrichis tricolor Dujardin, 1845

Subulura suctoria (Molin, 1860)
Syncuaria contorta (Molin, 1858)
Tetrameres fissispina (Diesing, 1861)

Genus *Geronticus* Wagler, 1832
G. eremita L., 1758
(No nematodes yet recorded.)

Genus *Nipponia* Reichenbach, 1852
N. nippon (Temm., 1835)
(No nematodes yet recorded.)

Genus *Threskiornis* Gray, 1842
T. aethiopicus (Lath., 1790)
(No nematodes yet recorded.)

T. melanocephalus (Lath., 1790)
(No nematodes yet recorded.)

Family *Ciconiidae*
Genus *Ciconia* Brisson, 1760
C. boyciana Swinh., 1873
(No nematodes yet recorded.)

C. ciconia (L., 1758)

Syngamus trachea (Montagu, 1811)
Cyathostoma verrucosum (Hovorka & Macko, 1959)
Hovorkonema variegatum (Creplin, 1849)

Heterakis gallinarum (Schrank, 1788)
Excisa excisa (Molin, 1860)
Desportesius brevicaudatus (Dujardin, 1845)
Skrjabinocara parvepapillata Macko, 1962

Syncuaria ciconiae (Gilbert, 1927)
Skrjabinocta ciconiae Morozov, 1958

Dicheilonema ciconiae (Schrank, 1788)
Paronchocerca ciconiarum Peters, 1936

C. nigra (L., 1758)

Eustrongylides mergorum (Rudolphi, 1809)
Syngamus trachea (Montagu, 1811)
Cyathostoma trifurcatum (Hovorka & Macko, 1959)
Hovorkonema variegatum (Creplin, 1849)

Contracaecum microcephalum (Rudolphi, 1809)
Desportesius sagittatus (Rudolphi, 1809)
Skrjabinocara parvepapillata Macko, 1962
Syncuaria ciconia (Gilbert, 1927)
Dicheilonema ciconiae (Schrank, 1788)

Ciconia sp.
Dispharynx nasuta (Rudolphi, 1819)

Order *Anseriformes*
Family *Anatidae*
Genus *Mergus* L., 1758
M. albellus (L., 1758)

Capillaria mergi Madsen, 1945
Eustrongylides mergorum (Rudolphi, 1809)
E. tubifex (Nitzsch, 1819)
Amidostomum acutum (Lundahl, 1848)
Contracaecum rudolphii Hartwich, 1964

Echinuria uncinata (Rudolphi, 1819)
Skrjabinoclava decorata (Solonitsin, 1928)
Streptocara crassicauda (Creplin, 1829)
Tetrameres crami Swales, 1933
T. fissispina (Diesing, 1861)

M. merganser L., 1758

Capillaria mergi Madsen, 1945
Thominx anatis (Schrank, 1790)
Eustrongylides mergorum (Rudolphi, 1809)
E. tubifex (Nitzsch, 1819)
Hystrichis coronatus Molin, 1861
H. tricolor Dujardin, 1845
Amidostomum orientale Ryzhikov & Pavlov, 1959

A. acutum (Lundahl, 1848)
Contracaecum rudolphii Hartwich, 1964
C. yamaguti Mawson, 1956
Paracuaria formosensis (Sugimoto, 1930)
Streptocara crassicauda (Creplin, 1829)
Ingliseria cirrohamata (Linstow, 1888)
Tetrameres fissispina (Diesing, 1861)

M. serrator L., 1758

Capillaria mergi Madsen, 1945
Eustrongylides mergorum (Rudolphi, 1809)
E. tubifex (Nitzsch, 1819)
Hystrichis tricolor Dujardin, 1845
Contracaecum rudolphii Hartwich, 1964
Cosmocephalus obvelatus (Creplin, 1825)

Paracuaria formosensis (Sugimoto, 1930)
Echinuria uncinata (Rudolphi, 1819)
Streptocara crassicauda (Creplin, 1829)
Ingliseria cirrohamata (Linstow, 1888)
Tetrameres crami Swales, 1933
T. fissispina (Diesing, 1861)

M. squamatus Gould., 1864

Capillaria mergi Madsen, 1945
Contracaecum micropapillatum (Stossich, 1890)

C. rudolphii Hartwich, 1964
Paracuaria formosensis (Sugimoto, 1930)

Order *Falconiformes*
Family *Pandionidae*
Genus *Pandion* Savigny, 1809
P. haliaetus (L., 1758)

Contracaecum pandioni Sobolev & Sudarikov, 1939
Porrocaecum angusticolle (Molin, 1860)
Sexansocara skrjabini Sobolev & Sudarikov, 1939
Physaloptera alata Rudolphi, 1819
Thelaziella aquillina (Baylis, 1934)
Pelecitus fulicaeatrae (Diesing, 1861)
Robertdollfusa longimicrofilaria (Parukhin & Oshmarin, 1960)

Family *Accipitridae*
Genus *Haliaetus* Savigny, 1809
H. albicilla (L., 1758)

Contracaecum milviensis Karokhin, 1937
Porrocaecum angusticolle (Molin, 1860)
P. depressum (Zeder, 1800)

H. leucocephalus (L., 1766)
(No nematodes yet recorded.)

H. leucoryphus (Pall., 1711)

Contracaecum haliaeti Baylis & Daubney, 1923
Porrocaecum depressum (Zeder, 1800)

H. pelagicus (Pall. 1811)
(No nematodes yet recorded.)

Order *Charadriiformes*
Suborder *Lari*
Family *Stercorariidae*
Genus *Stercorarius* Brisson, 1760
S. longicaudatus Vieill., 1819

Contracaecum rudolphii Hartwich, 1964
Porrocaecum ensicaudatum (Zeder, 1800)
P. semiteres (Zeder, 1800)
Cosmocephalus obvelatus (Creplin, 1825)
Stegophorus stellaepolaris (Parona, 1901)
S. stercorarii Leonov, Sergeeva & Tsimbalyuk, 1966
Seuratia shipleyi (Stossisch, 1900)
Tetrameres skrjabini Panova, 1926
Schistorophus skrjabini (Vasilkova, 1926)
Cordonema longifuniculata (Sobolev, 1952)

S. parasiticus (L., 1758)

Contracaecum rudolphii Hartwich, 1964
Cosmocephalus obvelatus (Creplin, 1825)
Paracuaria tridentata (Linstow, 1877)
Stegophorus stellaepolaris (Parona, 1901)
S. stercorarii Leonov, Sergeeva & Tsimbalyuk, 1966
Serautia shipleyi (Stossisch, 1900)
Tetrameres skrjabini Panova, 1926

S. pomarinus (Temm., 1815)

Contracaecum micropapillatum (Stossich, 1890)
Porrocaecum semiteres (Zeder, 1800)
Stegophorus stellaepolaris (Parona, 1901)

S. stercorarii Leonov, Sergeeva & Tsimbalyuk, 1966

S. skua Brünn., 1764
(No nematodes yet recorded.)

Family *Laridae*
Genus *Larus* L., 1758
L. argentatus Pontopp., 1763

Capillaria carbonis (Rudolphi, 1819)
Thominx anatis (Schrank, 1790)
Th. contorta (Creplin, 1839)
Syngamus trachea (Montagu, 1811)
Cyathostoma lari Blanchard, 1849
Contracaecum rudolphii Hartwich, 1964
Porrocaecum ensicaudatum (Zeder, 1800)
P. semiteres (Zeder, 1800)
Cosmocephalus obvelatus (Creplin, 1825)
C. aduncus (Creplin, 1846)
Paracuaria tridentata (Linstow, 1877)
Pectinospirura multidentata Sobolev, 1943
Rusguniella elongata (Rudolphi, 1819)

Streptocara crassicauda (Creplin, 1829)
Ingliseria cirrohamata (Linstow, 1888)
Stegophorus stellaepolaris (Parona, 1901)
S. stercorarii Leonov, Sergeeva & Tsimbalyuk, 1966
Seuratia shipleyi (Stossich, 1900)
Desmidocerca aerophila Skrjabin, 1915
Desmidocercella incognita Solonitsin, 1932
Schistorophus skrjabini (Vasilkova, 1926)
Tetrameres fissispina (Diesing, 1861)
T. skrjabini Panova, 1926
Aprocta turgida Stossich, 1902
Eufilaria lari Yamaguti, 1935

L. audouinii Payr., 1826
(No nematodes yet recorded.)

L. brunnicephalus Jerd., 1840
(No nematodes yet recorded.)

L. canus L., 1758

Capillaria carbonis (Rudolphi, 1819)
Thominx contorta (Creplin, 1839)
Strongyloides turkmenicus Kurtieva, 1953
Cyathostoma lari Blanchard, 1849
Contracaecum rudolphii Hartwich, 1964
C. variegatum (Rudolphi, 1809)
Porrocaecum ensicaudatum (Zeder, 1800)
Cosmocephalus aduncus (Creplin, 1846)
C. obvelatus (Creplin, 1825)

Rusguniella elongata (Rudolphi, 1819)
Paracuaria tridentata (Linstow, 1877)
Streptocara crassicauda (Creplin, 1829)
Seuratia shipleyi (Stossich, 1900)
Stegophorus stellaepolaris (Parona, 1901)
Schistorophus skrjabini (Vasilkova, 1926)
Tetrameres fissispina (Diesing, 1861)
T. skrjabini Panova, 1926
Eufilaria lari Yamaguti, 1935

L. cirrocephalus Vieill., 1818
(No nematodes yet recorded.)

L. crassirostris Vieill., 1818

Thominx contorta (Creplin, 1839)

Contracaecum microcephalum (Rudolphi, 1809)

C. rudolphii Hartwich, 1964
Cosmocephalus obvelatus (Creplin, 1825)
Paracuaria tridentata (Linstow, 1877)

Stegophorus stellaepolaris (Parona, 1901)
Viktorocara guschanscoi Leonov, 1958
Tetrameres skrjabini Panova, 1926

L. fuscus L., 1758

Thominx contorta (Creplin, 1839)
Syngamus trachea (Montagu, 1811)
Cyathostoma lari Blanchard, 1849
Contracaecum rudolphii Hartwich, 1964
Porrocaecum semiteres (Zeder, 1800)
Cosmocephalus obvelatus (Creplin, 1825)

C. aduncus (Creplin, 1846)
Paracuaria tridentata (Linstow, 1877)
Rusguniella elongata (Rudolphi, 1819)
Skrjabinoclava horrida (Rudolphi, 1809)
Tetrameres fissispina (Diesing, 1861)

L. genei Breme, 1840

Capillaria carbonis (Rudolphi, 1819)
Thominx anatis (Schrank, 1790)
Th. contorta (Creplin, 1839)
Contracaecum rudolphii Hartwich, 1964
Cosmocephalus obvelatus (Creplin, 1825)
C. aduncus (Creplin, 1846)
Dispharynx nasuta (Rudolphi, 1819)
Paracuaria tridentata (Linstow, 1877)

Rusguniella elongata (Rudolphi, 1819)
Echinuria heterobrachiata Wehr, 1937
Streptocara crassicauda (Creplin, 1829)
Desmidocercella numidica (Seurat, 1920)
Tetrameres skrjabini Panova, 1926
Sciadiocara umbellifera (Molin, 1860)
Aprocta turgida Stossich, 1902
Eufilaria lari Yamaguti, 1935

L. glaucescens Naum., 1840

Cosmocephalus obvelatus (Creplin, 1825)
Paracuaria tridentata (Linstow, 1877)
Stegophorus stellaepolaris (Parona, 1901)

Seuratia shipleyi (Stossich, 1900)
Eufilaria lari Yamaguti, 1935

L. glaucoides Meyer, 1822
(No nematodes yet recorded.)

L. hemprichii Bruch., 1853
(No nematodes yet recorded.)

L. hyperboreus Gunn., 1767

Thominx contorta (Creplin, 1839)
Contracaecum rudolphii Hartwich, 1964
Cosmocephalus obvelatus (Creplin, 1825)

C. aduncus (Creplin, 1846)
Paracuaria tridentata (Linstow, 1877)

L. ichthyaetus Pall., 1773

Thominx contorta (Creplin, 1839)
Contracaecum microcephalum (Rudolphi, 1809)
C. rudolphii Hartwich, 1964
C. variegatum (Rudolphi, 1809)
Porrocaecum ensicaudatum (Zeder, 1800)
Cosmocephalus obvelatus (Creplin, 1825)

C. aduncus (Creplin, 1846)
Paracuaria tridentata (Linstow, 1877)
Streptocara crassicauda (Creplin, 1829)
Tetrameres skrjabini Panova, 1926
Aprocta turgida Stossich, 1902

L. leucophthalmus Temm., 1825
(No nematodes yet recorded.)

L. marinus L., 1758

Thominx contorta (Creplin, 1839)
Contracaecum rudolphii Hartwich, 1964
Cosmocephalus obvelatus (Creplin, 1825)

C. aduncus (Creplin, 1846)
Paracuaria tridentata (Linstow, 1877)
Eufilaria lari Yamaguti, 1935

L. melanocephalus Temm., 1820

Cosmocephalus obvelatus (Creplin, 1825)

Paracuaria tridentata (Linstow, 1877)

L. minutus Pall., 1776

Thominx anatis (Schrank, 1790)
Th. contorta (Creplin, 1839)
Contracaecum rudolphii Hartwich, 1964
Cosmocephalus obvelatus (Creplin, 1825)
Paracuaria tridentata (Linstow, 1877)

Rusguniella elongata (Rudolphi, 1819)
Echinuria uncinata (Rudolphi, 1819)
Streptocara crassicauda (Creplin, 1829)
Tetrameres skrjabini Panova, 1926

L. relictus Lönn., 1931
(No nematodes yet recorded.)

L. ridibundus L., 1766

Thominx anatis (Schrank, 1790)
Th. contorta (Creplin, 1839)
Eustrongylides mergorum (Rudolphi, 1809)
Amidostomum fulicae (Rudolphi, 1819)
Cyathostoma lari Blanchard, 1849
Porrocaecum ensicaudatum (Zeder, 1800)
Contracaecum rudolphii (Hartwich, 1964)
Cosmocephalus obvelatus (Creplin, 1825)
C. aduncus (Creplin, 1846)
Dispharynx nasuta (Rudolphi, 1809)
Chevreuxia revoluta (Rudolphi, 1819)

Paracuaria tridentata (Linstow, 1877)
Rusguniella elongata (Rudolphi, 1819)
Streptocara crassicauda (Creplin, 1829)
Desmidocercella numidica (Seurat, 1920)
Tetrameres fissispina (Diesing, 1861)
T. pavonis Chertkova, 1953
T. skrjabini Panova, 1926
Sciadiocara umbellifera (Molin, 1860)
Aprocta turgida Stossich, 1902
Eufilaria lari Yamaguti, 1935
Pelecitus fulicaeatrae (Diesing, 1861)

L. saundersi (Swinh., 1871)
(No nematodes yet recorded.)

L. schistisagus Stejn., 1884

Thominx contorta (Creplin, 1839)
Cosmocephalus obvelatus (Creplin, 1825)
Paracuaria tridentata (Linstow, 1877)

Streptocara crassicauda (Creplin, 1829)
Tetrameres skrjabini Panova, 1926
Viktorocara schejkini Gushanskaya, 1950

Larus sp.

Porrocaecum ensicaudatum (Zeder, 1800)

Aprocta turgida Stossich, 1902

Genus *Rhynchops* L., 1758
R. flavirostris Vieill., 1816
(No nematodes yet recorded.)

Genus *Xema* Leach, 1819
X. sabini (Sabine, 1818)

Cosmocephalus obvelatus (Creplin, 1825)
Rusguniella elongata (Rudolphi, 1819)

Schistorophus laciniatus (Molin, 1860)
Eufilaria lari Yamaguti, 1935

Genus *Rissa* Stephens, 1826
R. brevirostris (Bruch, 1853)

Contracaecum rudolphii Hartwich, 1964
Paracauria tridentata (Linstow, 1877)

Schistorophus skrjabini (Vasilkova, 1926)

R. tridactyla (L., 1758)

Thominx contorta (Creplin, 1839)
Contracaecum rudolphii Hartwich, 1964
C. variegatum (Rudolphi, 1809)
Cosmocephalus obvelatus (Creplin, 1825)

Paracuaria tridentata (Linstow, 1877)
Streptocara crassicauda (Creplin, 1829)
Schistorophus skrjabini (Vasilkova, 1926)
Eufilaria lari Yamaguti, 1935

Genus *Rhodostethia* Mac Gillivray, 1824
R. rosea (Mc Gill., 1824)

Contracaecum rudolphii Hartwich, 1964

Paracauria tridentata (Linstow, 1877)

Genus *Pagophila* Kaup, 1829
P. eburnea (Phipps, 1774)

Paracuaria tridentata (Linstow, 1877)

Genus *Chlidonias* Rafinesque, 1822
Ch. hybrida (Pall., 1811)

Contracaecum rudolphii Hartwich, 1964
Cosmocephalus obvelatus (Creplin, 1825)
C. aduncus (Creplin, 1846)

Tetrameres skrjabini Panova, 1926
Sciadiocara umbellifera (Molin, 1860)

Ch. leucoptera (Temm., 1815)

Thominx contorta (Creplin, 1839)
Contracaecum rudolphii Hartwich, 1964
Cosmocephalus obvelatus (Creplin, 1825)
Paracuaria tridentata (Linstow, 1877)
Tetrameres fissispina (Diesing, 1861)

T. skrjabini Panova, 1926
Stellocaronema skrjabini Gilbert, 1930
Aprocta turgida Stossich, 1902
Eufilaria lari Yamaguti, 1935

Ch. nigra (L., 1758)

Capillaria carbonis (Rudolphi, 1819)
Thominx contorta (Creplin, 1839)
Porrocaecum ensicaudatum (Zeder, 1800)
Cosmocephalus obvelatus (Creplin, 1825)
Rusguniella elongata (Rudolphi, 1819)

Tetrameres fissispina (Diesing, 1861)
T. skrjabini Panova, 1926
Sciadiocara umbellifera (Molin, 1860)
Stellocaronema skrjabini Gilbert, 1930
Pelecitus fulicaeatrae (Diesing, 1861)

Genus Gelochelidon Brehm, 1830
G. nilotica (Gm., 1789)

Capillaria carbonis (Rudolphi, 1819)
Thominx contorta (Creplin, 1839)
Contracaecum rudolphii Hartwich, 1964
Cosmocephalus obvelatus (Creplin, 1825)
C. aduncus (Creplin, 1846)
Synhimantus niloticus Leonov, 1958

Skrjabinoclava horrida (Rudolphi, 1809)
Streptocara crassicauda (Creplin, 1829)
Tetrameres skrjabini Panova 1926
Schistorophus bihamatus (Mueller, 1897)
Sciadiocara umbellifera (Molin, 1860)
Aprocta turgida Stossich, 1902

Genus Hydroprogne Kaup, 1829
H. tschegrava (Lep., 1770)

Thominx anatis ((Schrank, 1790)
Th. contorta (Creplin, 1839)
Contracaecum rudolphii Hartwich, 1964
Cosmocephalus obvelatus (Creplin, 1825)
C. aduncus (Creplin, 1846)

Paracuaria tridentata (Linstow, 1877)
Streptocara crassicauda (Creplin, 1829)
Schistorophus acanthocephalicus (Molin, 1860)
Viktorocara guschanscoi Leonov, 1958

Genus Sterna L., 1758
S. albifrons Pall., 1764

Thominx anatis (Schrank, 1790)
Th. contorta (Creplin, 1839)
Cosmocephalus obvelatus (Creplin, 1825)

Paracuaria tridentata (Linstow, 1877)
Rusguniella elongata (Rudolphi, 1819)
Sciadiocara umbellifera (Molin, 1860)

S. aleutica Baird, 1869

Thominx contorta (Creplin, 1839)

Contracaecum rudolphii Hartwich, 1964

S. anaethetus Scop., 1786
(No nematodes yet recorded.)

S. bengalensis Less., 1831
(No nematodes yet recorded.)

S. bergii Licht., 1823
(No nematodes yet recorded.)

S. fuscata L., 1766
(No nematodes yet recorded.)

S. dougallii Mont., 1813
(No nematodes yet recorded.)

S. hirundo L., 1758

Capillaria carbonis (Rudolphi, 1819)
Thominx contorta (Creplin, 1839)
Syngamus trachea (Montagu, 1811)
Contracaecum microcephalum (Rudolphi, 1809)
C. rudolphii Hartwich, 1964
Porrocaecum ensicaudatum (Zeder, 1800)
Cosmocephalus obvelatus (Creplin, 1825)
C. aduncus (Creplin, 1846)
Paracuaria tridentata (Linstow, 1877)

Rusguniella elongata (Rudolphi, 1819)
Streptocara crassicauda (Creplin, 1829)
Stegophorus stercorarii Leonov, Sergeeva & Tsimbalyuk, 1966
Tetrameres skrjabini Panova, 1926
Schistorophus acanthocephalicus (Molin, 1860)
Sciadiocara umbellifera (Molin, 1860)
Viktorocara guschanscoi Leonov, 1958

S. maxima Bodd., 1783
(No nematodes yet recorded.)

S. paradisea Pontopp., 1763

Syngamus trachea (Montagu, 1811)
Contracaecum matwejewi Layman & Mudretsova, 1926
C. rudolphii Hartwich, 1964

Porrocaecum ardeae (Frölich, 1802)
P. ensicaudatum (Zeder, 1800)
P. semiteres (Zeder, 1800)
Cosmocephalus obvelatus (Creplin, 1825)

S. repressa Mart., 1916
(No nematodes yet recorded.)

S. sandvicensis Lath., 1787

Thominx anatis (Schrank, 1790)
Th. contorta (Creplin, 1839)
Contracaecum rudolphii Hartwich, 1964
Porrocaecum ensicaudatum (Zeder, 1800)
osmocephalus obvelatus (Creplin, 1825)

C. aduncus (Creplin, 1846)
Paracuaria tridentata (Linstow, 1877)
Streptocara crassicauda (Creplin, 1829)
Tetrameres skrjabini Panova, 1926
Viktorocara guschanscoi Leonov, 1958

S. saundersi Hume, 1877
(No nematodes yet recorded.)

S. sumatrana Raff., 1822
(No nematodes yet recorded.)

S. zimmermanni Reich., 1903
(No nematodes yet recorded.)

Suborder Alcae

Family *Alcidae*
Genus *Plotus* Gunnerus, 1761
P. alle (L., 1758)
(No nematodes yet recorded.)

Genus *Alca* L., 1758
A. torda L., 1758

Eustrongylides mergorum (Rudolphi, 1809)
E. tubifex (Nitzsch, 1819)
Contracaecum rudolphii Hartwich, 1964

C. variegatum (Rudolphi, 1809)
Cosmocephalus obvelatus (Creplin, 1825)
Streptocara crassicauda (Creplin, 1829)

Genus *Uria* Brisson, 1760
U. aalge (Pontopp., 1763)

Eustrongylides tubifex (Nitzsch, 1819)
E. mergorum (Rudolphi, 1809)
Contracaecum rudolphii Hartwich, 1964
C. variegatum (Rudolphi, 1809)
Cosmocephalus obvelatus (Creplin, 1825)

C. imperialis Morishita, 1930
Skrjabinocerca prima Shikhobalova, 1930
Streptocara crassicauda (Creplin, 1829)
Stegophorus stellaepolaris (Parona, 1901)
Eufilaria lari Yamaguti, 1935

U. lomvia (L., 1758)

Thominx contorta (Creplin, 1839))
Eustrongylides mergorum (Rudolphi, 1809)
Contracaecum rudolphii Hartwich, 1964
Streptocara crassicauda (Creplin, 1829)

Stegophorus stellaepolaris (Parona, 1901)
S. stercorarii Leonov, Sergeeva & Tsimbalyuk, 1966
Eufilaria lari Yamaguti, 1935

Genus *Cepphus* Pallas, 1769
C. carbo Pall., 1811

Contracaecum rudolphii Hartwich, 1964
Cosmocephalus obvelatus (Creplin, 1825)

Stegophorus stellaepolaris (Parona, 1901)

C. columba Pall., 1811
(No nematodes yet recorded.)

C. grylle (L., 1758)

Thominx contorta (Creplin, 1839)
Eustrongylides mergorum (Rudolphi, 1809)
Contracaecum rudolphii Hartwich, 1964

Stegophorus stercorarii Leonov, Sergeeva & Tsimbalyuk, 1966
Eufilaria lari Yamaguti, 1935

Genus *Brachyramphus* Brandt, 1837
B. brevirostris (Vigors, 1829)
(No nematodes yet recorded.)

B. marmoratus (Gm., 1789)
(No nematodes yet recorded.)

Genus *Synthliboramphus* Brandt, 1837
S. antiqus (Gm., 1789)

Contracaecum rudolphii Hartwich, 1964

Stegophorus stellaepolaris (Parona, 1901)

S. wumizusume (Temm., 1835)
(No nematodes yet recorded.)

Genus *Aethia* Merrem, 1788
A. cristatella (Pall., 1769)

Heterakis kurilensis Oshmarin, 1950   *Stegophorus stercorarii* Leonov, Sergeeva & Tsimbalyuk, 1966

A. pusilla (Pall., 1811)
Stegophorus stercorarii Leonov, Sergeeva & Tsimbalyuk, 1966

A. pygmaea (Gm., 1789)

Paracuaria tridentata (Linstow, 1877)   *Stegophorus stellaepolaris* (Parona, 1901)

Genus *Cyclorrhynchus* Kaup, 1829
C. psittacula (Pall., 1769)

Paracuaria tridentata (Linstow, 1877)   *Stegophorus stellaepolaris* (Parona, 1901)

Genus *Cerorhinca* Bonaparte, 1828
C. monocerata (Pall., 1811)
Cosmocephalus obvelatus (Creplin, 1825)

Genus *Fratercula* Brisson, 1760
F. arctica (L., 1758)
(No nematodes yet recorded.)

F. corniculata (Naum, 1821)

Contracaecum rudolphii Hartwich, 1964   *Stegophorus stercorarii* Leonov, Sergeeva & Tsimbalyuk, 1966
Streptocara crassicauda (Creplin, 1829)   *Seuratia shipleyi* (Stossich, 1900)

Genus *Lunda* Pallas, 1811
L. cirrhata (Pall., 1769)

Contracaecum rudolphii Hartwich, 1964   *Stegophorus stellaepolaris* (Parona, 1901)
Skrjabinocerca prima Shikhobalova, 1930   *S. stercorarii* Leonov, Sergeeva & Tsimbalyuk, 1966

Oder *Strigiformes*
Family *Strigidae*
Genus *Ketupa* Lesson, 1830
K. blakistoni (Seeb., 1884)
(No nematodes yet recorded.)

K. flavipes Hodg., 1836
(No nematodes yet recorded.)

K. zeylonensis (Gm., 1788)
(No nematodes yet recorded.)

Order *Coraciiformes*
Family *Alcedinidae*
Genus *Ceryle* Boie, 1828
C. lugubris (Temm., 1834)
(No nematodes yet recorded.)

C. rudis (L., 1758)
(No nematodes yet recorded.)

(?) *C. torquata* (L., 1766)
Monopetalonema alcedinis (Rudolphi, 1819)

Genus *Halcyon* Swainson, 1821
H. coromanda (Lath., 1790)
(No nematodes yet recorded.)

H. pileata (Bodd., 1783)

Skrjabinoclava halcyoni Ryzhikov & Khokhlova, 1964
Sobolevicephalus halcyonis Parukhin, 1964
Skrjabinocerca prima Shikhobalova, 1930
Viktorocara tenuis (Maplestone, 1932)

H. smyrnensis (L., 1758)
Skrjabinoclava halcyoni Ryzhikov & Khokhlova, 1964

Genus *Alcedo* L., 1758
A. atthis L., 1758

Aviculariella alcedonis (Yamaguti & Mitunaga, 1943)
Proyseria decora (Dujardin, 1845)

References

Abdou A. M. & Selim M. K., 1957: On the life-cycle of *Subulura suctoria*, a caecal nematode of poultry in Egypt. Z. ParasitKde, 18: 20—23.
Abdou A. M. & Selim M. K., 1963: The life-cycle of *Subulura suctoria* (Mol., 1860) Rail. and Henry, 1912 in the domestic fowl. Z. ParasitKde, 23: 45—49.
Ablasov N. A., 1957: (Helminth fauna of water birds of Kirghizia.) Trudy Inst. Zool.Parazit., Frunze, No. 6: 121—144 (in Russian).
Ablasov N. A. & Chibichenko N. T., 1961: (Helminth fauna of wild birds of Kirghizia.) In: Ptitsy Kirghizii, Tom. III., Frunze, pp. 187—279 (in Russian).
Ablasov N. A. & Chibichenko N. T., 1962: (Nematode fauna of wild birds of Kirghizia.) Izv. Akad. Nauk Kirgiz. SSR, Ser. biol. Nauk, 4: 113—130 (in Russian).
Ablasov N. A., Iksanov K. I. & Chibichenko N. T., 1960: (A short note on the helminths of pelicans of Lake Balkhash.) Izv. Akad. Nauk Kirgiz. SSR, Ser. biol. Nauk, 2: 181—182 (in Russian).
Adams J. R. & Gibson G. G., 1969: *Ancyracanthopsis bendelli* n. sp. *(Acuariidae: Schistorophinae)* from Pacific Coast grouse with observations on related nematode genera. Can. J. Zool., 47: 619—626.
Agapova A. I. & Zhatkanbaeva D., 1971: (Helminths of fish-eating birds of Markakol' Lake.) Mater. nauch. Konf. Vses. Obshch. Gel'mint., 1969—1970, pp. 3—6 (in Russian).
Akhumyan K. S., 1966: (Species composition of helminths of game and other wild birds of the Armenian SSR.) Biol. Zhurn. Armenii, 19: 97—104 (in Russian).
Alekseev V. M., 1970: (Zoogeographical characteristics of the helminth fauna of *Anseriformes* in Primorsk Territory.) Uchen. Zap. Dal'nevost. gos. Univ. 16: 3—11 (in Russian).
Alekseev V. M. & Smetanina Z. B., 1968: (Nematodes of fish-eating birds of Rimskij-Korsakov islands.) In: Skryabin K. I. & Mamaev Yu. L. (Editors), Gel'minty zhivotnykh Tikhogo okeana. Moscow: Izdat. „Nauka", pp. 97—104 (in Russian).
Ali M. M., 1970: A review and revision of the subfamily *Cyathostominae* Nicoll, 1927 (Nematoda, Syngamidae). Acta parasit. pol., 17: 237—246.
Ali M. M., 1971: A review and revision of the subfamily *Epomidiostomatinae* Skrjabin et Schulz, 1937. Riv. Parassit., 32: 179—192.
Alicata J. E., 1939: Preliminary note on the life history of *Subulura brumpti*, a common caecal nematode of poultry in Hawaii. J. Parasit., 25: 179—180.
Anderson R. C., 1959: The morphology of *Monopetalonema alcedinis* (Rudolphi, 1819) (Nematoda: *Filarioidea)* including its first-stage larva. Can. J. Zool., 37: 609—614.
Anderson R. C. & Chabaud A. G., 1959: Remarques sur la classification des *Splendidofilariinae*. Annls Parasit. hum. comp., 34: 53—63.
Anderson R. C. & Díaz-Ungría C., 1959: Revisión preliminar de las especies de *Thelazia* Bosc *(Spiruroidea: Thelaziidae)*, parásitas de aves. Mems. Soc. Cienc. nat. "La Salle", 19: 37—75.
Babaev Ya., 1970: (Study of helminth fauna of wild vertebrates in Turkmenia.) In: Tashliev A. O. & Shagalina L. M. (Editors), Akademik K. I. Skrjabin i razvitie helmintologicheskoi nauki v Turkmenii. Ashkabad: Izdat. "ILIM", pp. 16—40 (in Russian).

Babič I., 1936: Entoparaziti ptica zapadnog dijela drzave. Vet. Arh., 6: 297—302.
Baer J. G., 1956: Parasitic helminths collected in West Greenland. Meddr Grønland, 124: 1—55.
Baird W., 1853: Catalogue of the species of *Entozoa*, or intestinal worms, contained in the collection of the British Museum. London, 132 pp.
Bakke T. A., 1969: Fiskemåke som vert for *Syngamus Lari* i Norge. Fauna, Oslo 22: 153—157.
Bakke T. A., 1970: Tetrameriasis i Norge. Nord. Vet. Med. 22: 8—12.
Bakke T. A., 1972: Check list of helminth parasites of the common gull *(Larus canus L.)*. Rhizocrinus, 1: 1—20.
Bakke T. A., 1973: Studies of the helminth fauna of Norway XXVII: Syngamiasis in Norway. Norw. J. Zool., 21: 299—303.
Bakke T. A. & Baruš V., 1976: Studies of the helminth fauna of Norway XXXVIII: The common gull, *Larus canus L.*, as final host for Nematoda. III. Qualitative and quantitative data on species of *Capillariidae* Neveu-Lemaire, 1936, *Strongyloididae* Chitwood et McIntosh, 1934 and *Syngamidae* Leiper, 1912. Norw. J. Zool, 23: 183—191.
Bakke T. A. & Baruš V., 1976: Studies of the helminth fauna of Norway XXXVII: The common gull, *Larus canus* L., as final host for Nematoda. II. Qualitative and quantitative data on species of *Acuariidae, Capillariidae, Strongyloididae, Syngamidae*, and *Tetrameridae;* with notes on host-parasite relationship. Norw. J. Zool., 24: 7—31.
Baruš V., 1964 a: The morphological and biometrical variability of the nematode *Syngamus (Syngamus) trachea* (Montagu, 1811) Chapin, 1925 and a revision of the species composition of the subgenus *Syngamus*. Věst. čs. Spol. zool., 28: 290—304.
Baruš V., 1964 b: Studien über die exogene Phase des Entwicklungs-Zyklus von *Amidostomum fulicae* (Rudolphi, 1819) *(Nematoda, Amidostomatidae)*. Z. ParasitKde., 24: 112—120.
Baruš V., 1964 c: The species of *Capillaria* Zeder 1800 and *Thominx* Dujardin 1845 (Nematoda, *Trichocephaloidea*) in *Strigiformes* and *Falconiformes (Aves)* in Czechoslovakia. Čslka. parasit. (Praha), 11: 51—64.
Baruš V., 1967: Redescription and synonymy of the species *Paracuaria tridentata* (Linstow, 1877) *Nematoda, Acuariidae*. Folia parasit., Praha, 14: 281—286.
Baruš V., 1969: Studies on the nematode *Subulura suctoria*. I. Morphological and metrical variability. Folia parasit., Praha, 16: 303—311.
Baruš V., 1970 a: Studies of the nematode *Subulura suctoria*. II. Development in the intermediate host. Folia parasit., Praha, 17: 49—59.
Baruš V., 1970 b: Studies on the nematode *Subulura suctoria*. IV. Intermediate hosts. Folia parasit., Praha, 17: 191—199.
Baruš V., 1970 c: Beitrag zur Morphologie der Larven der Nematodenart *Cyathostoma lari* Blanchard, 1849, während der exogenen Entwicklungsphase. Z. ParasitKde., 34: 151 to 157.
Baruš V., 1974: Redescription and taxonomic position of *Trichosoma pachyderma* Linstow, 1877 *(Capillariidae)*. Folia parasit., Praha, 21: 381—384.
Baruš V. & Blažek K., 1965: Revision der exogenen und endogenen Phase des Entwicklungszyklus und der Pathogenität von *Syngamus (Syngamus) trachea* (Montagu, 1811) Chapin, 1928 in Organismus des Endwirtes. Čslka. Parasit. (Praha), 12: 47—70.
Baruš V. & Blažek K., 1970: Studies on the nematode *Subulura suctoria*, III. Development in the definitive host. Folia parasit., Praha, 17: 141—151.
Baruš V., Kullmann E. & Tenora F., 1972: Parasitische Nematoden aus Wirbeltieren Afghanistans. Acta Sci. Nat. Acad. Sci. Bohemoslovacae Brno, 6: 1—46.
Baruš V. & Lorenzo Hernández N., 1971: Nemátodos parásitos de aves en Cuba. Parte IV. Poeyana, No. 88: 1—15.
Baruš V., Ryšavý B., Groschaft J. & Folk Č., 1972: The helminth fauna of *Corvus frugilegus* L.

(*Aves, Passeriformes*) in Czechoslovakia and its ecological analysis. Acta Sci. Nat. Acad. Sci. Bohemoslovacae Brno, 3: 1—53.

Baruš V. & Tenora F., 1972: Notes on the systematics and taxonomy of the nematodes belonging to the family *Syngamidae* Leiper, 1912. Acta Univ. agric., fac. agronom., Brno 20: 275—286

Baruš V. & Zajíček D., 1967: Parasitic nematodes of birds of the order *Colymbiformes* in Czechoslovakia. Folia parasit., Praha, 14: 73—85.

Bashkirova E. Ya., 1960: (Nematode fauna of birds of Primorye Territory.) Trudy gelmint. Lab., 10: 46—57 (in Russian).

Baylis H. A., 1920: On the classification of the *Ascaridae*. I. The systematic value of certain characters of the alimentary canal. Parasitology, 12: 253—264.

Baylis H. A., 1923: Report on a collection of parasitic Nematodes, mainly from Egypt. Pt. 1. *Ascaridae* and *Heterakidae*. Parasitology, 15: 1—13.

Baylis H. A., 1928: Records of some parasitic worms from British vertebrates. Annls Mag. nat. Hist., Ser. 10, 1: 329—343.

Baylis H. A., 1934: Miscellaneous notes on parasitic worms. Annls Mag. nat. Hist., Ser. 10, 13: 223—228.

Baylis H. A., 1936: The fauna of British India, including Ceylon and Burma. Nematoda. Vol. I. *Ascaroidea* and *Strongyloidea*. London, XXXVI+1—408.

Baylis H. A., 1939: Further records of parasitic worms from British veterbrates. Annls Mag. nat. Hist., 4: 473—498.

Baylis H. A., 1944: Notes on some parasitic nematodes. Annls Mag. nat. Hist., 11: 793—804.

Baylis H. A. & Daubney R., 1922: Report on the parasitic nematodes in the collection of the Zoological Survey of India. Mems Ind. Mus. Calcutta, 7: 266—347.

Baylis H. A. & Daubney R., 1923: A further report on parasitic nematodes in the collection of the Zoological Survey of India. Rec. Indian Mus., 25: 551—578.

Belogurov O. I., 1965: (Nematode fauna of wild *Anseriformes* of continent border of the Sea of Okhotsk.) In: (Parasitic worms of domestic and wild animals.) Vladivostok, pp. 43 to 47 (in Russian).

Belogurov O. I., Leonov V. A. & Zueva L. S., 1968: (Helminth fauna of fish-eating bird. (gulls and guillemots) from the coast of the Sea of Okhotsk.) In: Skryabin K. I. & Mamaev Yu. L. (Editors), (Helminths of animals of the Pacific Ocean), Moscow: Izdats "Nauka", pp. 105—124 (in Russian).

Belogurov O. I. & Smetanina Z. B., 1965: (Short analysis of the helminth fauna of fish-eating birds on the continental coast of the Sea of Okhotsk.) Mat. nauch. Konf. vses. Obshch. Gelmint., Part I., pp. 12—16 (in Russian).

Belopolskaya M. M., 1952: (Parasite fauna of sea water birds.) Uchen. Zap. leningr. gos. Univ., Ser. biol. nauk (Zoologiya), 28: 127—180 (in Russian).

Belopolskaya M. M., 1959: (Parasite fauna of birds of Sudzukhin reservation (Primorye). 3. Roundworms (Nematoda).) In: Polyanski Y. I. (Editor), (Ekologicheskaya Parazitologyia.) Leningrad: Izdat. Leningr. Univ., pp. 3—21 (in Russian).

Belopolskaya M. M., 1963: (Survey of parasite fauna of birds of Sudzukhin reserve (Primorye). Parazit. Sb., 21: 221—244 (in Russian).

Belous E. V., 1971: (Helminth fauna of birds of the Kuril Islands.) In: Parazity zhivotnykh i rastenii Dalnogo Vostoka, Vladivostok: Dalnevost. knizhnoe Izdat., pp. 11—14 (in Russian).

Bezubik B., 1956: Materiały do helmintofauny ptaków wodnych Polski. Acta parasit. pol., 4: 59—88.

Blanchard R., 1848: Les vers du sang. Les hematozoaires de l'homme et des animaux. Paris, pp. 1—181.

Blanchard E., 1848—1849: Recherches sur l'organisation de vers. Ann. Sci. Nat. Zool., 10: 321—364; 11: 106—202; 12: 267—276.

Borgarenko L. F., 1960: (Nematodes of game-birds of Tadzhikistan.) Izv. Otd. sel-khoz. biol. nauk AN Tadzhik. SSR 1: 119—133 (in Russian).

Borgarenko L. F., 1970: (Helminths of birds of the order *Podicipediformes* in Tadzhikistan.) In: Vop. zool. Tadzhik., Dushanbe, pp. 39—47 (in Russian).

Borgarenko L. F. & Daiya G. G., 1972: (New data on nematodes of the genus *Capillaria* (*C. podicipitis* Yamaguti, 1941 and *C. ryjikovi* Daija sp. n.) parasitizing grebes.) Izv. AN Latv. SSR, 5: 44—49 (in Russian).

Boulenger C. L., 1928: Report on a collection of parasitic nematodes, mainly from Egypt. Part V., *Filaroidea*. Parasitology, 20: 32—55.

Boyd E. M., 1966: Observations on nematodes of herons in North America including three new species and new host and state records. J. Parasit., 52: 503—511.

Braun M., 1892: Verzeichnis von Eingeweidewürmern aus Mecklenburg. Arch. Ver. Fr. Naturg. Mecklenburg (1891), No. 5, pp. 97—117.

Bremser J. G., 1824: Icones helminthum systema Rudolphii entozoologicum illustrantes. Vienna, 12 pp., 18 plates.

Broek E. van den, 1968: Bird parasites collected by G. J. van Oordt on Spitsbergen. Ardea, 56: 286—287.

Broek E. van den & Jansen J. Jr., 1964: Parasites of animals in the Netherlands. Supplement I.: Parasites of wild birds. Ardea, 52: 111—116.

Broek E. van den & Jansen J., 1971: Parasites of animals in the Netherlands. Supplement V. Endoparasites of wild birds. Ardea, 59: 28—33.

Burt D. R. R. & Eadie J. M., 1958: *Cyathostoma lari* Blanchard, 1849 *(Nematoda, Strongyloidea)*: its anatomy, intra-specific variation and hosts, with a re-definition of the genus. J. Linn. Soc. (Zoology), 43: 575—586.

Campana Y., 1949: Une filaire nouvelle du héron cendré *(Ardea cinerea* L.) *Lemdana urbaini* n. sp. Annls Parasit. hum. comp., 24: 443—446.

Chabaud A. G., 1950: Cycle évolutif de *Synhimantus (Desportesius) spinulatus (Nematoda, Acuariidae)*. Annls Parasit. hum. comp., 25: 150—166.

Chabaud A. G., 1952: Identité de *Petroviprocta vigissi* Schachtachtinskaja, 1951 et d'*Avioserpens gailliardi* Chabaud et Campana, 1949. Annls Parasit. hum. comp., 26: 482.

Chabaud A. G., 1953: Sur un nématode *Acuariidae* parasite du martin-pêcheur *Alcedo atthis* (L.). Annls Parasit. hum. comp., 28: 365—371.

Chabaud A. G., 1954: Sur le cycle évolutif des spirurides et de nématodes ayant une biologie comparable. Valeur systématique des caractères biologiques (suite et fin). Annls Parasit. hum. comp., 29; 358—425.

Chabaud A. G., 1957: Note sur les nématodes du genre *Desmidocercella*. Annls Parasit. hum. comp., 32: 343—347.

Chabaud A. G., 1958: Essai de classification des nématodes *Habronematinae*. Annls Parasit. hum. comp., 33: 445—509.

Chabaud A. G. & Campana Y., 1949a: *Avioserpens galliardi* n. sp., parasite de l'aigrette *Egretta garzetta* L. Annls Parasit. hum. comp., 24: 67—76.

Chabaud A. G. & Campana Y., 1949b: A propos d'une variété nouvelle de *Synthimantus equispiculatus* Wu & Liu 1943. Création d'une nouveau sous-genre *(Desportesius)* n. subgen. Annls Parasit. hum. comp., 24: 77—92.

Chabaud A. G. & Campana Y., 1950a: Note sur le genre *Hadjelia* Seurat, 1916 *(Nematoda: Spiruridae)*. Annls Parasit. hum. comp., 25: 435—440.

Chabaud A. G. & Campana Y., 1950b: Nouveau parasite remarquable par l'atrophie de ses

organes; *Robertdollfusa paradoxa (Nematoda, Incertae sedis)*. Annls Parasit. hum. comp., 25: 325—334.
Chabaud A. G. & Choquet M. T., 1953: Nouvel essai de classification des filaires (superfamille des *Filarioidea*). Annls Parasit. hum. comp., 28: 172—192.
Chabaud A. G. & Choquet M. T., 1955: Helminthes de la région de Banyuls. II. Deux filaires parasites d'oiseaux. Vie Milieu, 6: 93—100.
Chabaud A. G. & Czapliński B., 1961a: Le nématode parasite de mouettes, *Paracuaria macdonaldi* Rao, 1951, est une forme de passage entre Habronematinae et Acuariinae. Cah. Biol. mar., 2: 67—70.
Chabaud A. G. & Czapliński B., 1961b: *Paracuaria macdonaldi* Rao, 1951 — forma przejściowa między Habronematinae i Acuariinae. Wiad. parazyt., 7, Suppl.: 211—212.
Chapin E. A., 1925a: Review of the nematode genera *Syngamus* Sieb. and *Cyathostoma* E. Blanch. J. Agric. Res. 30: 557—570.
Chapin E. A., 1925b: Descriptions of new internal parasites. Proc. U. S. natn. Mus., 68 (Art. 2): 4 pp.
Chertkova A. N., 1953: (New nematode, *Tetrameres (Pterowimeres) pavonis* n. subg., n. sp. from the peacock.) Papers on helminthology presented to Acad. K. I. Skrjabin on his 75th birthday. Moscow: Izdat. Akad. Nauk SSSR, pp. 737—739 (in Russian).
Chertkova A. N., 1961: (Nematodes and acanthocephalans of domestic *Galliformes.*) In: Chertkova A. N. and Petrov A. M., (Gelminty domashnikh ptits i vyzyvaemye imi zabolevaniya, 2. Nematody i akantocefaly domashnikh kurinykh ptits i zabolevaniya, vyzyvaemye nematodami. Moscow, Vses. Inst. Gelmint., pp. 5—270 (in Russian).
Chibichenko N. T., 1960: (Helminth fauna of predatory birds of Kirgizia. Izv. Akad. Nauk Kirgiz. SSR, Ser. biol. Nauk, 2: 169—175 (in Russian).
Chitwood B. G. & Wehr E. E., 1934: The value of cephalic structures as characters in nematode classification, with special reference to the superfamily *Spiruroidea*. Z. ParasitKde., 7: 273—335.
Chow C. Y., 1939: Notes sur quelques nématodes de l'Indochine française. Annls Parasit. hum. comp., 17: 20—31.
Ciurea I., 1924: Die Eustrongylidae-Larven bei Donaufischen. Z. Fleisch. Milch-Hygiene, 34: 134—137.
Ciurea I., 1938: Sur une infestation massive des poissons rapaces du lac Greaca par les larves d'*Eustrongylides excisus* Jägerskiöld. In: Grigore Antipa Hommage à son oeuvre, Bucaresti, pp. 149—162.
Clapham P. A., 1945: Some bird helminths from Antigua. J. Helminth., 21: 93—99.
Condorelli F. M., 1895: Ricerche zoologiche ed anatomo-istologiche sulla *Filaria labiata* Creplin. Boll. Soc. Roma Studi Zool., 4: 95—108, 248—262.
Cram E. B., 1927: Bird parasites of the nematode suborders *Strongylata, Ascaridata* and *Spirurata*. Bull. U. S. Nat. Mus., No. 140: 465 pp.
Cram E. B., 1931: Developmental stages of some nematodes of the *Spiruroidea* parasitic in poultry and game birds. U. S. Dept. Agric. Washington, Techn. Bull. No. 227, 1—28.
Cram E. B., 1936: Species of *Capillaria* parasitic in the upper digestive tract of birds. Techn. Bull. U. S. Dept. Agric., No. 516, 27 pp.
Creplin F. C. H., 1825: Observationes de entozois. Gryphiswaldiae 86 pp.
Creplin F. C. H., 1829: Novae observationes de entozois. Berolini, VI+134 pp.
Creplin F. C. H., 1839: Eingeweidewürmer, Binnenwürmer, Thierwürmer. Allg. Encykl. d. Wissensch. u. Künste, 32: 277—302.
Creplin F. C. H., 1845—1846: Nachträge zu Grult's Verzeichniss der Thiere, bei welchen Entozoen gefunden worden sind. Arch. Naturg. Berlin, 11 Jahr, Vol. 1: 325—330, und 12 Jahr, Vol. 1: 129—160.

Creplin F. C. H., 1849: Nachträge von Creplin zu Gurlt's Verzeichnisse der Thiere, in welchen Endozoen gefunden worden sind. Arch. Naturg., Berlin, Jahr 15, Vol. 1, pp. 52—80.
Creutz G. & Gottschalk C., 1969: Endoparasitenbefall bei Lachmöven in Abhängigkeit vom Alter. Angew. Parasit., 10: 80—91.
Cuckler A. C. & Alicata J. E., 1944: The life history of *Subulura brumpti*, a caecal nematode of poultry in Hawaii. Trans. Amer. microsc. Soc., 63: 345—357.
Czaplinski B., 1962a: Nematodes and acanthocephalans of domestic and wild *Anseriformes* in Poland. I. Revision of the genus *Amidostomum* Railliet et Henry, 1909. Acta parasit. pol., 10: 125—164.
Czaplinski B., 1962b: Nematodes and acanthocephalans of domestic and wild *Anseriformes* in Poland. II. *Nematoda* (excl. *Amidostomum*) and *Acanthocephala*. Acta parasit. pol., 10: 277—319.
Daiya G. G., 1966b: (Redescription of *Capillaria mergi* and *Thominx skrjabini* (*Nematoda: Capillariidae*).) Trudy gel'mint. Lab., 17: 49—53 (in Russian).
Daiya G. G., 1967a: (Nematodes of fish-eating birds of the Lower Ob River (based on the materials from the Ob expedition in 1965).) In: Sbornik rabot po gel'mintofaune ryb i ptits. Moscow: Izdat. Akad. Nauk SSSR, No. 162—67, pp. 133—153 (in Russian).
Daiya G. G., 1968: (Nematode fauna of *Accipitriformes* and owls in Yakutia.) Trudy gel'mint. Lab., 19: 73—82 (in Russian).
Daiya G. G., 1971: (Catalogue of parasitic worms of Latvia.) Zool. Muzeja raksti, Latv. univ., 7: 97—126 (in Latvian).
Danzan G., 1964: (Helminths of domestic and wild birds of the Mongolian People's Republic.) Trudy vses. Inst. Gel'mint., 11: 42—44 (in Russian).
Deslongchamps E. E., 1824: Ascaride-Ascaris. Encycl. méthodique, Paris, 2, pp. 83—112.
Desportes C., 1947: Tridelphie chez une filaire nouvelle parasite du héron pourpré (*Ardea purpurea* L.) Annls Parasit. hum. comp., 22: 36—44.
Diesing K. M., 1851: Systema helminthum. Vol. II., Vindobonae, VI+588 pp.
Diesing K. M., 1861: Revision der Nematoden. Sitzungsb. Akad. Wiss. Math. Naturw. Cl., (1860) 42: 595—736.
Dogel V. A., Bykhovsky B. E., 1939: (Fish parasites from the Caspian Sea.) Tr. komissii po komplexnomu izucheniu Kaspiiskogo morya, 7, Moscow-Leningrad, 150 pp. (in Russian).
Dollfus R. P., 1964: Miscellanea helminthologica Marocana, 35. Trois espèces de filaire d'oiseaux à ajouter à la faune du Maroc. Archs Inst. Pasteur Maroc, 6: 392—407.
Dollfus R. P. et al., 1961: Station expérimentale de parasitologie de Richelieu (Indre-et-Loire). Contribution à la faune parasitaire régionale. Annls Parasit. hum. comp., 36: 1—355.
Drasche R., 1882—1883: Revision der in der Nematoden-Sammlung des k. k. zoologischen Hofcabinetes befindlichen Original-Exemplare Diesing's und Molin's. Verh. zool.—bot. Ges. Wien, 32: 117—139; 33: 107—118; 193—218.
Dubinin V. B., 1938: (Changes in the parasite fauna of glossy ibis (*Plegadis falcinellus*) caused by the growth and migration of the host.) Trudy astrakh. gos. Zapovedn., 2: 114—212 (in Russian).
Dubinin V. B., 1949: (Experimental study of the life cycles of some parasitic worms of animals in the delta of the Volga River.) Parazit. Sb. Zool. Inst. AN SSSR, 11: 126—160 (in Russian).
Dubinin V. B., 1952: (Fauna of larvae of parasitic worms of vertebrates in the delta of the Volga River.) Parazit. Sb. Zool. Inst. AN SSSR, 14: 213—265 (in Russian).
Dubinin V. B., 1954: (Dynamics of parasite fauna of the pelicans in the delta of the Volga River.) Uchen. Zap. Leningr. Gos. Univ., Ser. biol. nauk, No. 35: 203—243 (in Russian).

Dubinin V. B. & Dubinina M. N., 1940: (Parasite fauna of bird colonies of Astrakhan reservation.) Trudy Astrakh. gos. Zapovedn., 3: 190—298 (in Russian).
Dubinina (Gorbunova) M. N., 1937: (Parasite fauna of the heron *(Nycticorax nycticorax)* and its changes associated with the migration of the host.) Zool. Zh., 16: 547—573 (in Russian).
Dubinina M. N. & Serkova O. P., 1951: (Roundworms of birds overwintering in South Tadzhikistan.) Parazit. Sb. Zool. Inst. AN SSSR, 13: 75—95 (in Russian).
Dujardin F., 1845: Histoire naturelle des helminthes ou vers intestinaux. Paris, XVI+654 pp.
Eberth C. J., 1863: Untersuchungen über Nematoden. Leipzig, 77 pp.
Ellis E. & Williams J. C., 1973: The longevity of some species of helminth parasites in naturally acquired infections of the lesser black-headed gull, *Larus fuscus* L. in Britain. J. Helminth., 47: 329—338.
Enigk K. & Dey-Hazra A., 1968: Die perkutane Infection bei *Amidostomum anseris (Strongyloidea, Nematoda)*. Z. Parasitkde., 31: 155—165.
Feyzullaev N. A., 1963a: (Helminth fauna *(Nematoda, Cestoda* and *Acanthocephala)* of birds of the order *Ciconiiformes* from the lowlands of the Azerbaidzhan SSR.) Izv. Akad. Nauk Azerb. SSR, Ser. biol. med., No. 2: 61—68 (in Russian).
Feyzullaev N. A., 1963b: Principal results of helminthological investigations of *Ciconiiformes* in the Azerbaidzhan SSR during 7 years (1956—1962). (Abstract) Mater. nauch. Konf. vses. Obshch. Gel'mint., Part 2, pp. 152—154 (in Russian).
Feyzullaev N. A., 1965: (Dependence of the helminth fauna of *Ciconiiformes* in the Azerbaidzhan SSR on ecological factors.) Trudy Inst. Zool., Baku, 24: 109—127 (in Russian).
Freitas J. F. T. & Almeida J. L., 1935: O genero *Capillaria* Zeder 1800 *(Nematoda — Trichuroidea)* a os capillarioses nas aves domesticas. Revta Dep. Nac. Prod. Anim., 2: 311—363.
Freitas J. F. T. & Lent H., 1936: O genero *Monopetalonema* Diesing, 1861 *(Nematoda: Filarioidea)*. Mems. Inst. Oswaldo Cruz., 31: 747—757.
Frölich J. A., 1802: Beiträge zur Naturgeschichte der Eingeweidewürmer. Naturforscher, Halle, 29: 5—96.
Garkavi B. L., 1949a: (Study of the life-cycle of the nematode *Streptocara crassicauda* (Creplin, 1829) parasitizing domestic and wild ducks.) Dokl. Akad. Nauk SSSR, 65: 421—424 (in Russian).
Garkavi B. L., 1949b: (Life-cycle of the nematode *Tetrameres fissispina*, a parasite of domestic and wild ducks.) Dokl. Akad. Nauk SSSR, 66: 1215—1218 (in Russian).
Garkavi B. L., 1950: (Reservoir hosts of the nematode *Streptocara crassicauda* (Creplin, 1829) Skrjabin 1915, a parasite of domestic and wild ducks.) Trudy vses. Inst. Gel'mint., 4: 5—7 (in Russian).
Garkavi B. L., 1953: (Life-cycle of the nematode *Streptocara crassicauda*, diagnostics and epizootology of streptocariasis of ducks.) Trudy vses. Inst. Gel'mint., 5: 5—22 (in Russian).
Gedoelst L., 1919: Le genre *Histiocephalus* et les espèces qui y ont été rapportées. C. r. Séanc. Soc. Biol., 82: 901—904.
Gendre E., 1920: Un genre nouveau *d'Acuariinae*. Act. Soc. linn. Bordeaux, 72: 40—42.
Gendre E., 1922: Sur quelques espèces d'*Habronema*, parasites des oiseaux. Act. Soc. linn., Bordeaux, 74: 112—133.
Gendre E., 1928: Nématodes. In: Joyeux C. E., Gendre E., Baer J. G., Recherches sur les helminthes de l'Afrique occidentale française, Paris, pp. 55—81.
Gibson G. G., 1964: Taxonomic and biological observations on *Streptocara californica* (Gedoelst, 1919) Gedoelst and Liégeois, 1922 and the genus *Streptocara (Nematoda: Acuariidae)*. Can. J. Zool., 42: 773—783.
Gibson G. G., 1968: Species composition of the genus *Streptocara* Railliet et al., 1912 and the

283

occurrence of these avian nematodes *(Acuariidae)* on the Canadian Pacific coast. Can. J. Zool., 46: 629—645.

Gibson G. G., 1972: *Sciadiocara denticulata* n. sp. *(Acuariidae)* from *Actitis macularia* (L.) and other nematodes from spotted sandpiper and black-bellied plover. Can. J. Zool., 50: 131—136.

Gilbert L., 1927: (Characteristics of two nematodes of the western part of the U.S.S.R.) Sb. rab. po gelm. posv. K. I. Skrjabinu, Moscow, pp. 54—61 (in Russian).

Gilbert L. I., 1930: (Nematode fauna of birds of the western part of the U.S.S.R.) Wiss. Mitt. Univ. Smolensk, 6: 91—111 (in Russian).

Ginetsinskaya T. A., 1952: (Parasites of *Ralliformes* and *Podiceps* of the Astrakhan reservation.) Trudy Leningr. Obshch. Estest. Otdel. Zool., 7: 53—72 (in Russian).

Ginetsinskaya T. A., 1952: (Exchange of parasitic worms in birds of the delta of the Volga River.) Uchen. Zap. leningr. gos. Univ., Ser. Biol., 28: 181—189 (in Russian).

Gmelin J. F., 1790—1917: Systema naturae per regna tria naturae, secundum classes ordines, genera species cum characteribus differentiis, synonymis, locis. Vol. 1 (6). Vermes. Lipsiae, pp. 3021—3910.

Goble F. C. & Kutz H. L., 1945: The genus *Dispharynx (Nematoda: Acuraiidae)* in galliform and passeriform birds. J. Parasit., 31: 323—331.

Golikova M. N., 1959: (Ecological and parasitological study of the biocoenosis of some lakes of the Kaliningrad region. II. Parasite fauna of birds.) In: Polyanski Y. I. (Editor), Ekologicheskaya parazitologiya. Leningrad: Izdat. Leningr. Univ., pp. 150—194 (in Russian).

Golovin O. V., 1958: (Helminth fauna of white-tail eagle in Kalinin region.) Byull. mosk. Obshch. Ispit. Prir., Kalinsk. Otdel. No. 1: 93—96 (in Russian).

Golovin O. V., 1964: ((Nematode fauna of birds of Kalinin region.) Uchen. Zap. Kalinin. gos. pedagog. Inst., No. 31: 283—288 (in Russian).

Gubanov N. M., 1950: (Influence of the conditions of the environment on change in the morphology of nematodes of birds.) Dokl. Akad. Nauk SSSR, 70: 173—175 (in Russian).

Gubanov N. M., 1954: (Helminth fauna of game animals of the Sea of Okhotsk and the Pacific Ocean.) Trudy gelmint. Lab., 7: 380—381 (in Russian).

Gubanov N. M., 1971: (Helminth fauna of *Gaviidae* of Yakutia.) In: Vrednye nasekomye i gelminty Yakutii, Yakutsk: Yakut. Knizhn. Izdat., pp. 85—90 (in Russian).

Gubanov N. M. & Daiya G. G., 1967: (Nematode fauna of birds of North Yakutia.) In: Sbornik rabot po gelmintofaune ryb i ptits. Moscow: Mimeographed. Deposited in VINITI. No. 162—167, pp. 120—132 (in Russian).

Gubanov N. M. & Sergeeva T. P., 1968: (Helminth fauna of *Rhodosthethia rosea.)* Sb. rab. po gelm. posvyashch. 90letiu akad. K. I. Skrjabina, Moscow, pp. 157—158 (in Russian).

Gubanov N. M. & Sergeeva T. P., 1971: (Helminth fauna of gulls from Yakutia.) In: Ammosov Yu. N. (Editor), Vrednye nasekomye i gelminty Yakutii, Yakutsk: Yakut. Knizhn. Izdat., pp. 96—101 (in Russian).

Gubsky V. S., 1960: (Helminth fauna of heron *(Nycticorax nycticorax* L.) of the Lower Dniester River.) Nauch. Ezhegodn. biol. Fak. odessk. Univ., No. 2: 94—96 (in Russian).

Guildal J. A., 1964: Some qualitative and quantitative investigations on the endoparasitic fauna of the Scandinavian-Baltic population of the black headed gull *(Larus ridibundus)*. Contributions to the parasitic fauna of Denmark. 2. Aarsskr. K. Vet. Landbohøjsk., pp. 224—249.

Guildal J. A., 1966: Some qualitative and quantitative investigations on the endoparasitic fauna of the Scandinavian population of the black headed gull *(Larus ridibundus* L.). (Abstract) Proc. of the first Intern. Congress of Parasitol., Roma, Sept. 21—26, 1964, Vol. 1, pp. 517—519.

Guildal J. A., 1968: Investigations on the endoparasitic fauna of the Scandinavian—Baltic population of the herring gull *(Larus argentatus* Pontoppidan 1763). A qualitative and quantitative study. Contributions to the parasitic fauna of Denmark. 3. Medd. Hyg. Bakt. Lab. K. Vet. -og Landbohøjsk. Aarsskr., pp. 59—78.

Gundłach J. L., 1969: Contribution to the helminthofauna of storks *(Ciconia ciconia* L. and *C. nigra* L.) originating from the Lublin Palatinate. Acta parasit. pol., 16: 83—89.

Gurlt E. F., 1845: Verzeichniss der Thiere, bei welchen Entozoen gefunden worden sind. Arch. Naturg., Berlin, 11 Jahr, 1: 223—325.

Gushanskaya L. Kh., 1950a: (New nematodes of the genus *Skrjabinocara* Kuraschvili, 1941.) Trudy gel'mint. Lab., 3: 191—198 (in Russian).

Gushanskaya L. Kh., 1950b: (Invalidation of the tribe *Antennocarea* and the genus *Antennocara.)* Trudy gel'mint. Lab., 4: 53—54 (in Russian).

Gushanskaya L. Kh., 1950c: (Study of *Spirurata* of water birds of the U.S.S.R.) Trudy gel'mint. Lab., 4: 55—63 (in Russian).

Gushanskaya L. Kh., 1950d: (New *Spirurata* of birds.) Trudy gel'mint. Lab., 4: 40—52 (in Russian).

Gushanskaya L. Kh., 1953: (Family *Desmidocercidae* Cram, 1927 and its position in the classification of nematodes.) Sb. Rab. po gelm. posv. 75letiu K. I. Skrjabina, Moscow, pp. 188—204 (in Russian).

Gushanskaya L. Kh. & Krotov A. I., 1952: (Discovery of a male of *Schistorophus skrjabini (Nematoda: Schistorophidae).* Trudy gel'mint. Lab., 6: 225—228 (in Russian).

Gutierrez R. O., 1943: Sobre la morphologia de una nueva especies de *Contracaecum (Nematoda, Ascaroidea).* Revta bras. Biol., 3: 159—172.

Gvozdev E. V. & Kasimzhanova B. A., 1965: (Nematode fauna of South Kazakhstan.) In: Mater. nauch. Konf. vses. Obshch. Gel'mint., Part I., pp 54—58 (in Russian).

Hamann O., 1891: Monografie der Acanthocephalan (Echinorhynchen), ihre Entwicklungsgeschichte, Histogenie und Anatomie. Jenaische Ztschr. Naturw., V. 25, n. F. 18: 113—121.

Hartwich G., 1959: Revision der vogelparasitischen Nematoden Mitteleuropas. I. Die Gattung *Porrocaecum* Railliet & Henry, 1912 *(Ascaridoidea).* Mitt. zool. Mus. Berl., 35: 107—147.

Hartwich G., 1964: Revision der vogelparasitischen Nematoden Mitteleuropas. II. Die Gattung *Contracaecum* Railliet et Henry, 1912 *(Ascaridoidea).* Mitt. zool. Mus. Berl., 40: 15—53.

Hartwich G., 1966: Vogelparasitische Nematoden aus der Mongolischen Volksrepublik. Mitt. zool. Mus. Berl., 42: 281—306.

Hoeppli R., Hsü H. F. & Wu, H. W., 1929: Helminthologische Beiträge aus Fukien und Chekiang. Arch. Schiffs- u. Tropenhyg., 33: 1—44.

Hovorka J. & Macko J., 1959: *Calcaronema* gen. nov. a new genus of the subfamily *Cyathostominae* Nicoll, 1927 *(Syngamidae* Leiper, 1912) and the description of the new species *C. trifurcatum* sp. n. and *C. verrucosum* sp. n. Helminthologia, 1: 103—112.

Hsü H. F., 1933: Some species of *Porrocaecum (Nematoda)* from birds in China. J. Parasit., 19: 280—285.

Hsü W. N., 1957: Studies on nematodes parasitic from Canton. Acta zool. sin., 9: 77—84.

Huizinga H. W., 1966: Studies on the life cycle and development of *Contracaecum spiculigerum* (Rudolphi, 1809) *Ascaroidea: Heterocheilidae)* from marine piscivorous birds. J. Elisha Mitchell Scient. Soc., 82: 181—195.

Iksanov K. I. & Dikambaeva L. K., 1962: (Study of nematode infection in fish-eating birds of Kirghizia.) Izv. Akad. Nauk Kirgiz. SSR, Ser. biol. Nauk, 4: 131—137 (in Russian).

Inglis W. G., 1957: A review of the nematode superfamily *Heterakoidea.* Ann. Mag. nat. Hist., Ser. XII, 10: 905—912.

Iordachescu D., 1957: Observations à propos de quelques nematodes parasites des oiseaux indigènes. Trav. Mus. Hist. Nat. Gr. Antipa, 1: 245—253.

Iordachescu D., 1962: Contribution à l'Helminthofauna des oiseaux. Trav. Mus. Hist. Nat. Gr. Antipa, 3: 517—523.

Iordachescu-Lazarescu D., 1963: Données nouvelles concernant la helminthofauna des oiseaux de Roumanie. Trav. Mus. Hist. Nat. Gr. Antipa, 4: 175—180.

Ivanitsky S. V., 1940: (Contribution to the study of the helminth fauna of vertebrates of Ukraine. Fauna of cestodes, nematodes and acanthocephalans.) Sb. Trud. Kharkov. vet. Inst., 19: 129—155 (in Russian).

Ivashkin V. M., Sobolev A. A. & Khromova L. A., 1971: (Principles of Nematology XXII. *Camallanata.*) Moscow: Izdat. Nauka 388 pp. (in Russian).

Jägerskiöld L. A., 1894: Beiträge zur Kenntnis der Nematoden. Zool. Jb., Abt. Anat., 7: 449—532.

Jägerskiöld L. A., 1909: Zur Kenntnis der Nematoden-Gattungen *Eustrongylides* und *Hystrichis.* Nova acta R. Soc. Sci. Upsaliensis, 2: 48 pp.

Jančev J., 1958: Untersuchungen über einige Helminthen und Helminthosen bei weissen Störchen und Pelikanen. Bull. Inst. Zool. Acad. Sci. Bulgarie, Sofia, 7: 393—416.

Jansen J. & Broek E., van den, 1966: Parasites of zoo-animals in the Netherlands and of exotic animals II. Bijdr. Dierk., 36: 65—68.

Jennings A. R. & Soulsby E. J. L., 1956: Diseases in wild birds. Third report. Bird Study, 3: 270—272.

Jennings A. R. & Soulsby E. J. L., 1957: Diseases of wild birds. Fourth report. Bird Study, 4: 216—220.

Jennings A. R. & Soulsby E. J. L., 1958: Disease in a colony of Blackheaded Gulls, *Larus ridibundus.* Ibis, 100: 305—312.

Jögis V. A., 1963: (Fauna of cestodes, nematodes and acanthocephalans of water and littoral birds of Pukhta environs.) Ezheg. obshch. estestvoispyt. AN Eston. SSR, (Loodusuur. Seltsi Aastar. Year 1962) 55: 94—128 (in Estonian).

Jögis V. A., 1967: Life cycle of *Porrocaecum semiteres* (Zeder, 1800) *(Nematoda: Ascaridata)* Parazitologiya, 1: 213—218 (in Russian).

Johnston T. H. & Mawson P. M., 1941: Additional nematodes from Australian birds. Trans. R. Soc. S. Austr., 66: 66—70.

Kamburov P. & Vasilev V., 1972: (On the helminth fauna in certain wild aquatic birds *(Anseriformes)* in Bulgaria.) Izv. tsent. khelmint. Lab., Sof., 15: 109—133 (in Bulgarian).

Karmanova E. M., 1956: (Study of life cycle of the nematode *Hystrichis tricolor* Dujardin, 1845, parasite of wild and domestic ducks.) Dokl. Akad. Nauk SSSR, 111: 245—247 (in Russian).

Karmanova E. M., 1958: (Biology of the nematodes of the suborder *Dioctophymata* (Skrjabin, 1927). (Papers on helminthology presented to Academician K. I. Skrjabin on his 80th Birthday.) Moscow: Izdatel. Akad. Nauk SSSR, pp. 148—151 (in Russian).

Karmanova E. M., 1959: (Biology of the nematode *Hystrichis tricolor* Dujardin, 1845 and some data on epizootology of hystrichosis of ducks.) Trudy gel'mint. Lab., 9: 113—125. (in Russian).

Karmanova E. M., 1960b: (Revision of the genus *Hystrichis* (Dujardin, 1845) *(Dioctophymata, Nematoda).*) Trudy gelmint. Lab., 10: 112—116 (in Russian).

Karmanova E. M., 1960a: (Helminth fauna of *Criodrilus lacuum.*) Trudy gel'mint. Lab., 10: 117—123 (in Russian).

Karmanova E. M., 1961: (Larvae of helminths found in *Oligochaeta* in south-western part of the delta of the Volga River.) Trudy astrakh. gos. Zapovedn., No. 5: 330—335 (in Russian).

Karmanova E. M., 1962: *(Oligochaeta* as intermediate hosts of nematodes of the subgenus *Dioctophymata.)* Sb. Mezhvuzovskoy nauchn. konf. po probleme kraevoy parazit., Odessky Gos. Univ., pp. 35—36 (in Russian).

Karmanova E. M., 1965: (Finding of intermediate hosts of the nematode *Eustrongylides excisus*, parasite of water birds.) Trudy gefmint. Lab., 15: 86—87 (in Russian.)

Karmanova E. M., 1968a: (Principles of nematodology. Vol. XX. *Dioctophymata.)* Moscow: Izdat. Nauka, 262 pp. (in Russian).

Karmanova E. M., 1968b: (Some peculiarities of the biology of nematodes *Dioctophymidea.)* In: Parazity cheloveka, zhivotnykh i rastenii i meri borby s nimi. Moscow: Izdat. Nauka, pp. 193—199 (in Russian).

Karokhin V. I., 1935: (Haemorrhage and inflammation in the region of location of *Streptocara crassicauda* var. *charadrii* Skrjabin, 1915 and *Acuaria (Syncuaria) ciconiae* Gilbert, 1927.) Trudy uralsk. zoovet. Inst., No. 1: 137—140 (in Russian).

Karokhin V. I., 1937: *(Contracaecum milviensis* sp. nov. — new representative of the genus *Contracaecum (Nematoda)* from *Milvus lineatus.)* (Papers on helminthology in commemoration of the 30 year jubileum of K. I. Skrjabin and of the 15th anniversary of the All-Union Institute of Helminthology.) pp. 275—280 (in Russian).

Karokhin V. I., 1949: (New representative of nematodes of the genus *Contracaecum* (from *Colymbus nigricollis).* Trudy gefmint. Lab., 2: 91—93 (in Russian).

Kasimov G. B., 1956: (Helminth fauna of economic game birds of the order *Galliformes.)* Moscow, 554 pp. (in Russian).

Kasimov G. S. & Feyzullaev N. A., 1965: (Helminth fauna of birds *(Galliformes, Columbiformes, Otidiformes, Ciconiiformes)* from the Kura-Arak lowlands of Azerbaidzhan. Trudy Inst. Zool., Baku, 24: 85—98 (in Russian.)

Kasimov G. S., Feyzullaev N. A., 1969: (Helminths of birds of the Small Caucasus.) Sb. Voprosy Parazit., Baku, pp. 102—116 (in Russian).

Khalil M., 1931: On two new species of nematodes from *Pelecanus onocrotalus*. Ann. trop. Med. Parasit., 25: 455—460.

Kibakin V. V., 1965: (Helminths of the birds of Gasan-Kuliysk reservation.) Mater. nauch. Konf. Vses. Obshch. Gefmint., Part. I, pp. 108—111 (in Russian).

Kibakin V. V. & Babacv Ya., 1964: (Study of helminth fauna of birds of the order *Lariformes.)* Tezisy dokladov Pervoy resp. konf. molodykh zoologov Turkmenistana, Ashkhabad, pp. 102—103 (in Russian).

Kibakin V. V., Dobrynin M. I., Skladchikov R. V. & Zhuchenko L. Ya., 1963: (Fauna of *Spirurata* of limnophilous birds of Karakum Canal.) Mater nauchn. Konf. Vses. Obshch. Gefmint., Part 1, p. 134 (in Russian).

Kobuley T. & Ryzhikov K. M., 1968: (Revision of the genus *Amidostomum (Nematoda: Strongylata).* Parazitologiya, 2: 306—311 (in Russian).

Kontrimavichus V. L. & Bakhmeteva T. L., 1960: (The helminth fauna of *Gavia* sp. in the Lena River basin.) Trudy gefmint. Lab., 10: 124—133 (in Russian).

Korpaczewska W. & Sulgostowska T., 1967: Materiały do helmintofauny ptaków wodnych Mazur (jezioro Warnołty). Wiad. parazyt., 13: 737—744.

Kosupko G. A., 1962: (Parasitic worms of gulls of the Moscow region.) Trudy Vses. Inst. Gefmint., 9: 11—15 (in Russian).

Kosupko G. A., 1963: (Nematodes of fish-eating birds of the Astrakhan reservation.) Mater. nauch. Konf. Vses. Obshch. Gefmint., Part 1, pp. 158—159 (in Russian).

Kovalenko I. I., 1960: (Study of the life cycles of some helminths of domestic ducks bred in the poultry-farms of the Azov Sea coast.) Probl. parazit. Tr. 3-ey nauch. konf. parazit. Ukrain. SSR, Kiev, pp. 168—170, also in Dokl. Akad. Nauk SSSR, 133: 1259—1261 (in Russian).

Kovalenko I. I., 1963: (Endemic of mixed infections with *Streptocara, Tetrameres* and *Polymorphus* in fowls.) Trudy Ukr. respubl. nauchno-issled. Obshch. Parazit., 2: 137—140 (in Russian).
Kowalewski M. M., 1904: Materyały do fauny helmintologicznej pasorzyticzej Polskieij. Sprawoz. Kom. Fizyogr. Comm. Akad. Umiej, Krakow, 38: 18—26.
Krastin N. I., 1957: (Thelaziases and their causative agents.) Blagoveshchensk, 174 pp. (in Russian).
Kreis H. A., 1955: *Contracaecum septentrionale*, ein neuer Parasit aus dem Kormoran; sein Lebenslauf, sowie Angaben über die Entwicklung der *Anisakinae*. (Beiträge zur Kenntnis parasitischer Nematoden XVI). Z. ParasitKde., 17: 106—121.
Kreis H. A., 1958: Parasitic Nematoda. Zoology of Iceland, 2 (15b), 24 pp.
Krishna Rao N. S., 1951: *Paracuaria macdonaldi* n. g., n. sp. (family *Acuariidae*, subfamily *Acuariinae*) from the sea gull *(Larus argentatus)*. Can. J. Zool., 29: 167—172.
Krivonogova F. D., 1963: (Helminth fauna of fish-eating birds of the Lower Amur river.) Trudy gel'mint. Lab., 13: 220—226, (in Russian.)
Krotov A. I., 1952: (Nematodes of birds of the lower reach of the River Amu-Darya.) Trudy gel'mint. Lab., 6: 273—277 (in Russian).
Krotov A. I., 1959: (Helminth fauna of vertebrates on the island of Sakhalin.) Sb. Rab. po gel'm., Moscow, No. 1: 98—102 (in Russian).
Krotov A. I. & Delyamure S. L., 1952: (Fauna of parasitic worms of mammals and birds of the U.S.S.R.) Trudy gel'mint. Lab., 6: 278—292 (in Russian).
Kulachkova V. G., 1950: (Parasite fauna of gulls and terns in the delta of the Danube River.) Uchen. Zap. leningr. gos. Univ., 23: 123—128 (in Russian).
Kulachkova V. G. & Kochetova I. V., 1964: (Characteristic features of helminth fauna of gulls of Kandalak Bay.) In: Prirodnaya ochagovost parazitarnykh i transmissionykh bolezney v Karelii, Moskva-Leningrad, Izdat. "Nauka", pp. 48—57 (in Russian).
Kurashvili B. E., 1941: (To the knowledge of helminth fauna of birds of Georgia.) Trudy zool. Inst., Tbilisi, 4: 53—100 (in Russian).
Kurashvili B. E., 1950: (Helminth fauna of game-birds of Georgia and some laws of its dynamics.) Trudy zool. Inst. Tbilisi, 9: 37—80 (in Russian).
Kurashvili B. E., 1953: (Helminth fauna of game-birds of Georgia.) Papers on helminthology presented to Acad. K. I. Skrjabin on his 75th birthday.) Moscow: Izdat. Akad. Nauk SSSR, pp. 340—346 (in Russian).
Kurashvili B. E., 1956: (Helminth fauna of birds of Lagodekh reservation.) Trudy zool. Inst. Tbilisi, 14: 105—145 (in Russian).
Kurashvili B. E., 1957: (Helminths of game birds of Georgia from the faunistical and ecological view.) Moscow: Izdat. Akad. Nauk SSSR, 434 pp. (in Russian).
Kurashvili B. E., 1961: (Study of helminth fauna of fish-eating birds in Georgia.) Soobshch. Akad. Nauk gruz. SSR, 26: 73—77 (in Russian).
Kurochkin Yu. V., 1954a: Scientific results of the 315th All-Union helminthological expedition. Trudy astrakh. gos. Zapovedn., No. 9: 8—31 (in Russian).
Kurochkin Yu. V., 1954b: (Life-cycle of the nematode — a pathogen of epomidiostomosis of ducks.) Dokl. Akad. Nauk SSSR, 98: 509—511 (in Russian).
Kurochkin Yu. V., 1958: (Study of nematodes of the genus *Skrjabinocara* Kuraschwili, 1941.) Trudy astrakh. gos. Zapovedn., No. 4: 325—336 (in Russian).
Kurochkin Yu. V. & Ryzhikov K. M., 1964: (Species composition of the genus *Paracuaria* Rao, 1951 *(Nematoda: Spirurata)*. Trudy astrakh. gos. Zapovedn., No. 9: 182—191 (in Russian).
Kurochkin Yu. V., Ryzhikov K. M. & Gubanov N. M., 1961: (Nematode fauna of *Anseriformes* of Vierkhoyaniye.) Trudy astrakh. gos. Zapovedn., No. 5: 326—329 (in Russian).

Kurochkin Yu. V. & Zablotsky V. I., 1961: (Helminth fauna of gulls of the Caspian Sea.) Trudy astrakh. gos. Zapovedn., No. 5: 296—318 (in Russian).

Kurtieva L., 1953: (A new nematode from the intestine of birds of the Turkmenian S. S. R. — *Strongyloides turkmenica* n. sp.) (Papers on helminthology presented to Acad. K. I. Skrjabin on his 75th birthday.) Moscow: Izdat. Akad. Nauk SSSR, pp. 347—348 (in Russian).

Lapage G., 1961: A list of the parasitic protozoa, helminths and *Arthropoda* recorded from species of the family *Anatidae* (ducks, geese and swans). Parasitology, 5: 1—109.

Layman E. M. & Andronova E. V., 1926: (New nematode of herons.) Rab. parazit. Lab. mosk. gos. Univ., pp. 47—49 (in Russian).

Layman E. M. & Mudretsova K. A., 1926: (Fauna of parasitic worms of the birds from Murman.) Rab. parazit. Lab. mosk. gos. Univ., pp. 38—46 (in Russian).

Lengy J., 1969: Notes on the classification of *Syngamidae (Nematoda)* with new data on some of the species. Israel J. Zool., 18: 9—23.

Lent H. & Freitas J. F. T., 1948: Uma caleção de Nematódeos, parasitos do vertebrados, do Museu de Historia Natural de Montevideo. Mems Inst. Oswaldo Cruz, 46: 1—71.

Leonov V. A., 1956: (Helminth-epizootological significance of fish-eating birds in the region of the Dnieper estuary of the Black Sea.) Problemy parazitologii, Tr. 2-y nauchn. konf. parazitol. Ukrain. SSR, Kiev, pp. 74—75 (in Russian).

Leonov V. A., 1958: (Helminth fauna of *Lariformes* in the animal reserve and adjoining territory in the Kherson area.) Uchen. Zap. gorkov. gos. pedagog. Inst., 20: 266—295 (in Russian).

Leonov V. A., 1960a (publ. 1961): (Helminth fauna of herons.) Uchen. Zap. gorkov. gos. pedagog. Inst., 27 (Gelm. Sb. No. 2): 29—37 (in Russian).

Leonov V. A., 1960b (publ. 1961): (Dynamics of the helminth fauna of the herring gull nesting in the territory of the Black Sea reserve.) Uchen. Zap. gorkov. gos. pedagog. Inst. (Gelm. Sb. No. 2) 27: 38—57 (in Russian).

Leonov V. A. & Belogurov O. I., 1963: (Nematodes of fish-eating birds of Koryaksk national district (Kamchatka). Mater. nauchn. Konf. vses. Obshch. Gelmint., Part I., pp. 179—181 (in Russian).

Leonov V. A., Belogurov O. I. & Tsimbalyuk A. K., 1964: *(Seuratia puffini* Yamaguti, 1941 — a genus and species of nematodes new to the U. S. S. R.) Zool. Zh., 43: 930—932 (in Russian).

Leonov V. A., Sergeeva T. P. & Tsimbalyuk A. K., 1966: (A new nematode — *Stegophorus stercorarii* n. sp. *(Nematoda: Acuariidae)).* Trudy gelmint. Lab., 17: 91—94 (in Russian).

Leonov V. A., Shevtsova L. S., 1970: (Nematodes of birds of the Wrangel Island.) Uchen. Zap. dalnevost. gos. Univ., 16: 46—56 (in Russian).

Leonov V. A., Tsimbalyuk A. K. & Belogurov O. I., 1963: To the species independence of the nematode *Paracuaria macdonaldi* Rao, 1951 and the position of the genus *Paracuaria* Rao, 1951 in the system of the family *Acuariidae (Spirurata: Acuariidae).* Vest. leningr. gos. Univ., 18 (15), Ser. Biol., 3: 155—157 (in Russian).

Levaschoff M. M., 1929: Beitrag zur Kenntnis der Fauna der parasitischen Nematoden des unteren Wolgagebietes. Z. ParazitKde., 2: 121—128.

Lewis E. A., 1926: Helminths of wild birds found in the Aberystwyth area. J. Helminth., 4: 7—12.

Lewis E. A., 1927: A survey of Welsh helminthology. J. Helminth., 5: 121—132.

Li H. C., 1934: Report on a collection of parasitic nematodes, mainly from North China. Part II. Spiruroidea. Trans. Am. microsc. Soc., 53: 174—195.

Linstow O. von, 1877a: Helminthologica. Arch. Naturg., Berlin, 43 Jahr, I, 1—18.

Linstow O. von, 1877b: Enthelminthologica. Arch. Naturg. Berlin, 43 Jahr, I, 173—198.

Linstow O., 1878: Compendium der Helminthologie. Hannover, XXII + 382 pp.

Linstow O., 1883: Nematoden, Trematoden und Acanthocephalen gesammelt von Prof. Fedtschenko in Turkestan. Arch. Naturg. Berlin, Jahr 49, I, 274—314.

Linstow O., 1884: Helminthologisches. Arch. Naturg., Berlin, Jahr 50, I, 125—145.

Linstow O., 1886: (Nematodes, trematodes and acanthocephalans collected by Fedchenko in Turkestan.) In: Journey to Turkestan, A. P. Fedchenko. Izv. imperat. obshch. lyubit. estest., antropol. i etnogr., 34: 13—15 (in Russian).

Linstow O., 1888: Report on the Entozoa collected by H. M. S. Challenger during the years 1873—76. Rept. Voyage H. M. S. Challenger, London, 23: 1—18.

Linstow O., 1889: Compendium der Helminthologie. Nachtrag. Hannover, 151 pp.

Linstow O., 1890: *Grus viridiorostris* getödtet durch den Parasitismus von *Syngamus sclerostomum* Molin. Zentbl. Bakt., 8: 259—261.

Linstow O., 1894: Helminthologische Studien. Jenaische Ztschr. Naturw., 28: 328—342.

Linstow O., 1899a: Nematoden aus dem Königlichen Zoologischen Museum in Berlin. Mitt. zool. Mus. Berl., 1—28.

Linstow O., 1899b: Zur Kenntnis der Genera *Hystrichis* und *Tropidocerca*. Arch. Naturg., 65: 155—164.

Linstow O., 1901a: Beobachtungen an Helminthen des senkenbergischen naturhistorischen Museums, des Breslauer zoologischen Instituts und anderen. Arch. Mikr. Anat., 58: 182—198.

Linstow O., 1903: Entozoa des zoologischen Museums der Kaiserlichen Akademie der Wissenschaften zu St. Petersburg II. Ann. Mus. Zool. Acad. imp. Sci. St. Petersburg, 8: 265—294.

Linstow O., 1906: Helminths from the collection of the Colombo Museum. Spolia zeylan., 3: 163—188.

Linstow O., 1907: Nematoden aus dem Königlichen Zoologischen Museum in Berlin. Mitt. zool. Mus. Berl., 3: 251—259.

Linstow O., 1909: Parasitische Nematoden. Nematodes, *Mermithidae* und *Gordidae*. II. In: Brauer's: Die Süsswasserfauna Deutschlands. Jena, Heft 15: 47—83.

Little M. D., 1966: Comparative morphology of six species of *Strongyloides (Nematoda)* and redefinition of the genus. J. Parasit., 52: 69—84.

López-Neyra C. R., 1947: Géneros y especies nuevas e mal conocidas de *Capillariinae*. Revta iber. Parasit., 7: 191—238.

López-Neyra C. R., 1948: Helmintos de los vertebrados Ibéricos. Vol. II. Granada; Instituto Nacional de Parasitología: 413—802.

López-Nreyra C. R., 1956: Revision de la super-familia *Filarioidea* (Weinland, 1858). Revta ibe . Parasit., 16: 3—212.

Lundahl C., 1847: Bemerkungen über zwei neue *Strongylus*-Arten. Notis. Sällsk. Fauna et Flora Fenn. Förh., 1: pp. 283—287.

Lyubimov M. P., 1937: *(Pharyngosetaria marcinowskyi* (Skrjabin, 1923) n. gen. from the gall bladder of common heron *Ardea cinerea.)* (Papers on helminthology in commemoration of the 30 year jubileum of K. I. Skrjabin.) Moscow, pp. 348—351 (in Russian).

McDonald M. E., 1969: Catalogue of Helminths of waterfowl *(Anatidae)*. Spec. scient. Rep. U. S. Fish Wildl. serv., Wildlife No. 126, Washington, VIII + 692 pp.

Machado de Mendonça J. & Oliveira Rodrigues H. de, 1968: Revisão do gênero *Seuratia* Skrjabin, 1916 e redescrição de espécie *Seuratia shipleyi* (Stossich, 1900) Skrjabin,1916 *(Nematoda, Spiruroidea)*. Mems Inst. Oswaldo Cruz, 66: 117—129.

Machida M., 1966: Helminths collected from a Pelagic Shag. Bull. Nat. Sci. Mus. Tokyo, 8: 447—450.

McIntosh A., 1927: Parasitic worms in the St. Andrews fauna in addition to the marine fauna of St. Andrews since 1874. Ann. Mag. nat. Hist., 9: 49—94.

Macko J. K., 1961a: Zistenie dvoch druhov červov *(Nematoda)* u volavky purpurovej *(Ardea purpurea* L.*)*. Biológia, Bratisl., 16: 845—847.

Macko J. K., 1961b: Niektoré pozorovania o premenlivosti druhu *Contracaecum (Contracaecum) nehli* Karokhin, 1949. Biológia, Bratisl., 16: 740—748.

Macko J. K., 1962a: Neue Art einer Nematode *Skrjabinocara parvepapillata* sp. n. aus dem Wirt *Ciconia ciconia* L. und *Ciconia nigra* L. Biológia, Bratisl., 17: 440—446.

Macko J. K., 1962b: Opis doteraz neznámeho samčeka druhu *Skrjabinocara parvepapillata* Macko, 1962 *(Nematoda)*. Biológia, Bratisl., 17: 837—840.

Macko J. K., 1963a: Auszug aus der Beschreibung neuer Helminthenarten bei freilebenden Vögeln in der Slowakei. Helminthologia, 4: 290—302.

Macko J. K., 1963b: K faune nematódov bociana bieleho, *Ciconia ciconia* L. Zool. Listy, 12: 337—342.

Macko J. K., 1964a: (Helminth fauna of the family *Laridae* and its relation to some ecological factors.) Studia Helminth., 1: 29—45 (in Russian).

Macko J. K., 1964b: (The nematode fauna and *Acanthocephala* in *Ardea cinerea* L. in Czechoslovakia.) Studia Helminth., 1: 21—28 (in Russian).

Macko J. K., 1964c: Zur Fauna der Nematoden des Schwarzstorches *(Ciconia nigra* L.*)* in der Slowakei (ČSSR). Helminthologia, 5: 21—32.

Macko J. K., 1964d: Príspevok k poznaniu fauny nematódov a pentastomíd čajkovitých vtákov Slovenska. Biológia, Bratisl., 19: 118—122.

Macko J. K. & Baruš V., 1973: The finding of *Victorocara guschanscoi (Nematoda, Schistorophidae)* in Czechoslovakia. Folia parasit, Praha, 20: 381.

Madsen H., 1945: The species of *Capillaria (Nematoda: Trichinelloidea)* parasitic in the digestive tract of Danish gallinaceous and anatine game birds with a revised list of species of *Capillaria* in birds. Dan. Rev. Game Biol., 1: 1—112.

Madsen H., 1950a: On the systematics of *Syngamus trachea* (Montagu, 1811) Chapin, 1925. J. Helminth., 24: 33—46.

Madsen H., 1950b: Studies on species of *Heterakis* (Nematodes) in birds. Dan. Rev. Game Biol., 1: 1—43.

Maksimova A. P., 1966: (Helminths of wild aquatic birds of water reservoirs in southern Kazakhstan.) In: Tokobaev, M. M. (Ed.) Gelminty zhivotnykh Kirgizii i sopredelnykh territoriy. Frunze; Izdat. „ILIM", pp. 47—56 (in Russian).

Maksimova A. P., 1967: (Helminths of wild *Anseriformes* of western Kazakhstan). (Trudy Inst. Zool. Alma-Ata, 27: 124—155 (in Russian).

Maplestone P. A., 1931: Parasitic nematodes obtained from animals dying in the Calcutta Zoological Gardens. Rec. Ind. Mus., 33: 71—171.

Maplestone P. A., 1932a: The genera *Heterakis* and *Pseudaspidodera* in Indian hosts. Indian J. med. Res., 20: 403—420.

Maplestone P. A., 1932b: Parasitic nematodes obtained from animals dying in the Calcutta Zoological Gardens. Pt. 9—11. Rec. Indian Mus., 34: 229—261.

Markov G., 1937: (Changes in the age of parasite fauna of *Uria lomvia lomvia*.) Trudy leningr. Obshch. Estest., 66: 456—666 (in Russian).

Markov G., 1941: (Parasitic worms of birds from nameless inlet (Novaya Zemlya).) Dokl. Akad. Nauk SSSR, 30: 573—576 (in Russian).

Mashtakov V. I., 1964: (Helminth fauna of gulls and terns of Azov-Sivash reservation.) Uchen. Zap. gorkov. gos. Inst., No. 62: 41—51 (in Russian).

Mawson P. M., 1956: Ascaroid nematodes from Canadian birds. Can. J. Zool., 34: 35—47.

Mégnin J. P., 1881: Sur le développement du *Syngamus trachealis:* Abstract of 1882 paper: In Gaz. Méd., Paris v. 52, 6. s, 3 (51) p. 726.

Mégnin J. P., 1882: Sur le développement du *Syngamus trachealis*. C. r. Séanc. Soc. Biol., v. 33, 7. s., 3: 348—350.

Mehlis E., 1831: Nova observationes de entozois. Auctore Dr. Fr. Chr. H. Creplin. Isis (Oken), Leipzig, (I), pp. 68—99.

Michelson V. K., 1968: (Materials of helminth fauna of wild birds from Latvia.) In: Ekologia vodoplavayushchikh ptits Latvii. Riga, pp. 243—251 (in Russian).

Molin R., 1858a: Versuch einer Monographie der Filarien. Sitzungsb. d. K. Akad. Wiss. Wien, Math.-naturw., 28: 365—461.

Molin R., 1858b: Prospectus helminthum, quae in prodromo faunae helminthologicae Venetiae continentur. Sitzungsb. d. K. Akad. Wiss. Wien, Math.-naturw., 30: 127—158.

Molin R., 1859: Prospectus helminthum, quae in parte secunda prodromi faunae helminthologicae venetae continentur. Sitzungsb. d. K. Akad. Wiss. Wien, Math.-naturw., 33: 287 to 302.

Molin R., 1860a: Una monografia del genere *Spiroptera*. Sitzungsb. d. K. Akad. Wiss. Wien, Math.-naturw., 38: 911—1005.

Molin K., 1860b: Una monografia del genere *Dispharagus* e una monografia del genere *Histiocephalus*. Sitzungsb. d. K. Akad. Wiss., Wien, Math.-naturw., 39: 479—516.

Molin R., 1860c: Trenta specie di nematoidi. Sitzungsb. d. K. Akad. Wiss. Wien, Math.-naturw., 40: 331—358.

Molin R., 1861a: Il sottordine degli acrofalli ordinato scientificamente secondo i risultamenti delle indagini anatomiche ed embriogeniche. Mem. R. Ist. Veneto Sci., Lett. ed Arti (1860), 9: 427—633.

Molin R., 1861b: Prodromus faunae helminthologicae venetae adjectis disquisitionibus anatomicis et criticis. Denkschr. d. K. Akad. d. Wissensch., Wien, Math.-Nat., 19: 189—338.

Mönnig H. O., 1923: South African parasitic nematodes. 9th and 10th Reports of the Director of Veterinary Education and Research. S. Africa Dept. Agric. Pretoria, pp. 435—478.

Montagu G., 1811: Account of a species of *Fasciola* which infests the trachea of poultry. Mem. Werner. Nat. Hist. Soc., 1: 194—198.

Morgan B. B., Schiller E. & Rausch R., 1949: The occurrence of *Contracaecum travassosi* (*Nematoda*) in North America. J. Parasit., 35: 541—542.

Morishita K., 1930: Two nematode parasites of the guillemot. Jap. J. Zool., 3: 67—72.

Morozov Yu. F., 1958: (Helminths of white stork in the Bialowieza virgin Forest.) Pervaya zoolog. konf. Bielorusk. SSR, Tezisy dokladov. pp. 168—169 (in Russian).

Mozgina A. A., 1967: (Helminth fauna of the black-headed gull in the Volgograd region.) Problemy Parasit. Tezisy dokladov V. nauch-konf. Ukr. resp. nauch. obshch. parazit., Kiev, pp. 175—176 (in Russian).

Mozgina A. A., 1969: (Epizootological significance of gulls and herons from Tsimlyansk water reservoir.) In: Markov G. S. et al., (Editors) (Parasitic animals in the Tsimlyansk water-reservoir). Volgograd, pp. 159—165 (in Russian).

Mozgovoy A. A., 1950: (New *Anisakidae* of birds.) Trudy gelmint. Lab., 3: 90—101 (in Russian).

Mozgovoy A. A., 1952a: (Biology of *Porrocaecum crassum* — nematode of water birds.) Trudy geľmint. Lab., 6: 114—125 (in Russian).

Mozgovoy A. A., 1952b: (Biological peculiarities of *Ascaridata*.) Trudy geľmint. Lab., 6: 126—130 (in Russian).

Mozgovoy A. A., 1953: (Principles of nematodology II. *Ascaridata* of animals and man and the diseases caused by them. Part II.) Moscow: Izdat. Akad. Nauk SSSR, 616 pp.(in Russian).

Mozgovoy A. A. & Bishaeva L., 1959: (Elucidation of the life cycle of *Porrocaecum heteroura* (*Ascaridata, Anisakidae*)). Helminthologia, 1: 195—197 (in Russian).

Mozgovoy A. A., Popova T. I. & Semenova M. K., 1965: (Elucidation of the life cycle of the nematode *Synhimantus brevicaudatus* (Dujardin, 1845) — the parasite of *Ciconiiformes* and fresh-water fishes.) Dokl. Akad. Nauk SSSR, 162: 719—721 (in Russian).

Mozgovoy A. A. & Romanova N. P., 1956: (Study of *Ascaridata* of birds and reptiles from the zoological garden of Moscow.) Trudy gelmint. Lab., 8: 77—84 (in Russian).

Mozgovoy A. A., Semenova M. K. & Shakhmatova V. I., 1965: (Life cycle of *Contracaecum microcephalum (Ascaridata: Anisakidae)*, nematode of water birds.) Mater. nauch. konf. vses. Obshch. Gelmint., Part I, pp. 154—159 (in Russian).

Mozgovoy A. A., Semenova M. K., & Shakhmatova V. I., 1968: (Life cycle of *Contracaecum microcephalum (Ascaridata: Anisakidae)* — parasite of fish-eating birds.) (Papers on helminthology presented to Acad. K. I. Skrjabin on his 90th birthday.) Moscow: Izdat. Akad. Nauk SSR, pp. 262—272 (in Russian).

Mozgovoy A. A., Shakhmatova V. I. & Semenova M. K., 1965a: (Life cycle of *Contracaecum spasskii* Mozgovoy, 1950 *(Ascaridata: Anisakidae)* — a parasite of fish-eating birds.) In: Raboty po parazitofaune Yugo-Zapada SSSR. Mat. zonaln. soveshch., Kishinev pp. 96—103 (in Russian).

Mozgovoy A. A., Shakhmatova V. I. & Semenova M. K., 1965b: (Study of life cycle of *Contracaecum spiculigerum (Ascaridata: Anisakidae)* — nematode of fish-eating birds.) Mater. nauchn. Konf. Vses. Obshch. Gelmint., Part IV, 169—174 (in Russian).

Mozgovoy A. A., Shakhmatova V. I. & Semenova M. K., 1968: (Life cycle of *Contracaecum spiculigerum (Ascaridata: Anisakidae)* — parasite of domestic and game birds.) Trudy gelmint. Lab., 19: 129—136 (in Russian).

Muehling P., 1898: Die Helminthen-Fauna der Wirbeltiere Ostpreussens. Arch. Naturg. Berlin, 64 Jahr, I: 1—118.

Mueller A., 1897: Helminthologische Mitteilungen. Arch. Naturg. Berlin, 63 Jahr, I: 1—26.

Mukhamadiev S. A., 1966: (Helminth fauna of herring-gull of Krym.) Izv. Akad. Nauk Tadzhik., Otd. biol. Nauk, 1: 119—121 (in Russian).

Myers B. J., Kuntz R. E. & Wells W. H., 1962: Helminth parasites of reptiles, birds, and mammals in Egypt. VII. Check list of nematodes collected from 1948 to 1955. Can. J. Zool., 40: 531—538.

Nathusius H., 1837: Helminthologische Beiträge. Über einige Eingeweidewürmer des Schwarzen Storches. Arch. Naturg. Berlin, 3 Jahr, v. I: 52—65.

Nicoll W., 1927: Parasitic worms in the St. Andrews fauna. In: McIntosh, W. C., Additions to the marine fauna of St. Andrews since 1874. Annls Mag. nat. Hist., 9: 31—50.

Nikolskaya N. P., 1939: (Parasite fauna of cormorant *(Phalacrocorax carbo)* of the Astrakhan reservation.) Uchen. zap. Leningr. Gos. Inst. (Sb. rab. po parazitol.) No. 11: 58—66 (in Russian).

Okorokov V. I., 1957: (Helminths of wild water birds of Chelyabinsk region and their seasonal dynamics.) Tez. Dokl. nauch. Konf. vses. Obshch. Gelmint., Part 1, pp. 228—230 (in Russian).

Okorokov V. I., 1964: (Helminth fauna of fish-eating birds of the order *Podicepites* in the Chelyabinsk region.) Mater. nauch. Konf. vses. Obshch. Gelmin., Part II, pp. 44—47 (in Russian).

Okorokov V. I. & Tkachev V. A., 1969: (The helminth fauna of wild aquatic birds of the order *Podicipitiformes* in the Southern Urals.) In: Problemy Parasit., Part I, Kiev, pp. 175—177 (in Russian).

Oliveira Rodrigues H. de & Machado da Mendonça J., 1967: Redescrição de *Stegophorus diomedeae* (Johnston e Mawson, 1942) Johnston e Mawson, 1945 *(Nematoda, Spiruroidea)*. Mems Inst. Oswaldo Cruz, 65: 149—152.

Ortlepp R. J., 1923: The life-history of *Syngamus trachealis* (Montagu) v. Siebold, the gape-worm of chicken. J. Helminth., 1: 119—140.

Ortlepp R. J., 1937: South African helminths. Part I. Onderst epoort J. vet. Sci. anim. Ind., 9: 311—336.

Osche G., 1955: Über Entwicklung, Zwischenwirt und Bau von *Porrocaecum talpae, Porrocaecum ensicaudatum* und *Habronema mansioni (Nematoda)*. Z. ParasitKde., 17: 144—164.

Osche G., 1959: Über Zwischenwirte, Fehlwirte und die Morphogenese der Lippenregion bei *Porrocaecum-* und *Contracaecum-* Arten *(Ascaridoidea, Nematoda)*. Z. ParasitKde., 19:458—484.

Oshmarin P. G., 1950: (Helminth fauna of birds of the Far East (Kamchatka, Zemlya Koryakov and the Kuril Islands). Trudy gelmint. Lab., 3: 166—179 (in Russian).

Oshmarin P. G., 1956: *(Spirurata, Tetrameridae* of domestic and wild birds of Primorye region.) Trudy Dalnevostochn. fil. AN SSSR, ser. zool., No. 3: 281—314 (in Russian).

Oshmarin P. G., 1959: *(Alcedospirura collaricephala* gen. et sp. n. — new species and genus of nematodes of birds of the Far East.) Zool. Zh., 38: 1310—1312 (in Russian).

Oshmarin P. G., 1963: (Helminths of mammals and birds in the Primorsk region.) Moscow: Izdat. Akad. Nauk SSSR, 323 pp. (in Russian).

Oshmarin P. G., 1965: (Helminth fauna of game-animals of Buryatia.) In: (Parasitic worms of domestic und wild animals.) Vladivostok, Dalnevost. gos. Univ., pp. 209—212 (in Russian).

Oshmarin P. G. & Belous E. V., 1951: (Filariid fauna of wild animals.) Trudy gelmint. Lab., 5: 121—127 (in Russian).

Oshmarin P. G. & Parukhin A. M., 1960: (The helminth fauna of *Pandion haliaetus* as an example of how the helminth fauna of animals is formed.) Zool. Zh., 39: 1303—1311 (in Russian).

Oshmarin P. G. & Parukhin A. M., 1963: (Trematodes and nematodes of birds and mammals of the Sikhote-Alinsk preserve.) Trudy Sikhote-Alianskogo Gos. zapoved. No. 3: 121—181 (in Russian).

Panin V. Ya., 1960: (Helminth fauna of birds of the Zaisan Valley.) Trudy Inst. Zool., Alma-Ata, 12: 166—172 (in Russian).

Panova L. G., 1926: (Study of nematodes of the gulls of the Don River region.) Trudy gos. Inst. eksper. Vet., 3: 82—85 (in Russian).

Panova L. G., 1927: (Helminthology in Kazakhstan.) Sb. Rabot po Gelmintol., pp. 121—137 (in Russian).

Parona C., 1887a: Elmintologia sarda. Ann. Mus. Civ. Storia Nat. Genova, Ser. 2, 4: 275 to 384.

Parona C., 1887b: Vermi parassiti in animali della Ligura. Ann. Mus. Civ. Genova, Ser. 2, 4: 483—501.

Parona C., 1894: L'elmintologia italiana da suoi primi tempi all'anno 1890. Storia, sistematica, corologia e bibliografia. Atti R. Univ. Genova, 13: 1-733.

Parona C., 1901: Diagnosi di una nuova specie di nematode, *Histiocephalus stellae-polaris* n. sp. Boll. Mus. Zool. Anat. Comp. R. Univ. Torino, 16: 1.

Parona C., 1902: Catalogo di elminti raccolti in vertebrati delle Isola d'Elba (second nota). Atti Soc. Ligust. Sci. Nat. Geogr., 13: 10—29.

Parona C., 1903: Elminti. Oss. Sc. Sped. polare di S.A.R. Luigi Amadeo di Savoira Duca degli Abruzzi, 1899—1900, Milano, 3 pp.

Parukhin A. M., 1964a: *(Sobolevicephalus chalcyonis* n. g., n. sp., a new nematode of the family Histiocephalidae Skrjabin, 1941.) Uchen. Zap. gorkov. gos. Univ., No. 62: 190—193 (in Russian).

Parukhin A. M., 1964b: (Occurrence of males of *Viktorocara halcyoni* Ryzhikov et Khokhlova

1964 *(Nematoda: Acuaridae)*. Mater. nauch. Konf. vses. Obshch. Gelmint., Part 2, Moscow, pp. 58—59 (in Russian).

Parukhin A. M., 1964c: (Study of helminth fauna of vertebrates from the Sikhote-Alinsk reservation.) Uchen. Zap. gorkow. gos. pedagog. Inst., 42: 141—159 (in Russian).

Parukhin A. M. & Oshmarin P. G., 1960: (Nematodes *Encephalonema longimicrofilaria* gen. et sp. n. from the brain of birds.) Zool. Zh., 6: 934—936 (in Russian).

Parukhin A. M. & Truskova G. M., 1963: (Results of helminthological investigations of fish-eating birds of Gorky water reservoir.) Uchen. Zap. gorkow. gos. Univ., No. 63: 37—42 (in Russian.)

Paskalskaya M. Yu., 1968: (Helminth fauna of wild water birds in the zone Kumund.) Sb. nauchn. Rab. Novosibirskoy nauchnoissled. vet. Stant., Novosibirsk, No. 3: 307—310 (in Russian).

Pavlov A. V., 1960: (Identity of the species *Amidostomum fulicae* (Rudolphi, 1819) and *Amidostomum raillieti* (Skrjabin, 1915).)Trudy geľmint. Lab., 10: 166—172.

Pemberton R. T., 1959: Life cycle of *Cyathostoma lari* Blanchard 1849 *(Nematoda, Strongyloidea* (Correspondence.) Nature, Lond., 184: 1423.

Pemberton R. T., 1960: Helminth parasites of some British birds. Ann. Mag. nat. Hist., Ser. 13: 455—463.

Pemberton R. T., 1963: Helminths of three species of British gulls, *Larus argentatus* Pont., *L. fuscus* L. and *L. ridibundus*. J. Helminth., 37: 57—88.

Peters B. G., 1936: *Paronchocerca ciconiarum* n. g., n. sp. from Saddlebilled Stork in West Africa. J. Helminth., 14: 1—10.

Petrov A. M. & Chertkova A. N., 1950: (Study of nematode fauna of birds of the southern Kirghizia.) Trudy geľmint. Lab. 4: 90—99 (in Russian).

Petter A. J., 1959: Redescription de *Paryseria adeliae* Johnston, 1938. Remarques sur le genre *Paryseria* et les genres voisins *Rusguniella, Aviculariella, Proyseria* (gen. nov.), *Seuratia*. Annls Parasit. hum. comp., 34: 322—330.

Radulescu J., 1966: Pelicani ca pradatori si raspinditori posibili de parazitore la pestii di pepinierele piscicole. Bul. Inst. de Cercetari si proieetari piscicole, 25: 75—78.

Radulescu A. & Lustun L., 1967: Contributii la cunoasterea parazitilor martinului rizator *(Larus ridibundus)* din Romania. Bul. Inst. Cercetari si proiectari piscicole, 26: 91—95.

Railliet A., 1895: Traité de zoologie médicale et agricole, 2nd (fasc. 2), Paris, pp. 423—424.

Railliet A., 1916: La Famille des *Thelaziidae*. J. Parasit., 2: 99—105.

Railliet A. & Henry A., 1909: Sur la classification des *Strongylidae*. 1. *Metastrongylinae*, 2. *Ankylostominae*. C. r. Séanc. Soc. Biol., 66: 85—88; 168—171.

Railliet A. & Henry A., 1912: Quelques nématodes parasites des reptiles. Bull. Soc. Path. exot., 5: 251—259.

Railliet A., Henry A. & Sisoff, 1912: Sur les affinités des dispharages *(Acuaria* Bremser), nématodes parasites des oiseaux. C. r. Séanc. Soc. Biol., 73: 622—624.

Reimer L., 1969: Helminthen von Kormoranen von Brutkolonien der Deutschen Demokratischen Republik. Wiss. Z. Ernst Moritz Arndt-Univ., Greifswald, Math.—naturwiss. Reihe, 18: 129—135.

Reimer L., 1973: Zur Helminthenfauna von *Larus canus* L. von Brutgebieten der Ostsee. Wiss. Z. Päd. Hochsch. „Liselotte Herrmann Güstrow", Math.-naturwiss. Fak., 2: 87—92.

Roman-Chiriac E., 1965: Cercetari asupra parazitofaunei pasarilor ihtiofage din delta Luňrii. An. Univ. Bucuresti, ser. stint. natur. biol. 14: 137—153.

Romanova N. P., 1947: (Study of the life cycle of *Echinuria uncinata* (Rud., 1819) — a nematode from the stomach of water birds.) Dokl. Akad. Nauk SSSR, 55: 375—376 (in Russian).

Romanova N. P., 1948: (Biology of *Cyathostoma* sp., parasitizing respiratory tract of the emu.

(Collected papers on helminthology dedicated to K. I. Skrjabin on the 40th Anniversary of his scientific achievement.) Moscow, pp. 189—194 (in Russian).

Rudolphi C. A., 1809: Entozoorum sive vermium intestinalium, historia naturalis. Amstelaedami, Vol. II: 457 pp.

Rudolphi C. A., 1819: Entozoorum synopsis cui accendunt mantissa duplex et indices locupletissimi. Berolini, 811 pp.

Ryšavý B., 1958: Helmintofauna kormoránů *(Phalacrocorax carbo* L.) hnízdících v Československu. Věst. Čs. Spol. zool., 22: 121—129.

Ryšavý B., 1959: Der Entwicklungszyklus von *Porrocaecum ensicaudatum* Zeder, 1800 *(Nematoda: Anisakidae)*. Acta vet. hung., 9: 317—323.

Ryšavý B., & Baruš V., 1965: (Distribution and forms of reservoir habitationism in nematodes.) Problemy Parazit., Kiev, No. 4, pp. 33—43 (in Russian).

Ryzhikov K. M., 1948: (Phylogenetic relationship of the nematodes of the family *Syngamidae* and an attempt to reconstruction their systematics.) Dokl. Akad. Nauk SSSR, 2: 733 to 736 (in Russian).

Ryzhikov K. M., 1949: (Principles of nematodology — 1. *Syngamidae* of domestic and wild animals.) Moscow-Leningrad, 164 pp. (in Russian).

Ryzhikov K. M., 1952: (New data on *Syngamidae*.) Trudy gel'mint. Lab., 6: 131—138 (in Russian).

Ryzhikov K. M., 1960: (Helminth fauna of *Somateria* spp.) Trudy gel'mint. Lab., 10: 173 to 188 (in Russian).

Ryzhikov K. M., 1963a: (Nematodes of *Anseriformes* from Kamchatka.) Trudy gel'mint. Lab., 13: 133—143 (in Russian).

Ryzhikov K. M., 1963b: (Helminth fauna of wild and domestic birds of the order *Anseriformes* from the Far East.) Trudy gel'mint. Lab., 13: 78—132 (in Russian).

Ryzhikov K. M., 1965: (New data about *Stegophorus stellae-polaris (Nematoda: Acuariidae)*. Helminthologia, 6: 173—180 (in Russian).

Ryzhikov K. M., 1966: (Species composition of the genus *Streptocara (Nematoda, Spirurata)*. Mater. nauchn. Konf. vses. obshch. Gel'mint., Part 3, pp. 240—244 (in Russian).

Ryzhikov K. M., 1967a: (Revision of the classification of the *Syngamidae.)* In: Problemy parazit. (Tezisy dokl. V. nauchn. konf. Ukr. resp. nauchn. obshch. parazitol.), Kiev, pp. 184—187 (in Russian).

Ryzhikov K. M., 1967b: Key to the helmints of domestic water birds.) Moscow: Izdat. "Nauka", 262 pp. ((in Russian).

Ryzhikov K. M. & Daiya G. G., 1967: (Nematodes of *Anseriformes* of Yakutia (based on materials of 290 SGE, 1953—1956). In: Sb. po gel'mintofaune ryb i ptits. Moscow. Mimeographed. Deposited in VINITI, No. 162—67, pp. 154—167 (in Russian).

Ryzhikov K. M. & Khokhlova I. G., 1964a: (Two new nematode species *(Skrjabinoclava halcyoni* n. sp. and *Cyrnea jubilarica* n. sp.) from wild birds in Vietnam.) Trudy gel'mint. Lab., 14: 187—193 (in Russian).

Ryzhikov K. M. & Khokhlova I. G., 1964b: (Contribution to the knowledge of nematodes of birds from Vietnam.) Mater. nauch. Konf. vses. Obshch. Gelmint., Part 2, Moscow, pp. 116—118 (in Russian).

Ryzhikov K. M. & Khokhlova I. G., 1964c: (Two new nematode species *(Schistorophus cirripedesmi* n. sp. and *Victorocara halcyoni* n. sp.) from wild birds in Vietnam.) Helminthologia, 5: 107—114 (in Russian).

Ryzhikov K. M. & Khokhlova I. G., 1967: (Nematode fauna of domestic and wild birds of Vietnam.) In: Sb. rab. po gelmintofaune ryb i ptits. Moscow. Mimeographed. Deposited in VINITI, No. 162—67, pp. 168—186 (in Russian).

Ryzhikov K. M. & Kozlov D., 1959: (Nematode fauna of wild birds of Turkmenistan.) Helminthologia, 1: 55—68 (in Russian).
Ryzhikov K. M. & Pavlov A., 1959: *(Amidostomum orientale* sp. n. — a new nematode of *Anseriformes* of Yakutia.) Helminthologia, 1: 69—73 (in Russian).
Ryzhikov K. M. & Zavadil R., 1958: Druhová příslušnost nematodů rodu *Cyathostoma* získaných ze pštrosů emu. Sb. Vys. šk. zeměd. Brno, Řada B, 2: 125—132.
Ryzhova A. A. & Dubov N. A., 1955: (Nematodes of game-birds of the Białowieża virgin forest.) Trudy Gork. selsk. gos. Inst., 7 (I): 223—235 (in Russian).
Saidov Yu. S., 1954: (New species of parasitic worms of fish-eating birds of Dagestan.) Trudy gelmint. Lab., 7: 265—274 (in Russian).
Sailov D. I., 1962: (Helminths of fish-eating birds of Kyzyl-Agach reservation and their ecology.) Trudy azerb. gos. pedagog. Inst., 21: 149—163 (in Russian).
Sailov D. I., 1963b: (Study of helminth fauna of fish-eating birds under conditions of C. M. Kirov national reservation in Kyzyl-Agach.) In: Problemy parazitologii. (Tr. 4—oy nauchn. konf. parazitologov USSR), Kiev, pp. 257—258 (in Russian).
Sailov D. I., 1965a: (Fauna of parasitic worms of *Steganopodes.)* Uchen. zap. Azerb. gos. Univ., ser. biol. nauk, No. 3: 35—42 (in Russian).
Sailov D. I., 1965b: (Helminths of fish-eating birds of C. M. Kirov national reservation in Kyzyl Agach from faunistical and ecological view.) Trudy zapovedn. Azerbaidzhana, Baku, No. 1: 130—146 (in Russian).
Sailov D. I., 1966: (Helminths of birds of the order *Podicepiformes* under conditions of southwestern Caspia.) Uchen. zap. Azerb. gos. Univ., ser. biol. nauk, No. 2: 35—40 (in Russian).
Sailov D. I., 1970: (Helminths of gulls in Azerbaidzhan.) In: Issled. po gelmintol. v Azerbaidzhane, Baku: Izdat. "ELM", pp. 166—172 (in Russian).
Samedov G. A., 1967a: (Acanthocephalans and nematodes of *Falconiformes* in the Lenkoran zone of Azerbaidzhan.) Izv. Akad. Nauk azerb. SSR, Ser. Biol. Nauk, No. 2: 44—50 (in Russian).
Samedov G. A., 1967b: *(Acanthocephala* and *Nematoda* of *Falconiformes* in the Lenkoran zone of Azerbaidzhan.) In: Mater. nauch. teoretich. konf. molodykh uchenykh AN Azerb. SSR, Baku, 4: 205—209 (in Russian).
Samedov G. A., 1969: (Helminths of fish-eating *Falconiformes* of south-eastern zone of Azerbaidzhan.) In: Problemy parazit., Part. I., Kiev, pp. 211—212 (in Russian).
Schmidt G. D. & Kuntz R. E., 1971: Nematode parasites of Oceanica. XV. *Acuariidae, Streptocaridae,* and *Seuratidae* of birds. Proc. helminth. Soc. Wash., 38: 217—223.
Schmidt G. D. & Kuntz R. E., 1972: Nematode parasites of Oceanica. XVII. *Schistorophidae, Spiruridae, Physalopteridae* and *Trichostrongylidae* of birds. Parasitology, 64: 269—278.
Schneider A. F., 1866: Monographie der Nematoden. Berlin, 357 pp.
Schrank F. P., 1788: Verzeichnisse der bisher hinlänglich bekannten Eingeweidewürmer, nebst einer Abhandlung über ihre Anverwandtschaften. München, 116 pp.
Semenov V. D., 1927: (Preliminary results of the study of helminth infection of birds of the West Region (according to data of the 30th SGE expedition in 1926.) Trudy Smolensk. obsch. estestvoisp. i vrachei, 2: 137—156 (in Russian).
Semenova M. K., 1971: (Life cycle of *Contracaecum micropapillatum* (Stossich, 1890) Baylis, 1920 — parasite of pelicans.) Trudy gelmint. Lab., 22: 148—152 (in Russian).
Semenova M. K., 1972: (Study of the biology of the nematode *Contracaecum micropapillatum* (Stossich, 1890) Baylis, 1920 *(Ascaridata).* In: Problemy parazitologii, Trudy VII nauch. Konf. Parasit. USSR. Part 2, Kiev, pp. 238—240 (in Russian).
Sergeeva T. P., 1968a: (Nematodes of gulls of the U.S.S.R.) Moscow, Mimeographed. Deposited in VINITI, No. 375—68, 31 pp. (in Russian).

Sergeeva T. P., 1968b: (On the systematic position and species composition of the genus *Rusguniella* (Seurat, 1919) *(Spirurata: Acuarioidea).*) Trudy gel'mint. Lab., 19: 163—168 (in Russian).

Sergeeva T. P., 1969: (Nematode fauna of gulls in the U.S.S.R.) Trudy gel'mint. Lab., 20: 146—155 (in Russian).

Sergienko M. I., 1963: (Nematode fauna of water birds from the valley of the Upper Dniester River.) In: Suchestna ta minula fauna zakhidnikh oblastey Ukrainy. Kiev, pp. 25—28 (in Russian).

Sergienko M. I., 1971: (Parasitic worms of gulls of the Upper Dniester basin.) Vest zool. AN Ukrainsk. SSR, No. 6: 43—48 (in Russian).

Sergienko M. I., 1972: (Parasitic worms of grebes of the Upper Dniester basin.) In: Parazity, parazitotsenozy i ikh likvidatsiya. Kiev, 1: 131—135 (in Russian).

Serkova O. P., 1948: (Nematodes of the birds of the Barabinsk Lakes.) Parazit. Sb. zool. Inst. AN SSSR, 18: 209—244 (in Russian).

Seurat L. G., 1912: Sur la quatrième mue des nématodes parasites. C. r. Séanc. Soc. Biol., 73: 279—281 (in Russian).

Seurat L. G., 1914a: Sur la morphologie de l'ovejecteur des *Tropidocerca*. C. r. Séanc. Soc. Biol., 76: 173—176.

Seurat L. G., 1914b: Sur un nouveau parasite de la Perdrix rouge. C. r. Séanc. Soc. Biol., 76: 390—393.

Seurat L. G., 1915: Sur deux *Tropidocerca* des *Ardeidae*. C. r. Séanc. Soc. Biol., 78: 279—282.

Seurat L. G., 1916a: Sur un nouveau type de *Spiruridae*. C. r. Séanc. Soc. Biol., 79: 517—519.

Seurat L. G., 1916b: Sur un nouvel *Habronema* du *Bubulcus lucidus* Raf. C. r. Séanc. Soc. Biol., 79: 295—297.

Seurat L. G., 1916c: Sur un nouveau dispharage des palmipèdes. C. r. Séanc. Soc. Biol., 79: 785—788.

Seurat L. G., 1918: Contribution à l'étude de la faune parasitaire de la Tunisie. Nématodes. Archs. Inst. Pasteur Tunis, 10: 243—275.

Seurat L. G., 1920: Sur une filaire du héron cendré. Bull. Soc. Hist. nat. Afr. N., 11: 142.

Shakhtakhtinskaya Z., 1951: (New nematode *Petroviprocta vigissi* nov. gen. nov. sp. from the thoracic cavity of heron.) Trudy gel'mint. Lab., 5: 162—164 (in Russian).

Shakhtakhtinskaya Z. M., 1953: (Fauna of parasitic worms of game birds in Azerbaidzhan.) Trudy Azerb. gos. pedagog. Inst., 1: 29—34 (in Russian).

Shakhtakhtinskaya Z. M., 1959a: (Helminths of domestic and game birds of Azerbaidzhan SSR.) Sb. Rab. po Gelm., Moscow, MSKH, 1: 197—202 (in Russian).

Shakhtakhtinskaya Z. M., 1959b: (Study of the fauna of parasitic worms of domestic- and game birds in Azerbaidzhan.) Trudy Azerb. gos. pedagog. Inst., 11: 35—48 (in Russian).

Shakhtakhtinskaya Z. M. & Sadykov R., 1967: (Study of helminth fauna of water and marsh birds of Nakhichevan ASSR.) Uchen. zap. Azerb. gos. Univ., Ser. Biol. Nauk No. 3: 33—37 (in Russian).

Shevtsova L. S., 1970: (Some peculiarities of the nematodes of birds in Chukotka.) Mat. 15 nauchn. kong. professorsko-prepod. sostava biologo-pochv. fak. Dalnevostochnogo gos. univ., Vladivostok, pp. 180—182 (in Russian).

Shigin A. A., 1951: (New filariid of heron.) Trudy gel'mint. Lab., 5: 168—172 (in Russian).

Shigin A. A., 1953: (New nematode of birds *Tetrameres ardeae* nov. sp.) Papers on helminthology presented to Acad. K. I. Skrjabin on his 75th birthday. Moscow: Izdat. Akad. Nauk SSSR, pp. 758—760 (in Russian).

Shigin A. A., 1954: (Results of helminthological investigation of fish-eating birds from Rybinsk water reservoir during three years (1949—1951). Trudy probl. temat. Soveshch. zool. Inst., 4: 57—60 (in Russian).

Shigin A. A., 1957: (Parasitic worms of herons and grebes of the Rybinsk water reservoir.) Trudy darv. gos. Zapov., No. 4: 245—289 (in Russian).
Shigin A. A., 1959: (Helminth fauna of fish-eating birds of the orders *Anseriformes* and *Accipitres* of Rybinsk water-reservoir.) Trudy darv. gos. Zapov., No. 5: 315—331 (in Russian).
Shigin A. A., 1961: (The helminth fauna of lariform birds in the Rybinsk water reservoir.) Trudy darv. gos. Zapov., No. 7: 309—362 (in Russian).
Shikhobalova N. P., 1930: On a new genus of the *Nematoda* fam. Acuariidae Seurat, 1913. J. Parasit., 16: 220—223.
Shmytova G. Y., 1967: (The life cycle of *Paracuaria macdonaldi* Rao, 1951 *(Nematoda: Spirurata)*.) Problemy Parasit. Tezisy dokl. V. nauchn. konf. Ukra. resp. nauchn. obsch. parasit., Kiev, pp. 219—220 (in Russian).
Siebold C. T. E., 1836: Helminthologische Beiträge. Zweiter Beitrag. *Syngamus trachealis*. Ein doppelleibiger Eingeweidewurm. Arch. Naturg. Berlin, 2 Jahr, vol. I: 105—116.
Siebold C. T. E., 1837: On a double bodied intestinal worm, the *Syngamus trachealis*. Lond. and Edinb. Phil. Mag., 10: 253—261.
Siebold C. T. E., 1838: Bericht über die Leistungen im Gebiete der Helminthologie. Arch. Naturg. Berlin, 4: 254—281.
Singh S. N., 1948a: Studies on the helminth parasites of birds in Hyderabad State. Nematoda I. J. Helminth., 22: 77—92.
Singh S. N., 1948b: Studies on the helminth parasites of birds in Hyderabad state. Nematoda II. J. Helminth., 22: 199—218.
Škarda J., 1964: Helmintofauna některých volně žijících ptáků v ČSSR. Sb. vys. Šk. zeměd., Brno, Řada B, 12: 269—293.
Skrjabin K. I., 1915a: (Nématodes des oiseaux du Turkestan-Russe.) Ezheg. zool. muz. Akad. Nauk. Also listed as (Ann. Muz. Zool. Acad. Imp. Sci. St. Petersb.), 20: 457—557 (in Russian).
Skrjabin K. I., 1915b: (Strongylids of muscular stomach of Turkestan birds. Species of the genus *Amidostomum* Raill. et Henry, 1909.) Vestn. Obshch. Vet. St. Petersb., 27: 693—700 (in Russian).
Skrjabin K. I., 1916a: (Characteristics of bird nematodes of the genus *Streptocara*.) Arkh. Vet. Nauk. St. Petersb., 46: 883—890 (in Russian).
Skrjabin K. I., 1916b: (Materials to the monographe on nematodes of birds. 1. Characteristics of the nematodes of the genus *Aprocta* Linstow, 1883). Ezheg. zool. muz. Akad. Nauk. Also listed as (Ann. Muz. Zool. Acad. Imp. Sci. St. Petersb.), 21: 117—129 (in Russian).
Skrjabin K. I., 1916c: *Seuratia* n. g. nouveau genre de nématodes d'oiseaux. C. r. Séanc. Soc. Biol., 79: 971—973.
Skrjabin K. I., 1917: Sur quelques nématodes des oiseaux de la Russie. Parasitology, 9: 460—481.
Skrjabin K. I., 1923: (Parasitic nematodes of fresh-water fauna of the European and, partly, of Asian Part of Russia.) In: Zernov et al., Presnovodnaya fauna Evropeiskoi Rossii. Moscow, 2: 1—98 (in Russian).
Skrjabin K. I., 1924: (Study of the changes of parasitic worms of the birds in Russia. 1. Parasitic worms of *Pelecanus onocrotalus* L.) Trudy gos. Inst. eksper. Vet., 2: 149—157 (in Russian).
Skrjabin K. I., 1926: (To the knowledge of round-worms (nematodes) from birds of the Palaearctic Region.) Ezheg. Zool. muz. Akad. Nauk, 27: 88—103 (in Russian).
Skrjabin K. I., & Shikhobalova N. P., 1948: *(Filariata* of animals and man.) Moscow: Orgiz. Selskhozgiz., 608 pp.
Skrjabin K. I., Shikhobalova N. P. & Lagodovskaya E. A., 1961: (Principles of nematodo-

logy. X. *Oxyurata* of animals and man.) Moscow: Izdat. Akad. Nauk SSSR, 499 pp. (in Russian).
Skrjabin K. I., Shikhobalova N. P. & Lagodovskaya E. A., 1964: (Principles of nematodology. XII. *Oxyurata* of animals and man. Part 3.) Moscow: Izdat. Nauka, 468 pp. (in Russian).
Skrjabin K. I., Shikhobalova N. P. & Orlov I. V., 1957: (Principles of nematodology. 6. *Trichocephalidae* and *Capillariidae* of animals and man and the diseases caused by them.) Moscow: Izdat. Akad. Nauk SSSR, 579 pp. (in Russian).
Skrjabin K. I., Shikhobalova N. P., Shultz R. S., Popova T. I., Boev S. N. & Delyamure S. L., 1952: (Key to the parasitic worms, nematodes. Part 3. *Strongylata*.) Moscow: Izdat. Akad. Nauk SSSR, 890 pp. (in Russian).
Skrjabin K. I., Shikhobalova N. P. & Sobolev A. A., 1949: (Key to the parasitic nematodes. Vol. I. *Spirurata* and *Filariata*.) Moscow: Izdat. Akad. Nauk SSSR, 519 pp. (in Russian).
Skrjabin K. I. & Sobolev A A., 1963: (Principles of nematodology. XI. *Spirurata* of animals and man and the diseases caused by them. Part I. *Spiruroidea*.) Moscow: Izdat. Akad. Nauk SSSR, 511 pp. (in Russian).
Skrjabin K. I. & Sobolev A. A., 1964: (Principles of nematodology, 12. *Spirurata* of animals and man and the diseases caused by them. Part 2. *Physalopteroidea*.) Moscow: Izdat. Akad. Nauk SSSR, 334 pp. (in Russian).
Skrjabin K. I., Sobolev A. A. & Ivashkin V. M., 1965: (Principles of nematodology, 14. *Spirurata* of animals and man and the diseases caused by them. Part 3. *Acuarioidea*.) Moscow: Izdat. Nauka, 570 pp. (in Russian).
Skrjabin K. I., Sobolev A. A. & Ivashkin V. M., 1967a: (Principles of nematodology. 16. *Spirurata* of animals and man and the diseases caused by them. Part 4. *Thelazioidea*.) Moscow: Izdat. Nauka, 624 pp. (in Russian).
Skrjabin K. I., Sobolev A. A. & Ivashkin V. M., 1967b: (Principles of nematodology. 19. *Spirurata* of animals and man and the diseases caused by them. Part 5. Supplement.) Moscow: Izdat. Nauka, 240 pp. (in Russian).
Smetanina Z. B., 1972: (Faunistic review of the nematodes of fish-eating birds of the Primorye Territory.) Problemy parazit. Trudy VII. Nauch. konf. Parazit. USSR, Part 2, Kiev, pp. 275—279 (in Russian).
Smetanina Z. B. & Alekseev V. M., 1967: (Nematode fauna of fish-eating birds of Rimskiy-Korsakov Island.) Tez. dokl. XI. nauch. konf. Dalnevost. gos. univ., Part 2, ser. biol. nauk, pp. 253—255 (in Russian).
Smetanina Z. B. & Alekseev V. M., 1968: *(Skrjabinobronema pileati* n. sp. from *Halcyon pileata.)* Parazitologiya, 2: 475—476 (in Russian).
Smith A. J., Fox. H. & White C. Y., 1908: Contributions to systematic helminthology. Univ. of Penn. Med. Bull., 20: 1—283.
Smogorzhevskaya L. A., 1955: (Helminth fauna of fish-eating birds of the Lower Dnieper River.) Tez. dokl. 8 soveshch. po parazit. probl., Zool. Inst. AN SSSR, Leningrad, pp. 140—141 (in Russian).
Smogorzhevskaya L. A., 1956: (Helminth fauna of the Lower Dnieper River.) Problemy Parazit. (Tr. 2 n. konf. parazitol. USSR), Kiev, pp. 111—112 (in Russian).
Smogorzhevskaya L. A., 1959: (Ecological characteristics of helminths of fish-eating birds of the Lower Dnieper River.) Vopr. ekologii, Kiev, 3: 222—231 (in Russian).
Smogorzhevskaya .L A., 1960: (Helminth fauna of red-necked grebe *(Colymbus griseigena* Bdd.) of the Krym Region.) Problemy parazit. (Tr. 3 nauch. konf. parazitol. USSR), Kiev, p. 68 (in Russian).
Smogorzhevskaya L. A., 1961: (The helminth fauna of *Phalacrocorax aristotelis* L. in the Crimea region.) Problemy parazit. (Tr. Ukr. Respubl. nauch. Obshch. Parazit.), Kiev, No. 1: 207—220 (in Russian).

Smogorzhevskaya L. A., 1962a: (Nematodes of fish-eating birds of the Dniester River lowland.) In: Mat. do vivchenya istor. ta prirody r-nu Kanivsk. zapovidn., Kiev, pp. 118 to 126 (in Russian).

Smogorzhevskaya L. A., 1962b: (Ecological characteristics of helminth fauna of fish-eating birds of Krym region.) Vopr. ekologii, Kiev, 8: 108—109 (in Russian).

Smogorzhevskaya L. A., 1964: (Review of helminth fauna of water birds on the territory of the Ukr. SSR.) In: Problemy parazit. (Tr. Ukr. resp. nauchn. obshch. parazit.) Kiev, No. 3: 125 — 188 (in Russian).

Smogorzhevskaya L. A., 1967: (Nematodes of aquatic and marsh birds on the Levoberezhnaya steppe of the Ukrainian SSR.) In: Problemy parazit. (Tezisy dokladov V. nauchn. konf. Ukr. resp. nauchn. obshch. parazit.), Kiev, pp. 193—194 (in Russian).

Smogorzhevskaya L. A., Kornyushin V. V., Iskova I. I. & Eminov A., 1965: (Helminth fauna of fish-eating birds of south-western Turkmenia.) Mater. nauch. Konf. vses. obshch. Gel'mint., Part 2, pp. 228—230 (in Russian).

Sobolev A. A., 1943a: (Revision of the family *Acuariidae (Nematoda)* Seurat, 1913 with description of the new subfamily *Echinuriinae* n. subf. and new genus *Skrjabinoclava* n. g.) Trudy Gorkov. gos. selsko-khoz. Inst., 4: 285—302 (in Russian).

Sobolev A. A., 1943b: (Evolution of nematodes of the family *Acuariidae.*) Dokl. Akad. Nauk SSSR, 39: 76—79 (in Russian).

Sobolev A. A., 1947: (Morphology of a peculiar nematode — parasite of Terek sandpiper.) Trudy Gork. ped. Inst., 12: 5—17 (in Russian).

Sobolev A. A., 1952: *(Skrjabinoclava longifuniculata* n. sp., a new nematode from birds.) Trudy gel'mint. Lab., 6: 293—295 (in Russian).

Sobolev A. A. & Sudarikov V. E., 1939: (New nematodes from osprey *Sexansocara skrjabini* n. g., n. sp. and *Contracaecum pandioni* n. sp.) Trudy Gork. selsko-khoz. inst., 3: 97—103 (in Russian).

Soloviev P. F., 1912: (Parasitic worms of birds of Turkestan.) Ezheg. zool. muz. Akad. Nauk, (Also listed as: Ann. Muz. Zool. Acad. Imp. Sci. St. Petersb.), 17: 86—115 (in Russian).

Solonitsin I. A., 1928a: (To the knowledge of helminth fauna of birds of the Volga—Kamsk Region (Nematodes and Trematodes of birds of the Chuvash and Tatar Republics.) Uchen. Zap. kazan. gos. ver. Inst., 38: 75—99 (in Russian).

Solonitsin I. A., 1928b: (On the helminth fauna of birds of the Volga—Kamsk Region (Nematodes and Trematodes). Trudy 3 vseross. sjezda zool., anat. i gistol., Leningrad, pp. 155—156 (in Russian).

Solonitsin I. A., 1932: Über zwei neue Arten von Nematoden in Vogel. Zentbl. Bakt. Abt. Orig., 124: 361—365.

Sonin M. D., 1959: (Filariids of birds of the Upper Yenisei River (the Tuva Autonomous Soviet Socialist Republic). Helminthologia, 1: 75—83 (in Russian).

Sonin M. D., 1963: (Filariids of birds of the Soviet Far East.) Trudy gel'mint. Lab., 13: 227—249 (in Russian).

Sonin M. D., 1965: (New genus *Parornithofilaria* nov. gen. *(Filariata, Splendidofilariidae)* and revision of the subfamily *Splendidofilariinae.*) Trudy gel'mint. Lab., 15: 140—144 (in Russian).

Sonin M. D., 1966: (Principles of nematodology. Vol. 17: *Filariata* of animals and man and the diseases caused by them. Part. I. *Aproctoidea.*) Moscow: Izdat. Nauka, 360 pp. (in Russian).

Sonin M. D., 1968: (Principles of nematodology. Vol. 21. *Filariata* of animals and man and the diseases caused by them. Prat. II. *Diplotriaenoidea.*) Moscow: Izdat. Nauka, 390 pp. (in Russian).

Sonin M. D. & Larchenko T. T., 1974: (Nematodes of birds of Tuva.) Trudy gel'mint. Lab., 24: 173—181 (in Russian).
Spasskaya L. P., 1949: (Nematodes of birds of West Siberia from the materials of the 25th Union Helminthological Expedition.) Trudy gel'mint. Lab., 2: 128—142 (in Russian).
Spassky A. A., 1952: (An investigation of the life-histories of *Anisakidae (Ascaridata: Anisakidae)*.) Trudy gelmint. Lab., 6: 72—73 (in Russian).
Sprehn C. E. W., 1932: Lehrbuch der Helminthologie. Berlin, 998 pp.
Sprehn C. E. W., 1962: Klasse Nematoda, Fadenwürmer. In: Brohmer P., Ehrmann P. & Ulmer G. (Editors), Die Tierwelt Mitteleuropas, 1 (5b), 191 pp.
Statirova N. A., 1946: (Helminth fauna of *Plegadis falcinellus* of Kazakhstan.) (Collected papers on helminthology dedicated by his pupils to K. I. Skrjabin in his 40th year of scientific teaching and administrative achievement.) Moscow, pp. 262—263 (in Russian).
Stossich M., 1889: Vermi parassiti in animali della Croazia. Glasn. Hrv. Nar. Družt., 4: 180—185.
Stossich M., 1890a: Elminti della Croazia. Glasn. Hrv. Nat. Družt., 5: 129—136.
Stossich M., 1890b: Elminti veneti raccolti dal Dr. Alessandro Conte de Ninni. Boll. Soc. adriat. Sci. nat., Trieste, 12: 49—56.
Stossich M., 1890c: Il genere *Trichosoma* Rudolfi. Bol. Soc. adriat. Sci. nat. Trieste, 12, 38 pp.
Stossich M., 1891a: Nuova serie di elminti veneti raccolti dal Dr. Alessandro Conte de Ninni. Secondo serio. Glasn. Hrv. Nar. Družt., 6: 216—219.
Stossich M., 1891b: Il genere *Dispharagus* Dujardin. Lavoro monografico. Boll. Soc. adriat. Sci. nat., Trieste, 13: 81—108.
Stossich M., 1892: Osservazioni elmintologiche. Glasn. Hrv. Nar. Družt., 7: 64—73.
Stossich M., 1893: Note elmintologiche. Boll. Soc. adriat. Sci. nat., Trieste, 14: 83—89.
Stossich M., 1895: Notizia elmintologiche. Boll. Soc. adriat. Sci. nat., Trieste, 16: 33—46.
Stossich M., 1896a: Il genere *Ascaris* Linné. Boll. Soc. adriat. Sci. nat., Trieste, 17: 9—120.
Stossich M., 1896b: Ricerche elmintologiche. Boll. Soc. adriat. Sci. nat., Trieste, 17: 121—136.
Stossich M., 1897: Filarie e Spiroptere. Lavoro monografico. Boll. Soc. adriat. Sci. nat., Trieste, 18: 13—162.
Stossich M., 1899: *Strongylidae*. Boll. Soc. adriat. Sci. nat., Trieste, 19: 55—152.
Stossich M., 1900: Contributo allo studio degli Elminti. Boll. Soc. adriat. Sci. nat., Trieste, 20: 1—9.
Stossich M., 1901: Osservazioni elmintologiche. Boll. Soc. adriat. Sci. nat., Trieste, 20: 89—104.
Stossich M., 1902: Sopra alcuni nematodi della collezione elmintologica del prof. dott. Corrado Parona. Boll. Mus. Zool. Anat. Comp., No. 116; 16 pp.
Sugimoto M., 1930: (On a new nematode parasite *(Streptocara formosensis* sp. nov.) in gizzard of Formosan domestic ducks.) J. Soc. trop. Agric., 2: 135—144 (in Japanese).
Sultanov M. A., 1958: (Role of game-birds in the distribution of helminthiases of domestic birds.) Uzbek. biol. Zh., 5: 17—21 (in Russian).
Sultanov M. A., 1959a: (Helminth fauna of domestic and game birds of Uzbekistan.) Trudy gel'mint. Lab., 9: 333—335 (in Russian).
Sultanov M. A., 1959b: (Helminth fauna of domestic and game birds of Uzbekistan.) Uzbek. biol. Zh., No. 2: 62—71 (in Russian).
Sultanov M. A., 1963: (Helminths of domestic and game birds of Uzbekistan.) Tashkent: Akad. Nauk Uzbek. SSR, 467 pp. (in Russian).
Sultanov M. A., Ryzhikov K. M. & Kozlov D. P., 1960: (Nematode fauna of wild birds of the mouth of the Amu Darya River.) Uzbek. biol. Zh., No. 1: 58—63 (in Russian).
Supryaga A. M., 1965a: (New nematode *Avioserpens mosgovoyi* n. sp. *(Camallanata: Dracunculidae)*

from coot *(Fulica atra)*. Mater. nauch. Konf. vses. Obshch. Gel'mint., Part 4, pp. 272 to 275 (in Russian).

Supryaga A. M., 1965b: (The life cycle of *Avioserpens mosgovoyi* n. sp. *(Camallanata: Dracunculidae)* — nematode of birds.) Materl. nauchn. Konf. vses. Obshch. Gel'mint., Part 4, pp. 275—277 (in Russian).

Supryaga A. M., 1968: (The question of reservoir parasitism of the nematode *Avioserpens mosgovoyi* Supryaga, 1965 — parasite of water birds.) Mater. nauchn. Konf. vses. Obshch. Gel'mint., Part. I, pp. 255—262 (in Russian).

Supryaga A. M., 1969a: *(Avioserpens mosgovoyi* of water birds and some questions of its problematics.) Tezisy dokl. IX. Mezhdunar. kongr. biologov-okhotovedov, Moscow, pp. 65—67 (in Russian).

Supryaga A. M., 1969b: (Lifespan of the nematode *Avioserpens mosgovoyi (Camallanata: Dracunculidae)* in the definitive host.) Tezisy dokl. 6 nauchn. konf. Ukr. resp. nauchn. obshch. parazit., Kiev, pp. 245—246 (in Russian).

Supryaga A. M., 1971a: (Revision of the genus *Avioserpens* Wehr et Chitwood, 1934 *(Camallanata, Dracunculidae)*. Trudy gel'mint. Lab., 22: 196—200 (in Russian).

Supryaga A. M., 1971b: (Life cycle of *Avioserpens mosgovoyi (Camallanata, Dracunculidae)* — nematode of water birds.) Sbornik Rabot po Gelmintologii posvyashchen. 90-letiyu so dnya rozhdeniya Akad. K. I. Skrjabina. Moscow: Izdat. Kolos, pp. 374—383 (in Russian).

Supryaga A. M., 1972: (Study of focus of *Avioserpens* of water birds.) Problemy parazit. (Trudy nauchn. konf. parazitol. Ukrain. SSR, Part 2) Kiev: Izdat. Naukova Dumka, pp. 304 to 306 (in Russian).

Swales W. E., 1933: *Tetrameres crami* sp. nov., a nematode parasitizing the proventriculus of a domestic duck in Canada. Can. J. Res., 8: 334—336.

Swales W. E., 1936a: *Tetrameres crami* Swales, 1933, a nematode parasite of dukcs in Canada. Morphological and biological studies. Can. J. Res., Sect. D, 14: 151—164.

Swales W. E., 1936b: Morphological and biological studies on *Tetrameres crami* Swales, 1933, an important nematode parasite of ducks. J. Parasit., 22: 528.

Swierstra D., Jansen J. Jr. & Broek E. van den, 1959: Parasites of animals in the Netherlands. Tijdschr. Diergeneesk., 84: 892—900.

Threlfall W., 1964: Factors concerned in the mortality of some birds which perished in Anglesey and northern Caernarvonshire during the winter of 1963 with special reference to parasitism by helminths. Ann. Mag. nat. Hist., 6: 721—737.

Threlfall W., 1965a: Helminth parasites and possible causes of death of some birds. Ibis, 107: 545—548.

Threlfall W., 1965b: Studies of the helminth parasites of herring gulls *(Larus argentatus* Pontopp) on the Newborough Warren Nature Reserve, Anglesey. Ph. D. Thesis Univ. of North Wales, 154 pp.

Threlfall W., 1965c: Life cycle of *Cyathostoma lari* Blanchard, 1849 *(Nematoda, Strongyloidea)*. (Correspondence.) Nature; Lond., 206: 1167—1168.

Threlfall W., 1966a: The method of attachment of *Cyathostoma lari* E. Blanchard, 1849 *(Nematoda: Strongyloidea)*. Can. J. Zool., 44: 1091—1092.

Threlfall W., 1966b: Studies on the blood of herring gulls and domestic fowl with some observations on changes brought about by infection with *Cyathostoma lari* Blanchard, 1849 *(Nematoda: Strongyloidea)*. Bull. Wildl. Dis. Ass., 2: 41—51.

Threlfall W., 1966c: The helminth parasites of the herring gull *(Larus argentatus* Pontopp). Tech. Commun. Commonw. Bur. Helminth., No. 37; 23 pp.

Threlfall W., 1966d: Experiments on eggs and adult specimens of *Cyathostoma lari* E. Blanchard, 1849. (Correspondence.) Nature, Lond., 212: 1063—1064.

Threlfall W., 1967: Studies on the helminth parasites of the herring gull *(Larus argentatus)* Pontopp., in Northern Caernarvonshire and Anglesey. Parasitology, 57: 431—453.

Tkachev V. A., 1971: (Growth dynamics of helminth infection of *Podiceps griseigena* and *Podiceps nigricollis* from the lakes of the Cheliabinsk region.) In: Voprosy zoologii, pp. 60 to 62 (in Russian).

Tolkacheva L. M., 1967: (Helminth fauna of water birds of the Lower Yenisei River and the Noril Lakes.) In: Sb. rabot po gelmintofaune ryb i ptits. Moscow. Mimeographed. Deposited in VINITI. No. 162—67, pp. 187—209 (in Russian).

Travassos L., 1914: Contribuições para o conhecimento da fauna helmintologica brazileira. 3. Sobre as especies brazileiras do genero *Tetrameres* Creplin, 1846. Mems Inst. Oswaldo Cruz, 6: 150—162.

Travassos L., 1915: Contribuicıões para o conhecimento da fauna helmintolojica brazileira. 5. Sobre as especies brazileiras do genero *Capillaria* Zeder, 1800. Mems Inst. Oswaldo Cruz, 7: 146—172.

Tsacheva-Petrova K., 1971: (Helminth fauna of free-living wild birds of western Stara Planina massif (Balkan Mountains, Bulgaria) *(Nematoda — Acanthocephala)*. Izv. zool. Inst., Sof., 33: 185—194 (in Bulgarian).

Tsimbalyuk A. K., 1965: (Nematode fauna of wild *Anseriformes* of the islands of the Bering Sea). In: Parasitic worms of domestic and wild animals: Papers on helminthology presented to Prof. A. A. Sobolev on the 40th anniversary of his scientific and teaching activity.) Vladivostok: Dalnevost. gos. Univ, pp. 340—342 (in Russian).

Tsimbalyuk A. K. & Belogurov O. I., 1964: (Nematode fauna of fish-eating birds of the islands of the Bering Sea.) Nauch. Dokl. vyssh. Shk., No. 4: 7—11 (in Russian).

Tsimbalyuk A. K. & Kulikov V. V., 1966: (The biology of *Skrjabinocerca prima* Schikhobalova, 1930 *(Nematoda, Acuariidae)* — parasite of birds.) Zool. Zh., 45: 1565—1569 (in Russian).

Tsimbalyuk A. K., Leonov V. A. & Belogurov O. I., 1963: (Species independence of the nematode *Paracuaria macdonaldi* Rao, 1951 and the position of the genus *Paracuaria* Rao, 1961 in the systematics of *Acuariidae (Spirirurata)*.) Vest. leningr. gos. Univ., 15: 155—157 (in Russian).

Tsimbalyuk A. K., Leonov V. A. & Belogurov O. I., 1964: *(Seuratia puffini* Yamaguti, 1941— a genus and species of nematodes new for the fauna of the USSR.) Zool. Zh., 43: 930 to 932 (in Russian).

Turemuratov A. T., 1962a: (Helminth fauna of herons and gulls of the delta of the Amu-Darya River.) Trudy gelmint. Lab., 12: 263—277 (in Russian).

Turemuratov A. T., 1962b: (Helminth fauna of pelicans of the Aral Sea basin.) Vest. karakalpaksk. Fil. Akad. Nauk Uzbek. SSR, No. 3: 44—49 (in Russian).

Turemuratov A. T., 1963a: (Helminth fauna of cormorants of the Aral Sea basin.) Mater. nauchn. Konf. vses. Obshch. Gelmint., Part 2, pp. 138—140 (in Russian).

Turemuratov A. T., 1963b: (Syngamids of *Pelecaniformes* from the Aral Sea and notes on the structure of this nematode family.) Vest. karakalpaksk. Fil. Akad. Nauk Uzbek. SSR, No. 3: 45—50 (in Russian).

Turemuratov A. T., 1964: (Some results of the study of helminth fauna of fish-eating birds of the Aral Sea basin.) In: Rybn. zapasy Aralskogo morya i puti ikh ratsionalnogo ispolzovanyia, Tashkent, pp. 127—132 (in Russian).

Turemuratov A. T., 1965a: (Helminth fauna of birds of the order *Colymbiformes* of the Aral Sea basin.) Vest. karakalpaksk. Fil. Akad. Nauk Uzbek. SSR, No. 2: 37—40 (in Russian).

Turemuratov A. T., 1965b: *(Pectinospirura sobolevi* n. sp. — a new species of nematode from herring-gull.) Helminthologia, 6: 121—124 (in Russian).

Turemuratov A. T., 1966: (Role of fish-eating birds of the Aral basin in the distribution of helminthiases of fishes.) In: Biol. osnovy rybn. khoz. na vodoemakh Srednei Azii i Kazakhstana, Izd. Nauka, Alma-Ata, pp. 966—967 (in Russian).

Vaidova S. M., 1963: (Helminth fauna of fish-eating birds of Azerbaidzhan.) In: Mat. nauchn. sessii Gel'mintol. resp. Zakavkazya (28.—30. X. 1961), Tbilisi, pp. 32—36 (in Russian).

Vaidova S. M., 1964: (Fauna and distribution of nematode in birds of Lenkoransk group of regions and Mugansk steppe of Azerbaidzhan.) Izv. Akad. Nauk Azerb. SSR, Ser. biol. Nauk, No. 4: 29—35 (in Russian).

Vaidova S. M., 1965: (Helminth fauna of fish-eating birds in water reservoirs of Kura-Araksinsk lowland of Azerbaidzhan.) Trudy Inst. Zool. Baku, 24: 99—108 (in Russian).

Vaidova S. M., 1969: (Helminths of fish-eating birds from waters of the Greater Caucasus Mountains in Azerbaidzhan.) In: Musaev M. A. et al. (Editors), Problemy parazitologii, Baku: Izdat. ELM., pp. 138—145 (in Russian).

Vaidova S. M., 1970: (Helminth fauna of fish-eating birds of Azerbaidzhan and the damage caused by it to the fish and game animal industry.) In: Issledovania po gelmintologii v Azerbaidzhane. Izdat. ELM, Baku, p. 43 (in Russian).

Vasilev I., 1968: (About taxonomic position of streptocarids of domestic birds in Bulgaria.) Izv. tsentr. khelmint. Lab., Sofia, 12: 5—21 (in Bulgarian).

Vasilkova Z. G., 1926: (Nematode fauna of gulls and terns of Kazakhstan (based on materials from 5th SGE).) Trudy gos. Inst. eksper. Vet., 4: 105—113 (in Russian).

Vasilkova Z. G., 1927: (Nematode fauna of gulls and terns of the USSR). Sb. Rab. Gel'mint., Moscow, pp. 44—49 (in Russian).

Vasilkova Z. G., 1930: (Nematodes of birds of North-Dvin province. In: Works of the 32nd and 38th SGE expedition on the territory of North-Dvin province in 1926—1927.) Izd. Ser.-Dvinsk. Gubzdrava i Gubvetotdela, Vyatka, pp. 68—86 (in Russian).

Vasilkova Z. G., & Gushanskaya L., 1930: Nématodes du genre *Eucoleus* Duj., 1845 des oiseaux. Annls Parasit. hum. comp., 8: 619—623.

Vassiliades G., 1970: Nématodes parasites d'Oiseaux malgaches. Annls Parasit. hum. comp., 45: 47—88.

Vevers G. M., 1920: Report on *Entozoa* collected from animals which died in the Zoological Gardens of London during eight months of 1919—1920. Proc. zool. Soc. Lond., 3: 405—410.

Vogel H., 1928: Zur Morphologie und Biologie von *Cyathostoma variegatum* (Creplin, 1849) *(Syngamus variegatus)*. Z. Infekt Krankh. Haustiere, 34: 97—117.

Vojtěchovská-Mayerová M., 1952: Nové nálezy parasitických červů u našich ptáků. Věst. Čs. Spol. zool., 16: 71—88.

Volskis R., 1966: Parasite fauna of nematodes in the birds of the Lithuanian SSR. Acta parasit. lith., 6: 47—56 (in Russian).

Volskis G., 1968: (Helminth fauna of birds of Zhuvintas reservation.) In: Zapovednik Zhuvintas, Vilnius, pp. 202—218 (in Russian).

Walter H., 1866: Helminthologische Studien. Ber. Vereins Naturk. Offenbach (1865—66), 7: 51—79.

Wang P. C., 1966: Notes on *Acuarioidea* of birds from Fukien, China. Acta Parasit. sin., 3: 2—29.

Wedl C., 1856: Über die Mundwerkzeuge von Nematoden. Sitzungsb. K. Acad. Wiss. Math.-naturw. Kl. Wien, 19: 33—69.

Wehr E. E., 1934: Descriptions of three bird nematodes including a new genus and a new species. J. Wash. Acad. Sci., 24: 341—347.

Wehr E. E., 1937: (Two new species of *Echinuria (Nematoda: Acuariidae)* from birds, with notes on other species of this genus.) (Papers on helminthology in commemoration of the

30 year Jubileum of K. I. Skrjabin and of the 15th anniversary of the All-Union Institute of Helminthology.) Moscow, pp. 763—768 (in Russian).

Williams I. C., 1961: (A list of parasitic worms, including twenty-two new records from British birds.) Ann. Mag. nat. Hist., 13: 467—480.

Williams O. L., 1929: Revision of the Nematode genus *Rusguniella* Seurat with a description of a new Central American species. Univ. Calif. Publs. Zool., 33: 1—12.

Witenberg G., 1925: Remarks on the anatomical structure and systematic position of the stork's lung filaria. J. Helminth, 3: 203—208.

Witenberg G., 1929: Parasitische Würmer von *Puffinus kuhli*. Ergebn. Sinai-Exped. 1927, Hebraisch. Univ. Jerusalem, pp. 118—124.

Wu H. W., 1933: Helminthological notes I. Sinensia, 4: 51—59.

Wu H. W. & Liu C. K., 1943: Helminthological notes III. Sinensia 14: 99—105.

Yamaguti S., 1935: Studies on the helminth fauna of Japan. Part 12. Avian Nematodes, I. Jap. J. Zool., 6: 403—431.

Yamaguti S., 1941: Studies of the helminth fauna of Japan. Part 36. Avian nematodes II. Jap. J. Zool., 9: 441—480.

Yamaguti S., 1961: Systema helminthum III. The nematodes of vertebrates. New York and London: Interscience Publishers Inc., pp. 1—679.

Yamaguti S. & Mitunaga Y., 1943: Nematode parasites of birds from Formosa I. Trans. nat. hist. Soc. Taiwan, 33: 300—311.

Yorke W. & Maplestone P. A., 1926: The nematode parasites of vertebrates. London, 536 pp.

Zablotsky V. I., 1962: (The helminth fauna of birds of prey on the border of the Caspian Sea.) Trudy Astrakh. gos. Zapovedn., 6: 91—114 (in Russian).

Zajíček D., 1961: Příspěvek k výskytu a vzájemnému vztahu cizopasných červů lysky černé *(Fulica atra* L.), racka chechtavého *(Larus ridibundus* L.*)* a kachny divoké *(Anas platyrhyncha* L.). Lesnictví, Sb. ČSAZV, 7: 495—514.

Zajíček D., 1964: The embryonal and postembryonal development of *Amidostomum boschadis* Petrow et Fediuschin, 1949 *(Nematoda)*. In: Ergens R. & Ryšavý B. (Editors), Parasitic worms and aquatic conditions. Proc. of symp., Prague, October 29th — November 2nd, pp. 137—143.

Zavadil R., 1957a: Das Vorkommen der Schmarotzer aus der Gattung *Cyathostoma* im Gebiete der ČSR. Práce z I. konf. čs. helmint., pp. 269—276.

Zavadil R., 1957b: Cyathostomosa ptáků, její původci a výskyt v Československu. Sb. vys. Šk. zemed., Brno, Řada B, 5: 105—121.

Zavadil R., 1961: Nález *Cyathostoma lari* E. Blanchard, 1894 u racka chechtavého *(Larus ridibundus* Linné 1766) z jižní Moravy. Čslká. Parasit., 8: 415—419.

Zeder J. G. H., 1800: Erster Nachtrag zur Naturgeschichte der Eingeweidewürmer mit Zufässen und Anmerkungen herausgegeben. Leipzig, 320 pp.

Zeder J. G. H., 1803: Anleitung zur Naturgeschichte der Eingeweidewürmer. Hamburg, 432 pp.

Zhatkanbaeva D., 1964: (Helminth fauna of aquatic birds in the Kazakh SSR.) Trudy Inst. Zool., Alma-Ata, 22: 110—125 (in Russian).

Zhatkanbaeva D., 1965: (Helminth fauna of grebes *(Podiceps griseigena, P. nigricollis)* in northern and central Kazakhstan.) Mater. nauch. Konf. vses. Obshch. Gel'mint., Part 1, pp. 79—81 (in Russian).

Zhatkanbaeva D., 1966a: (Helminth fauna of fish-eating birds of the Balkhash-Alakol basin.) Mater. nauch. Konf. vses. Obsch. Gel'mint., Part 3, Moscow, pp. 111—114 (in Russian).

Zhatkanbaeva D., 1966b: (Role of fish-eating birds in the distribution of helminthiases among fishes.) In: Bolezni ryb i mery borby s nimi. (Mater. nauch. proizv. Konf.) Alma-Ata, pp. 91—93 (in Russian).

Zhatkanbaeva D., 1971: (Helminths of *Larus ichthyaetus* Pall. from the lakes of Kurgaldzhin.) Trudy Inst. Zool., Alma-Ata, 31: 51—59 (in Russian).

Zhelyazkova-Paspaleva A., 1962a: (Helminth fauna of free-living birds of the Strandzhan region.) Izv. tsent. khelmint. Lab., Sofia, 7: 137—152 (in Bulgarian).

Zhelyazkova-Paspaleva A., 1962b: (Contribution to the knowledge of helminth fauna of free-living birds. I. Helminths of *Ciconia nigra* L.) Izv. zool. Inst., Sof., 11: 207—209 (in Bulgarian).

Index to Nematode Taxa

Descriptions on pages printed in boldface type

acanthocephalica, Spiroptera 182
acanthocephalicus, Schistorophus **182**
Acuaria 121, **122**
 anthuris 122
 ciconiae 161
 contorta 161
 orientalis 140
 pelagica 208
 phalacrocoracis **122**
 raillieti 141
Acuariidae 113, **120**, 196
Acuariinae 120, **121**
acuticauda, Physaloptera 180
acutissima, Subulura 110
acutum, Amidostomum **46**, 50, 51
acutum, Dicheilonema 230
acutus, Strongylus 47
Adenophorea 17, **18**
adunca, Spiroptera 130
aduncus, Cosmocephalus 127, **129**
africanus, Eustrongylides 33, **35**
alata, Physaloptera **179**, 180
alcedinis, Filaria 233
 Monopetalonema **232**
alcedonis, Aviculariella **123**, 125
 Rusguniella 125
Alcedospirura collaricephala 125
alcyona, Aviculariella 123
Amidostomidae **46**
Amidostomum **46**, 50, 51
 acutum 47
 anseris 47, **48**
 boschadis 47
 fulicae 47, **50**
 orientale 47, **52**
 raillieti 51
anatinum, Epomidiostomum 55
anatis, Strongylus 41
 Thominx **27**
 Trichocephalus 28

Trichosoma 23
Ancyracanthopsis bihamatus 183
Ancyracanthus bihamatus 183
andersoni, Contracaecum 69, **72**
Anenteronema skrjabini 186
angusticolle, Ascaris 94
 Porrocaecum 91, **93**
Anisakidae **68**
Anisakoidea 68
anseris, Amidostomum 47, **48**
 Strongylus 50
anthuris, Acuaria 122
Aphasmidia 17
Aprocta **225**
 cylindrica 225
 matronensis 226
 milinskii 226
 turgida **225**, 226
Aproctidae **224**
Aproctinae 225
aquillae, Ascaris, Gmelin, 1790 74
 Ascaris, Smith, Fox & White, 1908 74
aquillina, Thelazia 223
 Thelaziella **222**
arctica, Rusguniella 152
arcticus, Syngamus 56, **58**
ardea, Tetrameres **210**
 Tetrameres (Gynaecophila) 209
ardeae, Ascaris 96
 Filaria, Nawrotzky, 1914 (in part) 177
 nigrae, Filaria 232
 Pelecitus 244
 Porrocaecum 91, **95**, 96
argentata, Pectinospirura 149
Ascaridata 18, 43, **68**
Ascaridida 17, 43
Ascaridoidea 68
Ascaris angusticolle 94
 aquillae Gmelin, 1790 74
 aquillae Smith, Fox & White, 1908 74

309

ardeae 96
crassa 93
fissicollis 74
gallinarum 107
microcephala 71
micropapillata 76
ovalis 80
praelonga 101
reticulata 102
sagittatus 142
variegata 89
zeylanica 74
Aviculariella 121, **123**
   alcedonis **123**, 125
   alcyona 123
   collaricephala 125
Avioserpens **246**
   galliardi **247**
   mosgovoyi **248**
   multipapillosa 248
   nana 248
   taiwana 246
Avioserpentinae 246
bancrofti, Contracaecum 76
behningi, Lemdana **235**
bihamatus, Ancyracanthopsis 183
   Ancyracanthus 183
   Schistorophus 181, **182**
bilabiatum, Dicheilonema 230
bodenheimeri, Contracaecum 70, **73**
boschadis, Amidostomum 47
boularti, Cyathostoma 68
brantae, Cyathostoma 68
brevicaudatus, Desportesius 135, **137**
   Dispharagus 137
brevicole, Trichosoma 23
bronchialis, Cyathostoma 68
brumpti, Subulura 112
buckleyi, Skrjabinocara 155, 157
butoridi, Pharyngosetaria 177
Calcaronema 65
   trifurcatum 64, 65
californica, Streptocara 196, 197, 198
Camallanata 17, 43, **245**
Capillaria 15, **19**
   carbonis 19, **20**
   herodiae 19, **21**
   mergi 19, **22**
   obsignata 19
   obsignata of Babaev, 1970 25

podicipitis 19, **23**
ryjikovi 19, **25**
venteli 31
Capillariidae **19**
capillaris, Spiroptera 182
californica, Streptocara 196, 197, 198
carbonis, Capillaria 19, **20**
   Trichosomum 21
Chandlerella shaldybini 229
Cheilospirura phalacrocoracis 122
Chevreuxia 121, **125**
   revoluta **125**
chui, Thelazia 223
ciconiae, Acuaria 161
   Contortospiculum 232
   Dicheilonema **231**
   Filaria 232
   Skrjabinocta **229**
   Syncuaria 135, **160**
ciconiarum, Paronchocerca **240**
cincli, Skrjabinoclava 173
cirrohamata, Filaria 200
   Ingliseria 196, **200**, 201
   Streptocara 200
clausa, Physaloptera 179
coccinea, Tetrameres 209, **211**, 212
   Tropidocerca 212
collaricephala, Alcedospirura 125
   Aviculariella 125
contorta, Acuaria 161
   Syncuaria **161**, 162
   Thominx 26, **28**, 29
Contortospiculum 232
   ciconiae 232
contortum, Trichosoma 29
Contracaecum **69**
   andersoni 69, **72**
   bancrofti 79
   bodenheimeri 70, **73**
   haliaeti 70, **73**, 74, 82
   himeu 85
   magnicollare 89
   magnipapillatum 89
   matwejewi 69, **75**
   microcephalum 69, **70**
   micropapillatum 69, **76**, 77
   milviensis 70, **77**
   nehli 80
   oschmarini 89
   ovale 69, **79**, 80, 83

pandioni 69, **80**, 82
podicipitis 80
praestriatum 69, **82**, 83
rudolphii 70, **83**, 85
ruficolle 80
septentrionale 69, **86**
sp. 91
spasskii 80
spiculigerum 85
torquatum 89
travassosi 69, **87**
umiu 85
variegatum 69, **88**, 89
yamaguti 69, **90**, 91
Cordonema 165, **168**
   longifuniculata **168**, 174
   solonitzini 168
   venusta 168
coronatus, Hystrichis 40, **41**, 42
Cosmocephalus 122, **126**
   sp. of Ryzhikov & Khokhlova, 1965, 143
   aduncus 127, **129**
   diesingi 127
   faridi 127, **130**
   imperialis 127, **131**
   jaenschi 127, **132**
   obvelatus 126, **127**
crami, Tetrameres 210, **212**
   Tetrameres (Petrowimeres) 210
crassa, Ascaris 93
crassicauda, Dispharagus 199
   Spiroptera 199
   Streptocara 196, 197, **198**, 199
crassum, Porrocaecum 91, **92**
cruzi, Microtetrameres 219
Cyathostoma 55, **59**, 65
  boularti 68
  brantae 68
  bronchialis 68
  (Cyathostoma) 65
  (Hovorkonema) 65
  lari 59, **60**
  microspiculum 59, **61**, 62
  trifurcatum 59, **62**
  turemuratovi 62
  variegatus 68
  verrucosum 59, **64**
Cyathostomum variegatum 64
cylindrica, Aprocta 225
Cyrnea **114**, 120

(Cyrnea) 114
eurycerca 114
excisa 114
ficheuri 114
leptoptera 114, **116**
monoptera 114, **117**
(Procyrnea) 114, **117**
Cyrneinae 114
decora, Dispharagus 203
  Proyseria **201**
decorata, Decorataria 133, **134**, 135
  Echinuria 134, 171
  Skrjabinoclava **170**, 171
Decorataria 121, **133**
  decorata 133, **134**, 135
denticulata, Sciadiocara 188
depressa, Fusaria 98
depressum, Porrocaecum 91, **96**
Desmidocerca **174**
  aerophila 174, **175**
  numidica 177
Desmidocercella 174, **175**
  incognita 176, **178**, 179
  leiperi 177
  lubimovi 177
  numidica 175, **176**, 177
  skrjabini 178
Desmidocercidae 113, **174**
Desportesius 122, **135**
  brevicaudatus 135, **137**
  equispiculatus 135, **138**
  groffi 135, **138**
  invaginatus 135, **136**, 137
  orientalis 135, **139**, 140
  raillieti 135, **140**
  sagittatus 135, **141**
  spinulatus 135, **142**, 143
Dicheilonema **230**, 232
  acutum 230
  bilabiatum 230
  ciconiae **231**
Dicheilonematinae 230
diesingi, Cosmocephalus 127
differens, Subulura 112
Dioctophmyata 17, **32**
Dioctophymida 17
Dioctophymidae **32**
Dioctophyminae 32
Diplotiaenidae **230**
Diplotriaeninae 224, 230

Dispharagus sp. of Wedl, 1856 153
   brevicaudatus 137
   crassicauda 199
   decora 203
   invaginatus 137
Dispharynx 121, **144**
   nasuta **144**, 145
   spiralis 145
dogieli, Streptocara 197
Dracunculidae **246**
Echinuria 165
   decorata 134, 171
   heterobrachiata 165, **167**
   horrida of Singh, 1943 173
   uncinata **165**, 167
Echinuriinae 120, 121, **165**
elegans, Hystrichis, of Stossich, 1899 37
   Strongylus 39
elongata, Rusguniella **151**, 152, 153
   Spiroptera 152
Encephalonema longimicrofilaria 251
ensicaudata, Fusaria 99
ensicaudatum, Porrocaecum 91, **98**
Epomidiostominae 53
Epomidiostomum **53**
   anatinum 55
   uncinatum **54**, 55
equispiculatus, Desportesius 138
equispiculatus spinulatus, Synhimantus 143
   Synhimatus 138
Eucoleus laricola 29
   pachyderma 29
Eufilaria **227**
   lari **227**
   sergenti 227
Eufilariinae 226
eurycerca, Cyrnea 114
Eustrongylides **32**, 34, 39
   africanus 33, **35**
   excisus 33, 34, **36**, 37
   mergorum 33, 35, **38**
   sinicus 33, **39**
   tubifex **33**, 34
Eustrongylinae 32
Excisa **118**
excisa, Cyrnea 120
Excisa excisa **118**, 120
excisa, Excisa **118**, 120
excisa, Hadjelia 120
   Seurocyrnea 120

Spiroptera 120
   excisus, Eustrongylides 33, 34, **36**, 37
faridi, Cosmocephalus 127, **130**
Fasciola trachea 57
ficheuri, Cyrnea **114**
   Habronema 115
Filaria alcedinis 233
   ardeae Nawrotzky, 1914 (in part) 177
   ardeae nigrae 232
   ciconiae 232
   cirrohamata 200
   labiata 232
   marcinowskyi 177
   physalura 233
   recta 197
   squamata 157
   tridentata 147
   uncinata 167
Filariata 17, 43, **224**
fissicollis, Ascaris 74
fissispina, Tetrameres **213**, 214
   Tetrameres (Petrowimerea) 210
   Tropidocerca 214
formosensis, Paracuaria 145, **148**
   streptocara 149
fulicae, Amidostomum 47, **50**
   Spiroptera 51
fulicaeatrae, Pelecitus 244
   Spirofilaria 244
Fusaria depressa 98
   ensicaudata 99
galliardi, Avioserpens **247**
gallinarum, Ascaris 107
   Heterakis 105, **106**
gigantissima Placentonema 13
glareolae, Stellocaronema 191
Gnathostoma shipleyi 208
groffi, Desportesius 135, **138**
   Synthimantus 138
gubanovi, Tetrameres 214
   Tetrameres (Tetrameres) 210
guschanscoi, Viktorocara 192, **193**
gusi, Koriakinema 196
(Gynaecophila) Tetrameres 209
gynaecophlia, Tetrameres 209, **216**
   Tropidocerca 216
Habronema ficheuri 115
   monoptera 117
   sobolevi 120
Hadjelia excisa 120

halcyoni, Skrjabinoclava 170, **171**, 173
    Viktorocara 195
halcyonis, Sobolevicephalus **188**, 189
halieti, Contracaecum 70, **73**, 74, 82
Hamatospiculum 235
helix, Spirofilaria 244
    Spiroptera 244
herodiae, Capillaria 19, **21**
Heterakidae **105**
Heterakinae 105
Heterakis **105**
    gallinarum 105, **106**
    kurilensis 105, **107**
    pavonis 105, **108**, 109
    suctoria 112
    yamadori 109
heterobrachiata, Echinuria 165, **167**
Heterospiculoides 234, **237**
    skrjabini 237, **238**
Heterospiculum 234, **238**
    sobolevi 237, **239**
heteroura, Porrocaecum 104
himeu, Contracaecum 85
Histiocephalus laciniatus 184
    stellaepolaris 205
horrida, Echinuria, of Singh, 1943 173
    Skrjabinoclava 170, **173**
    Skrjabinoclava, of Sergeeva, 1969 169
horridus, Strongylus 174
Houdemeres tonkinensis 242
Hovorkonema 55, 65, **66**, 68
    tadornae 66
    variegatum **66**
Hystrichis 32, **40**, 42
    coronatus 40, **41**, 42
    elegans of Stossich, 1899 37
    mergi-merganseris 42
    neglectus 41
    orispinus 41
    tricolor **40**, 41
    varispinosus 41
    wedli 41
Icosiellinae 234
imperialis, Cosmocephalus 127, **131**
incognita, Desmidocercella 176, **178**, 179
Ingliseria 196, **200**
    cirrohamata 196, **200**, 201
invaginatus, Desportesius 135, **136**, 137
    Dispharagus 137
jaenschi, Cosmocephalus 127, **132**

Koriakinema 196
    gusi 196
kurilensis, Heterakis 105, **107**
labiata, Filaria 232
laciniatus, Histiocephalus 184
    Schistorophus **183**, 184
lari, Cyathostoma **59**, 60
    Eufilaria **227**
laricola, Eucoleus 29
    Thominx 29
laticeps, Synhimantus 162
leiperi, Desmidocercella 177
Leipoanematinae 110
legendrei, Sciadiocara 188
Lemdana **234**
    behningi **235**
    limbookengi **235**
    lomonti **237**
    marthae 235
    urbaini 177
Lemdaninae 234
leptoptera, Cyrnea 114, **116**
    Spiroptera 117
limbookengi, Lemdana **235**
lomonti, Lemdana **237**
longicornis, Schistorophus 181
longifuniculata, Cordonema 168
    Skrjabinoclava 168
longimicrofilaria, Encephalonema 251
lubimovi, Desmidocercella 177
macdonaldi, Paracuaria 147
magnicollare, Contracaecum 89
magnipapillatum, Contracaecum 89
manica, Thominx 26
marcinowskyi, Filaria 177
matronensis, Aprocta 226
matwejewi, Contracaecum 69, **75**
mergi, Capillaria 19, **22**
mergi-merganseris, Hystrichis 42
mergorum, Eustrongylides 33, 35, **38**
    Strongylus 39
Meteterakinae 105
microcephala, Ascaris 71
microcephalum, Contracaecum 69, **70**
micropapillata, Ascaris 76
micropapillatum, Contracaecum 69, **76**, 77
microspiculum, Cyathostoma 59, **61**, 62
    Syngamus (Ornithogamus) 62
Microtetrameres 209, **219**
    cruzi 219

pelecani **220**
spiralis 220, **221**
milinskii, Aprocta 226
milviensis, Contracaecum 70, **77**
Monopetalonema 230, **232**
　alcedinis **232**, 234
monoptera, Cyrnea 114, **117**
　Habronema 117
mosgovoyi, Avioserpens **248**
multidentata, Pectinospirura **149**, 151
multipapillosa, Avioserpens 248
nana, Avioserpens 248
nasuta, Dispharynx 145
　Spiroptera 145
neglectus, Hystrichis 41
nehli, Contracaecum 80
niloticus, Synhimantus **162**
numidica, Desmidocerca 177
　Desmidocercella 175, **176**, 177
nyctardeae, Thelazia 224
　Thelaziella 222, **223**, 224
obsignata, Capillaria 19
　Capillaria, of Babaev, 1970 25
obvelata, Spiroptera 128
obvelatus, Cosmocephalus 128
orientale, Amidostomum 47, **52**
orientalis, Acuaria 140
　Desportesius 135, **139**, 140
orispinus, Hystrichis 41
oschmarini, Contracaecum 89
Oshimaia 246
Oswaldofilariidae 224, **234**
Oswaldofilariinae 234
ovale, Contracaecum 69, **79**, 80, 83
ovalis, Ascaris 80
Oxyspirurinae 221
Oxyurata 18, 43, **104**, 221
pachyderma, Eucoleus 29
　Trichosoma 29
pandinoi, Contracaecum 69, **80**, 82
papillosus, Strongylus 39
Paracuaria 121, **145**
　formosensis 145, **148**
　macdonaldi 147
　tridentata 145, **146**, 147, 198
paradoxa, Rictularia 208
　Robertdollfusa **250**
　Tetrameres 209
Paronchocerca 234, **240**
　ciconiarum **240**

tonkinensis **242**
Parornithofilaria 227, **228**
　shaldybini **228**
　stantchinskyi 228
parvepapillata, Skrjabinocara 155, **157**
Paryseria 203
pavonis, Heterakis 105, **108**, 109
　Tetrameres **216**
　Tetrameres (Petrowimeres) 209
pectinifera, Streptocara 197, 199
Pectinospirura 121, **149**
　argentata 149
　multidentata **149**, 151
　sobolevi 150
pelagica, Acuaria 208
pelecani, Microtetrameres **220**
Pelecitus 234, **234**
　ardeae 244
　fulicaeatrae **243**, 245
　helicinus 243
　podicipitis **245**
penihamata, Streptocara 197
Petroviprocta 246
　vigissi 248
petrowi, Skrjabinocta 229
　(Petrowimeres) Tetrameres 210
phalacrocoracis, Acuaria **122**
　Cheilospirura 122
　Porrocaecum 91, 99
Pharyngosetaria butoridi 177
Phasmidia 17
Physaloptera **179**
　acuticauda 180
　alata **179**, 180
　clausa 179
　striata 120
　tenuicollis 180
Physalopteridae 131, **179**
Physalopterinae 179
physalura, Filaria 233
pileati, Skrjabinobronema 189
Placentonema gigantissima 13
platyptera, Thelazia 223
podicipitis, Capillaria 19, **23**
　Contracaecum 80
　Pelecitus **245**
　Spirofilaria 245
Porrocaecum 69, **91**
　angusticolle 91, **93**
　ardeae 91, **95**, 96

crassum 91, **92**
depressum 91, **96**
ensicaudatum 91, **98**
heteroura 104
phalacrocoracis 91, **99**
praelongum 91, **100**
reticulatum 91, **101**, 102
semiteres 91, **102**, 104
serpentulus 96
praelonga, Ascaris 101
praelongum, Porrocaecum 91, **100**
praestriatum, Contracaecum 69, **82**, 83
prima, Skrjabinocerca 158, **159**
procellariae, Seuratia 208
Proleptinae 179
Proyseria 196, **201**
    decora **201**
puffini, Seuratia 208
raillieti, Acuaria 141
    Amidostomum 51
    Desportesius 141
    Thominx 29
recta, Filaria 197
    Streptocara 197
reticulata, Ascaris 102
reticulatum, Porrocaecum 91, **101**, 102
revoluta, Chevreuxia 126
    Spiroptera 126
Rhabditida 17, **43**
Rhabditata 18, **43**
Rictularia paradoxa 208
rissae, Streptocara 147, 197
Robertdollfusa **250**
    longimicrofilaria 14, **251**
    paradoxa 251
Robertdollfusidae 246, **250**
rostombekovi, Skrjabinocara 155
rudolphii, Contracaecum 70, **83**, 85
ruficolle, Contracaecum 80
Rusguniella 120, 121, **151**, 196
    alcedonis 125
    arctica 152
    elongata **151**, 152, 153
    skrjabini 152
    tringae 152
    wedli 151, **153**
ryjikovi, Capillaria 19, **25**
sagittatus, Ascaris 142
    Desportesius 135, **141**, 142
schejkini, Viktorocara **192**

schigini, Tetrameres 210, **217**
    Tetrameres (Gynaecophila) 210
Sciadiocara 180, 181, **186**
    denticulata 188
    legendrei 188
    umbellifera **186**, 188
schikhobalovi, Skrjabinocara 155
Schistorophidae 133, **180**
Schistorophus 180, **181**
    acanthocephalicus **181**
    bihamatus **182**
    laciniatus **183**, 184
    longicornis 181
    skrjabini 182, 184, **185**, 186
Secernentea 17, **43**
semiteres, Porrocaecum 91, **102**, 104
septentrionale, Contracaecum 69, **86**
sergenti, Eufilaria 227
serpentulus, Porrocaecum 96
Seuratia 196, **206**
    procellariae 208
    puffini 208
    shipleyi 206, **207**
    yamaguti 208
Seuratinae 196
Seurocyrnea excisa 120
Sexansocara 122, **153**
    skrjabini **154**
shaldybini, Chandlerella 229
    Parornithofilaria **228**, 229
    Splendidofilaria 229
shipleyi, Gnathostoma 208
    Seuratia 206, **207**, 208
sinicus, Eustrongylides 33, **39**
sirry, Synhimantus 162, **164**
skrjabilina, Thelazia 223
skrjabini, Anenteronema 186
    Desmidocercella 178
    Heterospiculoides 237, **238**
    Rusguniella 152
    Schistorophus 182, 184, **185**, 186
    Sexansocara **154**
    Skrjabinocara 155
    Stellocaronema **190**, 191
    Tetrameres **218**
    Tetrameres (Tetrameres) 210
Skrjabinobronema pileati 189
Skrjabinocara 121, **155**
    buckleyi 155, 157
    parvepapillata 155, **157**

rostombekovi 155
schikhobalovi 155
skrjabini 155
squamata **155**
squamata of Johnston & Mawson, 1941 157
timofejevi 155
victori 155
Skrjabinocerca 121, **158**
  prima 158, **159**
Skrjabinoclava 165, **170**
  cincli 173
  decorata **170**, 171
  halcyoni 170, **171**, 173
  horrida 170, **173**
  horrida of Sergeeva, 1969 169
  longifuniculata 168
  solonitzini 168
Skrjabinocta 227, **229**
  ciconiae **229**
  petrowi 229
Smetaleksenema 189
sobolevi, Habronema 120
  Heterospiculum 237, **239**
  Pectinospirura 150
Sobolevicephalus 180, 181, **188**, 189
  halcyonis **188**, 189
solonitzini, Cordonema 168
  Skrjabinoclava 168
somateriae, Streptocara 149
spasskii, Contracaecum 80
spiculigerum, Contracaecum 85
spinulatus, Desportesius 135, **142**, 143
spirale, Thominx 26, **30**
spiralis, Dispharynx 145
  Microtetrameres **220**, 221
  Tropidocerca 221
Spirofilaria fulicaeatrae 245
  helix 244
  podicipitis 245
Spiroptera acanthocephalica 182
  adunca 130
  capillaris 182
  crassicauda 199
  elongata 152
  excisa 120
  fulicae 51
  helix 244
  leptoptera 117
  nasuta 145
  obvelata 128

revoluta 126
tadorna 41
tridentata 147
umbellifera 188
Spiruracercinae 114
Spirurata 17, 43, **112**
Spirurida 17, 43
Spiruridae **113**
Spirurinae 114
spizaeti, Thelazia 223
Splendidofilaira shaldybini 229
Splendidofilariidae 224, **226**
Splendidofilariinae 226
squamata, Filaria 157
  Skrjabinocara **155**
  Skrjabinocara, of Johnston & Mawson, 1941 157
stantchinskyi, Parornithofilaria 228
Stegophorus 196, **203**
  stercorarii 203, **205**
  stellaepolaris **203**, 205
stellaepolaris, Histiocephalus 205
  Stegophorus **203**, 205
Stellocaronema 180, 181, **190**
  fausti 190
  glareolae 191
  skrjabini **190**, 191
stercoralis, Strongyloides 44
stercorarii, Stegophorus 203, **205**
stereura, Thelaziella 222
Streptocara **196**
  californica 196, 197, 198
  cirrohamata 200
  crassicauda 196, 197, **198**, 199
  dogieli 197
  formosensis 149
  pectinifera 197, 199
  penihamata 197
  recta 197, 199
  rissae 147, 197
  somateriae 149
  transcaucasica 147, 197
  tridentata 147
Streptocaridae 113, 120, **195**
Streptocarinae 195
striata, Physaloptera 120
Strongylata 18, 43, **45**
Strongyloides **44**
  stercoralis 44
  turkmenicus **44**

Strongyloididae **43**
Strongylus acutus 47
  anatis 41
  anseris 50
  elegans 39
  horridus 174
  mergorum 39
  papillosus 39
  trachealis 67
  tubifex 34, 37
  tubifex, auctores 41
  tubifex of Rudolphi, 1819 39
  uncinatum 54
Subulura **110**
  acutissima 110
  brumpti 112
  differens 112
  suctoria **110**, 112
Subuluridae 105, **110**
Subulurinae 110
suctoria, Heterakis 112
  Subulura 105, **110**
Squamofilariinae 225
Syncuaria 121, **160**
  ciconiae 135, **160**
  contorta 160, **161**
  sp. of Škarda, 1964 135
Syngamidae 46, **55**
Syngamus 55, **56**, 57, 62, 65
  arcticus 56, **58**
  (Ornithogamus) microspiculum 62
  trachea **56**
Synhimantus 122, **162**
  equispiculatus 143
  equispiculatus spinulatus 143
  groffi 138
  laticeps 162
  niloticus **162**
  sirry 162, **164**
tadorna, Spiroptera 41
tadornae, Hovorkonema 66
taiwana, Avioserpens 246
tenuicollis, Physaloptera 180
tenuis, Viktorocara 192, **195**
Tetracheilonematinae 225
Tetrameres 13, 209
  ardea **210**
  coccinea 209, **211**, 212
  crami 210, **212**
  fissispina **213**, 214

  gubanovi **214**
Tetrameres (Gynaecophila) 209
  gynaecophila 209, **216**
  (Gynaecophila) ardea 210
  (Gynaecophila) schigini 210
  paradoxa 209
  pavonis 210, 216
  (Petrowimeres) 209
  (Petrowimeres) crami 210
  (Petrowimeres) fissispina 210
  (Petrowimeres) pavonis 210
  schigini **217**
  skrjabini **218**
  (Tetramares) 209
  (Tetrameres) gubanovi 210
  (Tetrameres) skrjabini 210
(Tetrameres), Tetrameres 209
Tetrameridae 113, **208**
Thelazia aquillina 223
  chui 223
  nyctardeae **223**
  platyptera 223
  skrjabilina 223
  spizaeti 223
Thelaziella 221, **222**
  aquillina **222**
  nyctardeae 222, **223**
  stereura 222
Thelaziidae 113, **221**
Thelaziinae 221
Thominx 19, **26**
  anatis 26, **27**
  contorta 26, **28**, 29
  laricola 29
  manica 26
  raillieti 29
  spirale 26, **30**
timofejevi, Skrjabinocara 155
tonkinensis, Houdemeres 242
  Paronchocerca **242**
torquatus, Contracaecum 89
trachea, Fasciola 57
  Syngamus **56**
trachealis, Strongylus 67
transcaucasica, Streptocara 147, 197
travassosi, Contracaecum 69, **87**
Trichocephalata 17, **18**
Trichocephalida 17
Trichocephalus anatis 28
Trichosoma anatis 23

brevicole 23
contortum 29
pachyderma 29
Trichosomum carbonis 21
Trychostrongylidae 46, **53**
tricolor, Hystrichis **40**, 41
tridentata, Filaria 147
    Paracuaria 145, **146**, 147, 198
    Spiroptera 147
    Streptocara 147
trifurcatum, Calcaronema 64, 65
    Cyathostoma 59, **62**
tringae, Rusguniella 152
Tropidocerca coccinea 212
    fissispina 214
    gynaecophila 216
    spiralis 221
Tropisuridae 209
Tropisurus 209
tubifex, Eustrongylides **33**, 34
    Strongylus 34, 37
    Strongylus, auctores 41
    Strongylus, of Rudolphi, 1819 39
turemuratovi, Cyathostoma 62
turgida, Aprocta **225**, 226
turkmenica, Strongyloides **44**
umbellifera, Sciadiocara **186**, 188
    Spiroptera 188

umiu, Contracaecum 85
uncinata, Echinuria **165**, 167
    Filaria 167
uncinatus, Epomidiostomum **54**, 55
    Strongylus 54
urbaini, Lemdana 177
variegata, Ascaris 89
variegatum, Contracaecum 69, **88**, 89
    Cyathostoma 55, **59**, 65
    Cyathostomum 64
    Hovorkonema 55, 65, **66**, 68
varispinosus, Hystrichis 41
venteli, Capillaria 31
venusta, Cordonema 168
verrucosum, Cyathostoma 59, **64**
victori, Skrjabinocara 155
vigissi, Petroviprocta 248
Viktorocara 180, 181, **191**
    guschanscoi 192, **193**
    halcyoni 195
    schejkini **192**
    tenuis 192, **195**
wedli, Hystrichis 41
    Rusguniella 151, **153**
yamadori, Heterakis 109
yamaguti, Contracaecum 69, **90**, 91
    Seuratia 208
zeylanica, Ascaris 74